IET ENERGY ENGINEERING SERIES 256

Wireless Power Transfer Technologies

Other volumes in this series:

Wireless Power Transfer Technologies

Theory and technologies,

2nd edition

Edited by
Naoki Shinohara

The Institution of Engineering and Technology

Published by The Institution of Engineering and Technology, London, United Kingdom

The Institution of Engineering and Technology is registered as a Charity in England & Wales (no. 211014) and Scotland (no. SC038698).

© The Institution of Engineering and Technology 2024

First published 2018, Second edition 2024

The Institution of Engineering and Technology
Futures Place
Kings Way, Stevenage
Hertfordshire, SG1 2UA, United Kingdom

www.theiet.org

British Library Cataloguing in Publication Data
A catalogue record for this product is available from the British Library

ISBN 978-1-83953-892-6 (hardback)
ISBN 978-1-83953-893-3 (PDF)

Typeset in India by MPS Limited

Contents

About the editor

Naoki Shinohara is a professor at the Research Institute for Sustainable Humanosphere, Kyoto University, Japan. He has been performing key research about wireless power transfer technologies for 30 years, including solar power satellites and microwave power transmission. He has published several books. He also served as IEEE Distinguished Microwave Lecturer, as IEEE MTT-S Technical Committee 25 Chair, and serves as IEEE WPT Conference Steering Committee Member and Wireless Power Transfer Consortium for Practical Applications Chair.

Chapter 1

Introduction

Naoki Shinohara[1]

Nikola Tesla dreamed of a "wirelessly powered world" at the end of the nineteenth century. He sometimes tried to transmit power wirelessly, believing that he could send it anywhere on the Earth using resonance phenomena [1]. Unfortunately, his dream has failed to materialize. Nevertheless, he is considered a pioneer of wireless power transfer (WPT) (Figure 1.1).

Figure 1.1   Nikola Tesla sitting in front of a spiral coil used in his wireless-power experiments

[1]Research Institute for Sustainable Humanosphere, Kyoto University, Japan

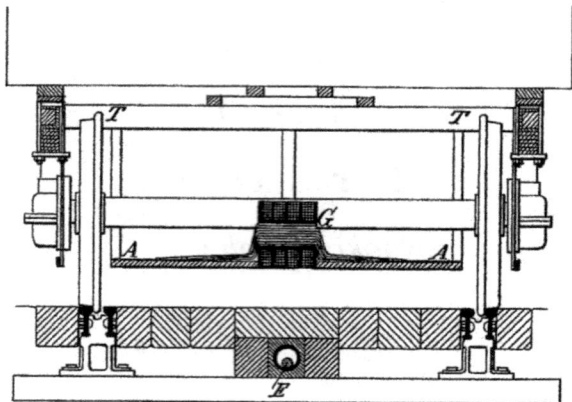

Figure 1.2   Diagram from the patent for Hutin and LeBlanc's first wireless charger for electric vehicles [2]

At around the same time, in 1894, Hutin and LeBlanc proposed an apparatus and a method for powering an electrical vehicle (EV) inductively using an AC generator of approximately 3-kHz frequency [2] (Figure 1.2). EVs were developed shortly after the steam engine, approximately 100 years ago. However, with the development of internal-combustion engine, the EV fell out of favor. As a result, after Hutin and LeBlanc, the EV inductive-coupling WPT charger was forgotten the same way Tesla's dream of WPT was forgotten after him.

There were unremitting effort to carry the electricity wirelessly after the Tesla's dream. In Japan in 1926, an interesting WPT experiment was carried out by Yagi and Uda, the inventors of the Yagi–Uda antenna. They placed nonfeed parasitic elements between a transmitting antenna and a receiving antenna with a frequency of 68 MHz to transmit wireless power (Figure 1.3) [3]. This device was named "wave canal," which was similar to the Yagi–Uda antenna. It was a WPT experiment via radio waves. But it can be considered a coupling WPT because the wireless power was transmitted through coupling antennas/directors as resonators. They successfully received approximately 200 mW by transmitting 2–3 W of electric power. In the same period in the United States, H. V. Noble demonstrated the WPT via 100-MHz radio wave at the Chicago World Fair III in 1933 (Figure 1.4). Distance between a transmitting antenna and a receiving antenna was 5–12 m, and transmitted wireless power was approximately 15 kW. The dream of the wireless power is taken over after the World War II by, e.g., Brown in the United States with microwaves in the 1960s [4], D. Otto in New Zeeland with inductive VLF in the 1970s [5], and a lot of researchers. Figure 1.5 indicates the first WPT-aided flying drone experiment via microwave by Brown in the 1960s. But except some commercial inductive wireless charger for a wireless telephone, an electric toothbrush, and a shaver, the WPT technology became a lost technology. In an IC card, the inductive WPT at 13.58 MHz is adopted. But it is considered near-field communication system, not the WPT.

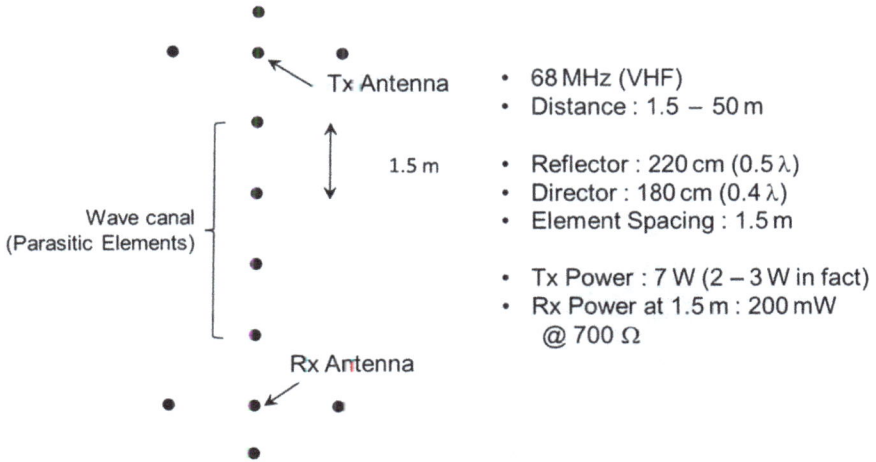

- 68 MHz (VHF)
- Distance : 1.5 – 50 m

- Reflector : 220 cm (0.5 λ)
- Director : 180 cm (0.4 λ)
- Element Spacing : 1.5 m

- Tx Power : 7 W (2 – 3 W in fact)
- Rx Power at 1.5 m : 200 mW
  @ 700 Ω

Tx Antenna

1.5 m

Wave canal
(Parasitic Elements)

Rx Antenna

Figure 1.3   Wave Canal by Yagi and Uda in Japan in 1926

*Figure 1.4   Demonstration of WPT via Radio Waves by Harrell Noble at the
Chicago World Fair III in 1933*

A century after Tesla, Hutin, and LeBlanc, we now have the same dream of WPT. Over the last hundred years, there have been many studies, developments, and commercial products based on WPT technology as described before. Its operational principles can be divided into "electromagnetic coupling WPT" and "uncoupled WPT" or "radiative WPT." The electromagnetic coupling WPT is called as near-field WPT, which uses high-frequency magnetic or electric fields. It is also called inductive-coupling WPT when high-frequency magnetic field is used.

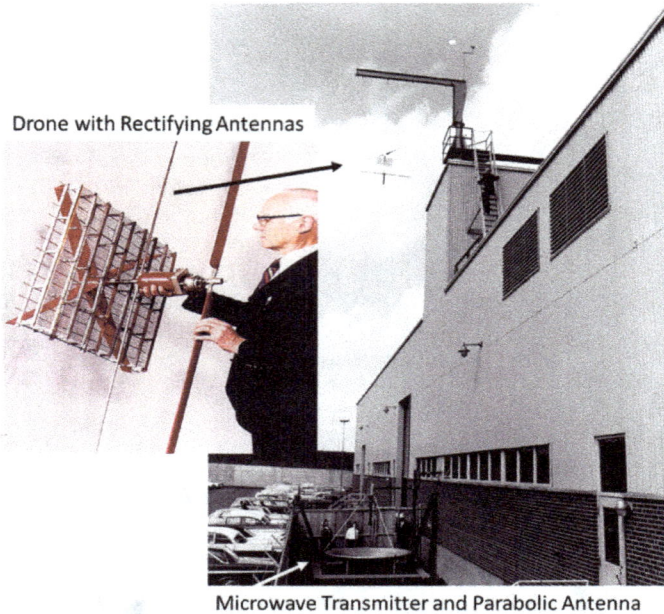

Figure 1.5   First WPT-aided flying drone experiment via microwave by Brown in the 1960s

Inductive-coupling WPT is explained by Ampere's law and Faraday's law. Radiative WPT is called far-field WPT, which is a WPT that happens via radio waves. The WPT via radio waves is explained by Maxwell's equations. Recently, resonance-coupling WPT has been developed as a revised inductive-coupling WPT. A research group at the Massachusetts Institute of Technology proposed a resonance-coupling WPT system in 2007 [6]. The resonance-coupling WPT is based on the inductive-coupling WPT. With the help of resonance phenomena, the distance between the transmitter and receiver can be increased more than that can be in the conventional inductive-coupling WPT. Figure 1.6 indicates the relationship between Ampere's law, Faraday's law, Maxwell's equations, the inductive-coupling WPT, the resonance-coupling WPT, and WPT via radio waves.

In this book, the theory, techniques, and applications of WPT are introduced. After the introduction, the theory of coupling WPT is described. Then, we discuss circuit technologies for the coupling WPT. Next, the theory and techniques of WPT via radio waves are explained, and numerous practical WPT applications are introduced. Finally, important issues in WPT, including safety problems, are discussed. After reading this book, you will also share the dream of WPT.

Inductive Coupling WPT (Near Field WPT)

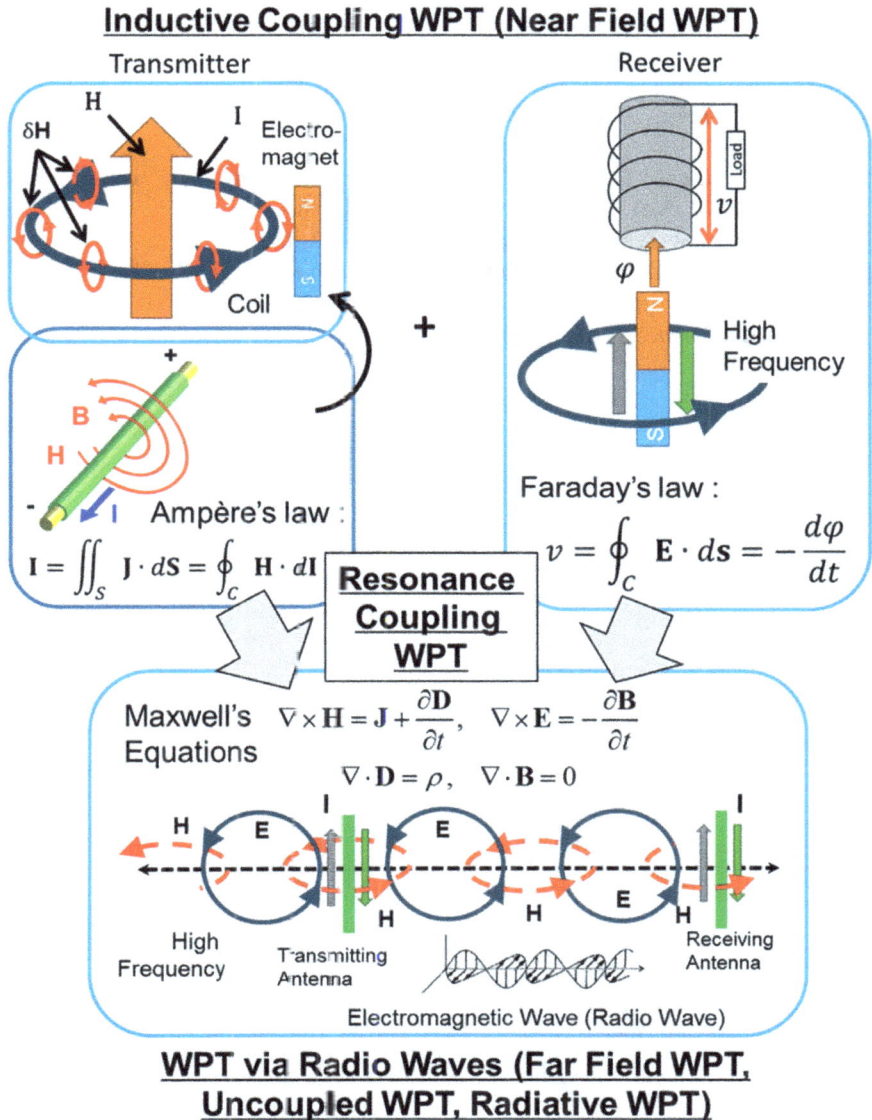

$$I = \iint_S \mathbf{J} \cdot d\mathbf{S} = \oint_c \mathbf{H} \cdot d\mathbf{I}$$

Faraday's law :
$$v = \oint_C \mathbf{E} \cdot d\mathbf{s} = -\frac{d\varphi}{dt}$$

Resonance Coupling WPT

Maxwell's Equations
$$\nabla \times \mathbf{H} = \mathbf{J} + \frac{\partial \mathbf{D}}{\partial t}, \quad \nabla \times \mathbf{E} = -\frac{\partial \mathbf{B}}{\partial t}$$
$$\nabla \cdot \mathbf{D} = \rho, \quad \nabla \cdot \mathbf{B} = 0$$

WPT via Radio Waves (Far Field WPT, Uncoupled WPT, Radiative WPT)

Figure 1.6   Relationship between Ampere's law, Faraday's law, Maxwell's equations, inductive-coupling WPT, resonance-coupling WPT, and WPT via radio waves

References

[1]   N. Tesla, "The transmission of electric energy without wires." The Thirteenth Anniversary Number of the Electrical World and Engineer, March 5, 1904.

[2]   M. Hutin and M. LeBlanc, "Transformer system for electric railways," US Patent Number 527,875, 1894.

[3]   H. Yagi and S. Uda, "On the feasibility of power transmission by electric waves." *Proc. 3rd Pan-Pacific Science Congress*, Vol. 2, 1926, pp. 1307–1313.

[4]   W. C. Brown, "The history of power transmission by radio waves", *IEEE Trans. MTT*, Vol. 32, No. 9, pp.1230–1242, 1984.

[5]   D. Otto, New Zealand Patent Number 167,422, 1974.

[6]   A. Kurs, A. Karalis, R. Moffatt, J. D. Joannopoulos, P. Fisher and M. Soljačić, "Wireless power transfer via strongly coupled magnetic resonances", *Science*, Vol. 317, pp. 83–86, 2007.

Chapter 2

Basic theory of inductive coupling

Hidetoshi Matsuki[1]

2.1 Introduction

The noncontact electricity transmission method for supplying electricity to an electrical apparatus is based on Faraday's law and is a recent (less than 20 years old) development in the field of household appliances. The inductive coupling method is based on electromagnetic induction and was developed by M. Faraday in 1831.

A recent application of the technology is the wireless power transfer (WPT) system. Based on the current energy situation, interest in electric vehicles (EVs) has increased, and WPT has attracted attention as a next-generation technology for charging of EV batteries. The noncontact electricity transmission technique has also attracted attention for medical applications. This chapter explains the basic characteristics of WPT systems and the basic theory of inductive coupling [1–7].

2.2 Wireless power transfer system (WPT system)

2.2.1 Basic theory of WPT system

When electrical energy from an AC power supply such as a commercial power supply or from a DC power supply such as solar cells is converted to high-frequency electrical energy by using a high-frequency inverter, the wireless feeding device (Tx.) releases electrical energy through a transmission device into space. Then, the receiving system (Rx.) converts the electrical power into DC in the recipient electrical appliance.

A key feature of noncontact electricity transmission is the distribution of electromagnetic energy over space by using an electromagnetic field or wave. Such transmission requires an electricity conversion apparatus such as a high-frequency inverter or converter. Therefore, improvement in the efficiency of the electrical power transmission of these converters is also important.

---

[1]Emeritus Professor of Tohoku University, Japan

Since electromagnetic energy is distributed over space, considering the medical and environmental influence of electromagnetic waves is important [8–10].

Furthermore, the improvement of the battery characteristics is essential; battery charging technology such as that for portable devices is thought to be one of the primary applications of WPT technology. For such improvement, facile high-speed charging, circuit-like manageability, safe security, and energy storage density are important.

WPT systems are primarily classified as microwave, evanescent wave, magnetic resonance, electrical resonance, or electromagnetic induction methods.

In this chapter, we discuss noncontact electricity transmission methods in an electrical power range of milliwatts to hundreds of kilowatts, using an electromagnetic wave or an electromagnetic field of a frequency band that is lower than that of a microwave. In addition, three methods, including magnetic field resonance, electric field resonance, and electromagnetic induction, are termed electromagnetic field resonance as discussed in a subsequent paragraph.

Researchers have recently suggested an electromagnetic induction method that is not premised on transformer coupling. It is revealed that the electric power transmission over a range including a magnetic field resonance method is enabled by adopting this method.

As shown in Figure 2.1, the electromagnetic field is distributed within the space between the electrical power transmission appliance (Tx.) and the power-receiving device (Rx.). The spatial distribution pattern varies in accordance with the distance from the Tx. Over approximately one wavelength, the mode of the spatial pattern of the electromagnetic field distribution changes.

Figure 2.2 illustrates the spatial distribution established in the space surrounding the power-releasing device (Tx.). In Figure 2.2, the zebra pattern represents the distance showing one-half wavelength. The spatial domain ranging (approximately) one wavelength is termed the near-field region. The range that encompasses more than one wavelength is termed the far-field region.

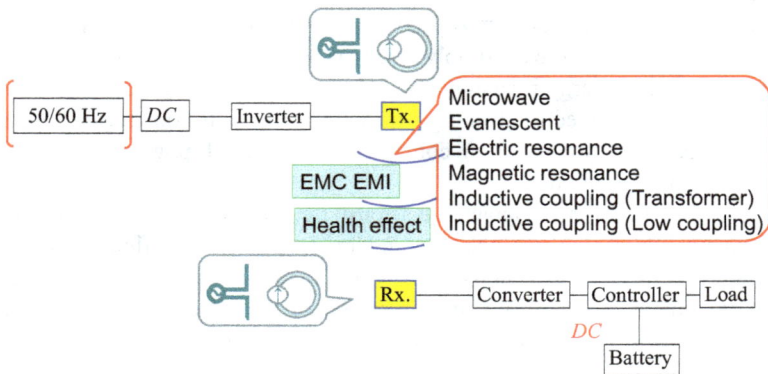

Figure 2.1   Block diagram of wireless power transfer system

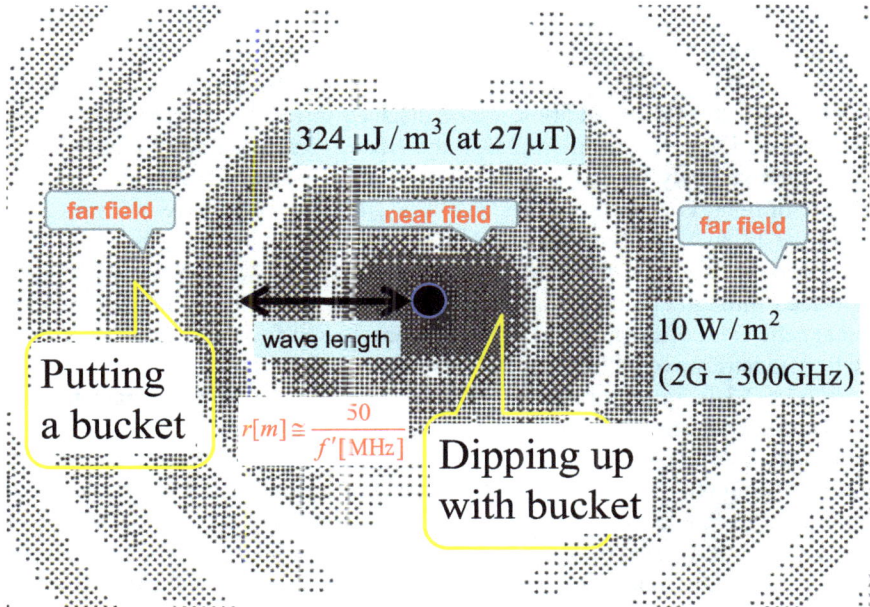

Figure 2.2   Spatial distribution of a magnetic field established in the space surrounding the power-releasing device

In the far-field region, the electromagnetic wave "flies away." In this region, the electric field and magnetic field vectors are at right angles, and these vectors constitute a Poynting vector, which shows the electromagnetic energy flow. In these domains, the energy radially propagates at high speed. A figurative high-pressure water current, in which the energy is compared to a water current, splashes in all directions in these domains.

Therefore, only a bucket (which catches water and becomes the electrical antenna) is required to figuratively receive water (energy) in a location separate from a transmission device. The water inherently jumps into a bucket. However, the bucket has to be devised such that water does not splash into the neighborhood. In other words, an antenna shape (circuitry design to reduce reflection) is important.

An energy density of 10 W/m² (equivalent to the aforementioned water current quantity) in the far field is the value of the Poynting vector that ICNIRP [8] regulates in a frequency range of more than 2 GHz to prevent adverse medical effects.

Because the far field does not grow up enough in the space regions from the transmission source to near one wavelength, energy to change in terms of time is accumulated in these spaces. In other words, there is negligible flow of electromagnetic energy in the near-field region. Figuratively, it is the domain where the quantity of water accumulated in space changes only in terms of time. The "mean volume of water available" is zero at time. This is because the stockpile fluctuates periodically with time. In other words, a mechanism is necessary for a bucket to draw up water. The electrical power transformer functions in such a manner.

A space energy density of 324 µJ/m³ is regulated by ICNIRP in a frequency range up to 100 kHz. In addition, space distance in meter, 50 over frequency in MHz, which is shown in Figure 2.2 is the size of the near-field region which is found approximately by supposing the propagation speed of the energy to be the velocity of light. For example, the size of the 2.45 GHz microwave becomes 2 cm, and of the 13.56 MHz RF wave is 3.7 m.

Based on this, WPT to the IC chip card using the RF wave is known to not be performed through an electromagnetic wave of the far field but through an electromagnetic field of the near field. The antenna design for an IC chip based on an electromagnetic wave will clearly not provide a useful result.

Figure 2.3 is an elaboration of Figure 2.2 using an electrical parameter. The WPT method can be classified by the electrical characteristics of the transmission source into two approximate types: (1) electrical field source transmission by a constant voltage (CV) source, where CV refers to the value of the root-mean-square amplitude in accordance with the sinusoidal alternating voltage and (2) magnetic field source transmission by a constant current source, where constant current refers to the value of the root-mean-square amplitude in accordance with the sinusoidal alternating current.

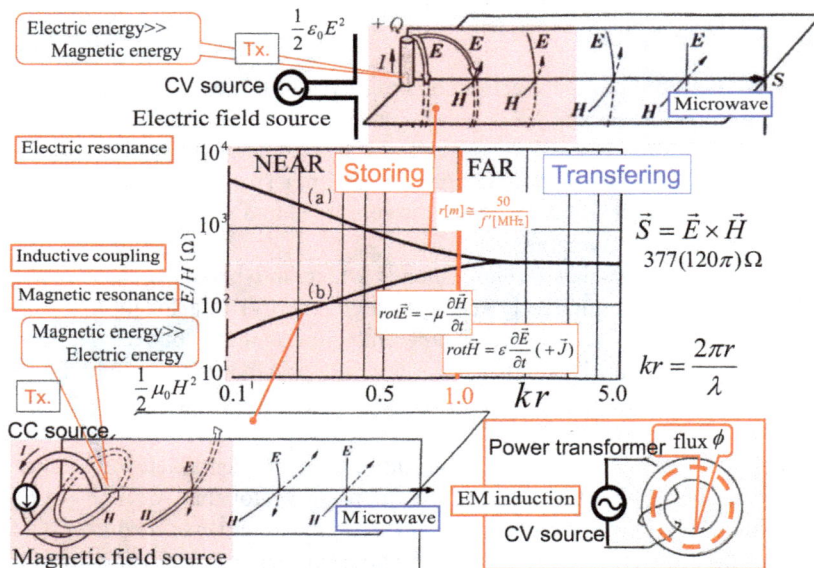

Figure 2.3 *Spatial distributions of the electric and magnetic fields established in the space surrounding the power-releasing device by using an electrical parameter*

An alternating electric field occurs in space because of a CV source in the electric field source transmission and only electrical field energy accumulates in the neighborhood of the transmission source. As a result, the spatial impedance becomes high (only if a displacement current is definitely ignored). Some of the energy accumulates as magnetic field energy in the place that distance left from the source, whereas the remainder decreases as electric field energy. On the other hand, the magnetic field energy gradually increases. Consequently, the impedance of the space decreases.

Conversely, an alternating magnetic field occurs in space because of a constant current source in the magnetic field source transmission, and only magnetic field energy accumulates in the neighborhood of the transmission source. As a result, the spatial impedance decreases. Some of the energy accumulates as electrical field energy in the place that distance left from the source due to Faraday's law, whereas the remainder decreases as magnetic field energy. Conversely, the electric field energy gradually increases. Consequently, the impedance of the space increases (when displacement current is considered it is restrained, and the rise of value gets closer to a constant value). The units of accumulation energy (i.e., the spatial density of the electrical or magnetic energy) are joules per cubic centimeter.

In both the cases, electric field source transmission and magnetic field source transmission, electromagnetic energy accumulates in space, and the quantity of accumulated energy changes with time. But both cases do not have the property as "the energy transmission." Therefore, an active mechanism for absorbing energy is needed in the "receiving side" to transfer the energy to the load.

In contrast, a magnetic field and an electric field following the Maxwell equations are inductively generated in space each other as the far distance from a power supply. As a result, both the electrical field and magnetic field distribution are induced only by the spatial distribution just before each and will be decided. Consequently, the spatial distribution of the electromagnetic field in the electric field source transmission and that in the magnetic field source transmission becomes similar.

It shows the same space impedance (377 Ω) to be decided at the dielectric constant ε_0 of the vacuum and magnetic permeability μ_0 in vacuum.

The accumulated energy in space transmits at the velocity of light in this region, and the quantity of energy transmission can be expressed in terms of the value $|\vec{E} \times \vec{H}|$ of the Poynting vector. The units are watts per square meter.

Representative methods for WPT are classified as being either the microwave, magnetic field resonance, electric field resonance, or the electromagnetic induction methods. The microwave method utilizes the electromagnetic far field for WPT, whereas the magnetic field resonance, electric field resonance, and electromagnetic induction methods are WPT methods that primarily utilize the near field. The evanescent wave method utilizes the near field for energy infusion and instead uses the far field for energy propagation.

The electromagnetic induction method represented by a power transformer in Figure 2.3 is included in neither the electric field source transmission nor the magnetic field source transmission. The power transformer works with a mode

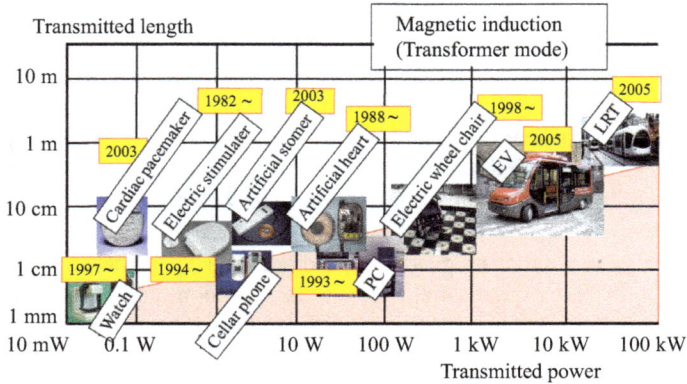

Figure 2.4   Classification of devices targeted for WPT applications

illustrated in the lower right-hand image of Figure 2.3. In other words, in a power transformer, the peak value of the primary alternating magnetic flux ϕ in a magnetic core is always kept constant by the CV source, even if the secondary reaction magnetic flux caused by the secondary load current flowing in the secondary winding interlinks with the primary winding in order to disturb a change of the primary magnetic flux.

Then, the new magnetomotive force flows in the primary winding to cancel the magnetomotive force generated by the secondary load current in the secondary winding. As a result, the electric power flowing into the primary terminal is conveyed to the load connected at the secondary terminal and is consumed as the secondary electrical power.

Therefore, representative examples of the noncontact electricity transmission method based on electromagnetic induction often include a transformer, but it may instead be said that the transformer propagates the electrical power by using a unique mechanism (i.e., noncontact electricity transmission).

Figure 2.4 shows devices selected for WPT that have been developed to date for electricity transmission. The transmission distances are also shown. Electrical power of 150 kW grade has already been reported with the thing of the trial manufacture stage by 10 mW grade.

Figure 2.5 shows suggested primary noncontact electricity transmission methods in accordance with WPT. Electrical power transmission using microwaves enables transmission over macroscale wavelengths. Electrical waves on the macroscale are of low power (0.1 W or less) and are tightly regulated due to health concerns. As a result, one must assign a strict upper limit to the energy density that can propagate.

2.2.2   Microwave method

The microwave method primarily uses a microwave (electrical wave) of 2.45 GHz and is a transmission method that uses the far-field region. In other words, without

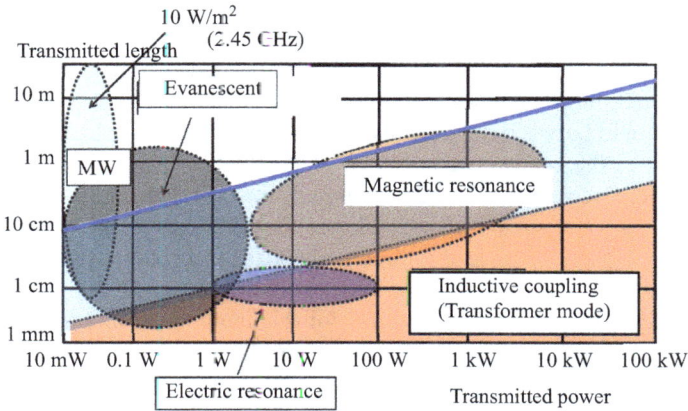

Figure 2.5   Range of suggested primary noncontact WPT methods

considering the consequences for human health, the microwave method is superior within this limit. A limiting level is the value of the Poynting vector (10 W/m^2).

A 10 km microwave transmission device that transmits energy of 1 million kW can be conceived. Figure 2.5 shows portable device sizes. In this method, antenna design and improvement of the rectifying circuits' efficiency are central to practical use.

2.2.3   Magnetic resonance method

The MIT proposed the magnetic field resonance method in 2007. This method can transmit the electrical energy and can postpone distance till the space by "the coil" of the shape that the electric field produces to use an opening electric field. It is a transmission method by the LC resonance using the inductance by the magnetic field and the capacitance by the electric field. A coil design and a resonance circuit design are central to practical use. The frequency band that is able to make use of the advantages provided by the magnetic field resonance method seems to be MHz obi, but "making it is desirable a low frequency" for the rise in efficiency, and a 100 kHz zone is pointed to when I intend by transmission electricity increase. As a result, a difference with low coupling electromagnetic induction method mentioned later becomes uncertain.

2.2.4   Electrical resonance method

This method essentially includes an electrical noncontact component in the interior of the condenser, and electrical energy transmits at the place by a displacement current. The electrical field resonance method utilizes this fact, and it is the transmission method that uses the energy accumulated in the electrical field for noncontact electricity transmission. In this method, the transmission efficiency improves by using the LC resonance phenomenon with inductance as the circuit element.

666666666666

As for the electric field resonance method, the energy transmission distance in the air is limited because of uses of electric field on a noncontact–contact transmission part. The electric field resonance method is, nevertheless, considered to be advantageous. A large pole plate area is necessary because of the need for a uniform electric field between the pole plates when a voltage is applied to one point of the pole plate of the condenser.

In addition, regarding the circuit impedance, the electrical field resonance method has characteristics that are suitable for modulating the large electricity transmission. This is because the circuit impedance decreases to a larger extent when one uses a higher frequency or larger pole area.

A resonance circuit design is central for this method. Such a design is advantageous in terms of electricity transmission of the narrow gap and for electrical field use turns for application such as the noncontact electrically even if the devices come into contact mechanically.

2.2.5 Electromagnetic induction method

The electromagnetic induction method involves dipping up the magnetic field energy using a coil. Here, the magnetic coupling between the coils is high. An interlinkage between the common magnetic flux with both coils for electrical energy transmission is important. In particular, it is a method to do the electrical power transmission mechanism of the power transformer for electricity.

In this manner, naturally bigger coils will be used if it becomes the larger electricity transmission. Accordingly, the transmitting distance increases in accordance with increasing power transmission. Recently, an LC type of electromagnetic induction (the LC booster method) to only use magnetic field energy formed in space for electricity transmission was developed without using a space spread electric field.

Unlike the magnetic field resonance method, the characteristics of this method have the following two features: the electromagnetic induction method not being a limit of the frequency and the limit of the coil shape and to work by low magnetic coupling.

Figure 2.5 shows the upper limit area, mentioned previously, in which the LC type of the magnetic induction method transmits electrical energy.

The electricity transmission distance when this method is used may be almost equal to that observed when the magnetic field resonance method is used. Since electrical power transmission is possible even if the coupling coefficient is several percent, the common perception is that these two methods perform equally well in practical use.

Because this electromagnetic induction method does not use an electric field of the space, the transmission distance falls by comparing with a magnetic field resonance method essentially. Both methods show equal transmission performance at frequencies below 100 kHz. Because a coil is put at the position bathing in a linkage magnetic field directly by this method, it is with a key that the choice of the wire rod and Ritz wire.

2.3   Magnetic induction

2.3.1   Power transformer

The electromagnetic induction method is the method of dipping up magnetic field energy by using a coil. Figure 2.5 assumes transformer coupling. In this manner, inherently larger coils will be used if it becomes the larger electricity transmission. Accordingly, the transmission distance increases in accordance with increasing power transmission. Furthermore, the area shown in Figure 2.5 is obtained from proof results shown in Figure 2.4.

Figure 2.6 illustrates an equivalent circuit for an electrical power transformer having a magnetic core. In the power transformer, an alternating magnetic flux in a magnetic core, which is always kept at a fixed maximum value relative to the alternating magnetic flux by a CV source, conveys electrical energy to a second side winding. Then, the electrical energy is transmitted. The load connected to a secondary coil consumes the electrical energy. It is desirable to maintain the coupling factor at 100%.

The condenser connected to the primary side mainly serves to compensate for reactive power or a power factor drop by a load connected to the second side. In the power transformer variant of the electromagnetic induction method, the circuitry

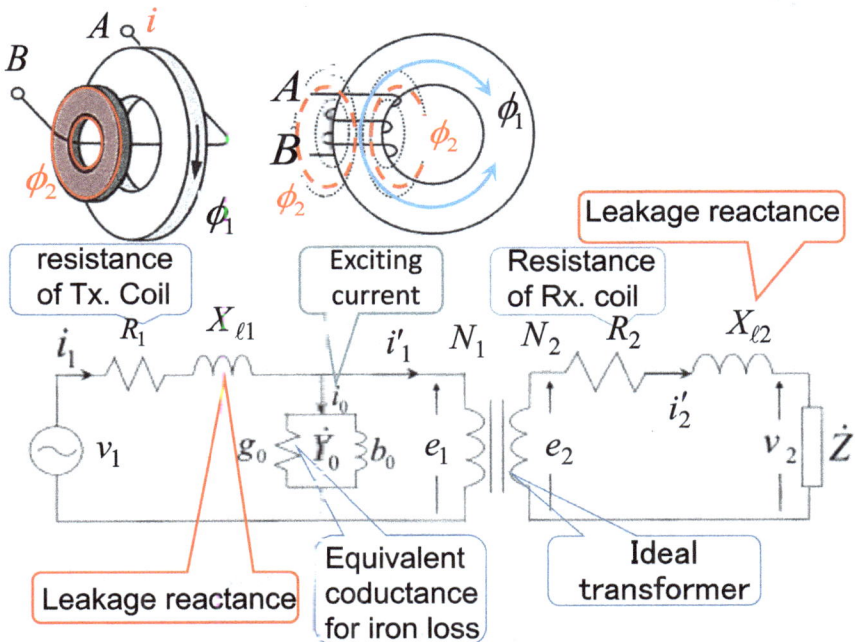

Figure 2.6   Equivalent circuit for an electrical power transformer comprising a magnetic core

design, including the condenser, has been performed from the point of view called the coil axis gap compensation between the primary and second coils.

In this manner, the transformer variant of the noncontact electricity transmission method points at a method to work under the circuit condition termed the CV drive in a method classified as an electromagnetic induction method. Of course, we will present a method that operates under other conditions in a subsequent section.

Consider the operation of the power transformer for electricity.

As mentioned above, Figure 2.6 depicts an equivalent circuit generally used for the analysis of the power transformer for electricity. In Figure 2.6, $X_{\ell 1}$ is the leakage reactance of the transmission side. This reactance represents the magnetic flux ϕ_2 which does not link with the power-receiving circuit. Similarly, $X_{\ell 2}$ is the leakage reactance of the power-receiving side. R_1 and R_2 are the winding resistances of the power transmission and receiving coils, respectively. Y_0 is the exciting admittance, and g_0 is the exciting conductance representing the iron loss of the core material.

The restraint of the winding resistance is clearly important. Since a magnetic core is used in the transformer for electricity, the coupling factor between the transmission and receiving coils becomes approximately 100%, and the leakage reactance becomes approximately zero. As a result, R, $X \cong 0$. It can supply a CV for any electrical load by connecting a CV power supply to the primary circuit of the transformer, and the electricity transmission efficiency can be nearly 100%. In the power transformer for electricity, winding resistance and the influence of the leakage reactance are dealt with as the voltage fluctuation that is slight (about 1%) in the load terminal.

Conversely, regarding the discussion on the efficiency of the power transformer for electricity, we did not consider the voltage change in the electrical load terminal.

Therefore, the copper loss that is equal to iron loss will produce the efficiency maximum point in load to occur, and the greatest efficiency's concept does not exist in the analysis of the transformer for electricity when using air core without iron loss.

It becomes the constitution such as Figure 2.7 by applying the concept of the transformer for electricity to WPT. However, the magnetic core often becomes the air core.

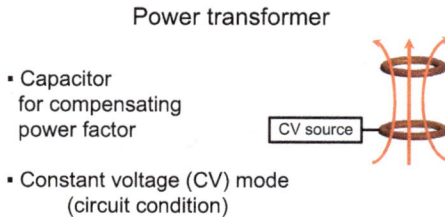

Figure 2.7   Schematic of a WPT based on a power transformer

In this circuit, the leakage reactance increases compared with a power transformer for electricity to realize a high coupling coefficient using a magnetic core. Even if the circuit is driven by a CV, the source voltage change at the load terminal is not avoided. As a result, the condition using a condenser for compensation is not considered in the concept such as the power factor compensation in a network analysis.

Consider an equivalent circuit of the transformer constitution having an air core as shown in Figure 2.8, including twin coils facing each other. The leakage reactance of the circuit under consideration increases as a result ratio of magnetic flux which does not link each other increase.

By assuming a circuit having an air core, the excitation conductance to express iron loss does not exist. It becomes only the coil resistance to bring a power loss.

Figure 2.9 shows the scenario in which the efficiency is maximized using the equivalent circuit shown in Figure 2.8. When this condition is attained, a terminal voltage change is produced by the load connected, which is a point to be considered. At this point, the load is optimal. This quantity becomes the capacitive impedance shown in Figure 2.9.

The main factor to reduce efficiency is inductive impedance to express leakage reactance to understand it from Figure 2.9. If this impedance is canceled by a condenser and if the compensated voltage descends in the leakage reactance, the applied voltage drop in load will be compensated. As a result, a drop in the efficiency of electricity transmission can be prevented.

Equivalent circuit for power transformer (High efficiency)

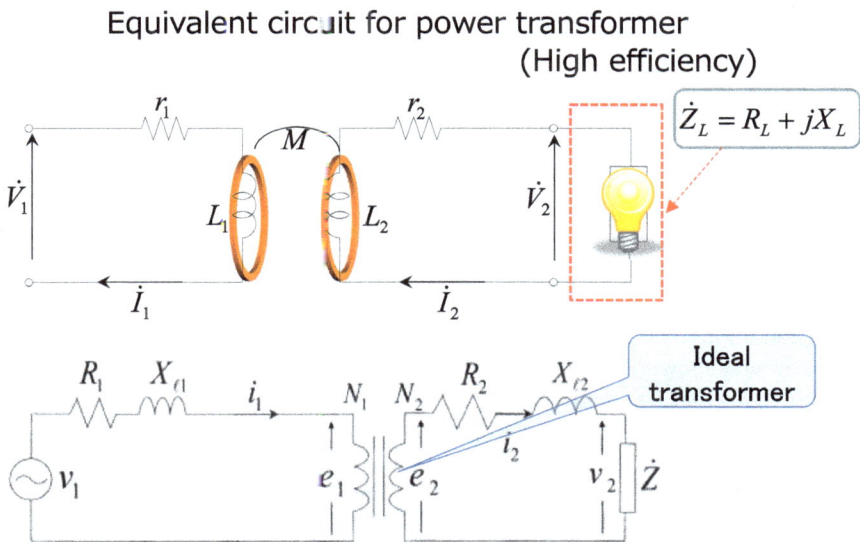

Figure 2.8   An equivalent circuit of the transformer comprising an air core

Optimal load impedance : $\dot{Z}_{Lm} = R_{Lm} - jX_{Lm}$ (no core)

Optimal resistance   : $R_{Lm} = \sqrt{r_2^2 + \dfrac{r_2}{r_1}(\omega M)^2} = r_2\sqrt{1+\alpha}$

Optimal reactance   : $-jX_{Lm} = -j\omega L_2 \to \dfrac{1}{j\omega C}$ $i.e.\ C \equiv 1/\omega^2 L_2$

Maximum efficiency: $\eta_{\max} = \left[1 + \dfrac{2(1+\sqrt{1+\alpha})}{\alpha}\right]^{-1}$
for optimal load

$\alpha \equiv \dfrac{(\omega M)^2}{r_1 r_2} = k^2 Q_1 Q_2$ $\begin{aligned}\dot{Z}_{Lm} &= R_{Lm} - j\omega L_2 \\ &= R_{Lm} + \dfrac{1}{j\omega C}\end{aligned}$

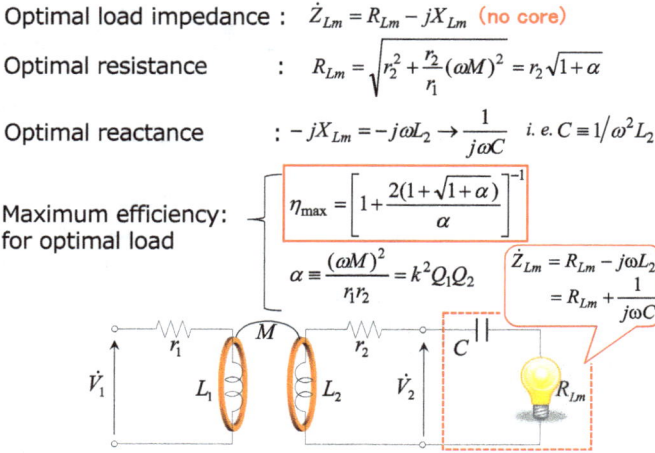

Figure 2.9   Maximal efficiency using an equivalent circuit

Furthermore, by introducing a circuit parameter α

$$\alpha = k^2 Q_1 Q_2 \tag{2.1}$$

where k is the coupling factor of a coil pair, Q_1 is the Q value of power transmitting coil, Q_2 is the Q value of power-receiving coil.

The relations between the maximum efficiency $\eta_{\max}$ and the parameter α can be expressed:

$$\eta_{\max} = \frac{1}{1 + \frac{2(1+\sqrt{1+\alpha})}{\alpha}} \tag{2.2}$$

Figure 2.10 shows an example indicating the relation between the optimal load to maximize the transmission efficiency and the maximum efficiency level along with the corresponding coil properties. The maximal efficiency can be expressed in terms of the parameter α, defined as each Q level of the transmission and power-receiving coil in the product of the square of the coupling factor.

A condition to be provided in Figure 2.10 is the condition that is common to all of the circuits shown in Figure 2.8, and it may be said that it applies to all of the WPT methods that use multicoils.

Therefore, these relations regulate both the electromagnetic induction and magnetic field resonance methods. It will be a reason to refer to electromagnetic induction method and magnetic resonance method as 'magnetic coupling method'.

2.3.2   Magnetic induction (LC mode)

In reference to the operation of the power transformer to generate electricity, consider a method to plan a technological advance in WPT. In a preceding

Maximum efficiency condition for magnetic induction

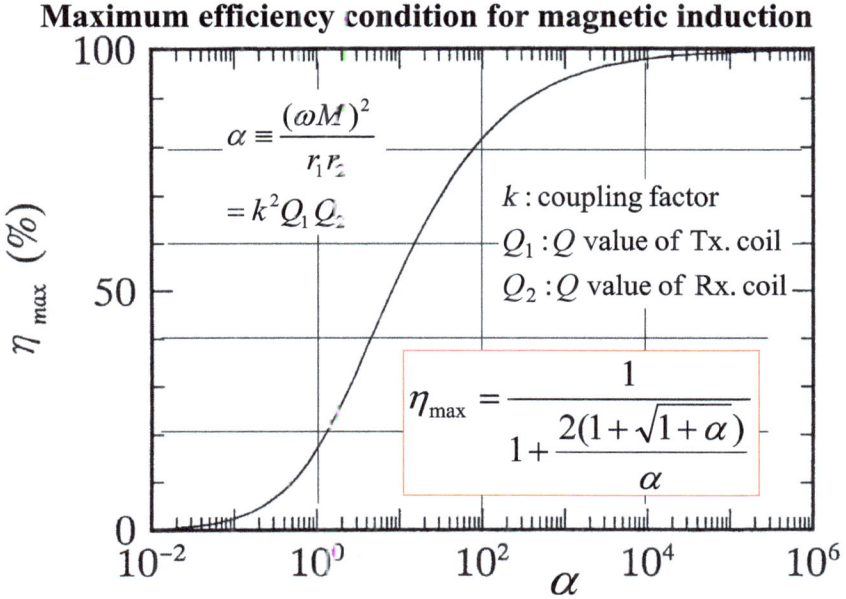

$$\alpha \equiv \frac{(\omega M)^2}{r_1 r_2}$$

$$= k^2 Q_1 Q_2$$

k : coupling factor
$Q_1 : Q$ value of Tx. coil
$Q_2 : Q$ value of Rx. coil

$$\eta_{\max} = \frac{1}{1 + \dfrac{2(1 + \sqrt{1 + \alpha})}{\alpha}}$$

Figure 2.10   An example indicating the relation between the optimal load to maximize the transmission efficiency and the maximal efficiency level, along with the corresponding coil properties

paragraph, it is discussed that handling of the leakage reactance is a focus with winding resistance in WPT.

Consider an example of doing so. For example, a method to insert a condenser in the equivalent circuit as shown in Figure 2.11 is considered without the influence of the leakage reactance. The circuit is known to become the short-circuit state in the series resonance circuit. In other words, the terminal voltage between terminals a and c in Figure 2.11 becomes zero. In this manner, the transmission voltage is directly applied to the power-receiving terminal without leakage reactance, where the winding resistance is neglected.

Furthermore, under the series resonance condition, the voltage between terminals b and c becomes Q times as large as the voltage e_2, and the voltage between terminals a and b becomes Q times the voltage having the opposite phase.

By using the Q times voltage v_2 occurring in terminal bc between the leakage reactance $X_{\ell 2}$, as shown in the upper hand image of Figure 2.12, electrical power can be efficiently transmitted. The lower image of Figure 2.12 shows the equivalent circuit. Here, we show the WPT method using three coils in total in conjunction with a transmission and reception power coil. We consider it to be a circuit to compensate for the decrease of the coupling factor in accordance with the resonance voltage. Even if the transmission device is one electric wire, the operation on the equivalent circuit occurs in the same manner as long as the value of the coupling factor between the transmission device and the receiver is constant.

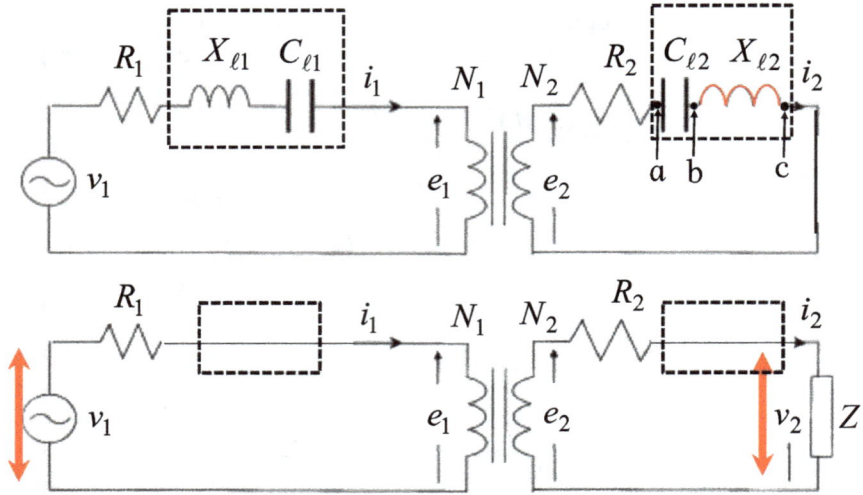

Figure 2.11   *Short-circuit state in the series resonance circuit in which the terminal voltage between terminals a and c becomes zero*

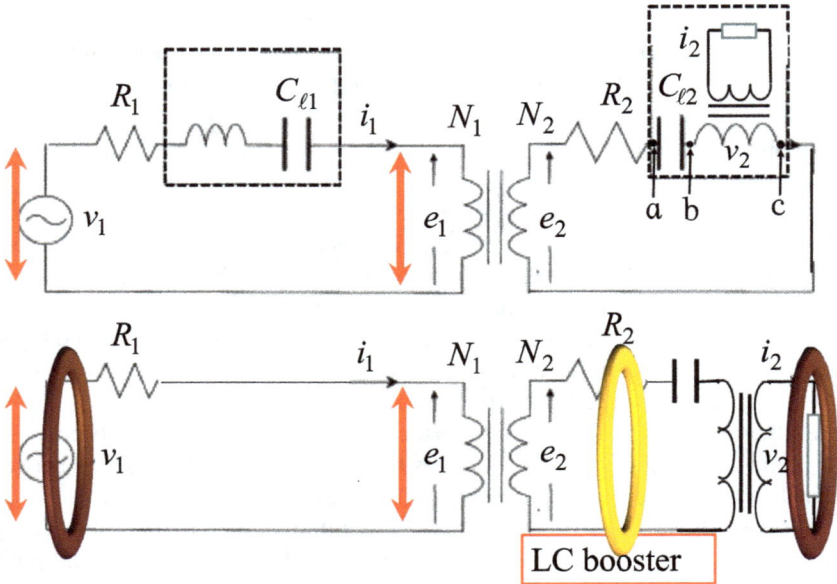

Figure 2.12   *Concept of compensating for the decrease of the coupling factor in accordance with the resonance voltage*

According to this analysis, the minimum value of the coupling factor that is necessary for WPT is approximately several percent. We term this WPT method the LC booster method.

As mentioned above, a voltage change occurring at the receiving side when load is fluctuated needs to be considered. When a noncontact energy transmission system shows the characteristics such as the general-purpose power supply.

In such a case, we have to supply electric power to a load unlike the load to realize the largest efficiency. Maintaining the maximum efficiency during operation is expected to be difficult. It is necessary to perform the design that takes efficiency, both the voltage changes into consideration by using capacitance for the transmission side.

Figure 2.13 compares the features of the high coupling type in the electromagnetic induction method with that of the characteristics of the LC type. The high coupling type assumes so-called transformer coupling, and it is clearly desirable for the primary magnetic flux to completely link to the secondary side coil. It is a noncontact electricity transmission method using magnetic flux, but a CV source is used for power supply. In the LC type of electricity transmission, the magnetic field space of the low energy density is established around the power transmission side by a constant current source and located LC resonator (LC booster) in the neighborhood of the second side (receive an electric side) coil. The combination with an LC booster and the second side coil is a high coupling type (i.e., transformer type). Several percent is enough for coupling factor between a primary coil and the secondary coil. A high Q level is necessary for increasing the power transmission efficiency.

Both the high coupling and LC types of electricity transmission are based on Faraday's electromagnetic induction law. Actually, transformer type power transmission is used in Faraday's law, but all of the electromagnetic induction method is not a transformer.

Figure 2.14 likens the energy flow in high coupling and LC power transmission to water flow. We compare the former to a system in which a small-capacity tank is virtually united with a large-capacity tank having a fundal pipe. Water collected in the large-capacity tank moves to a low-volume tank in a state of constant water

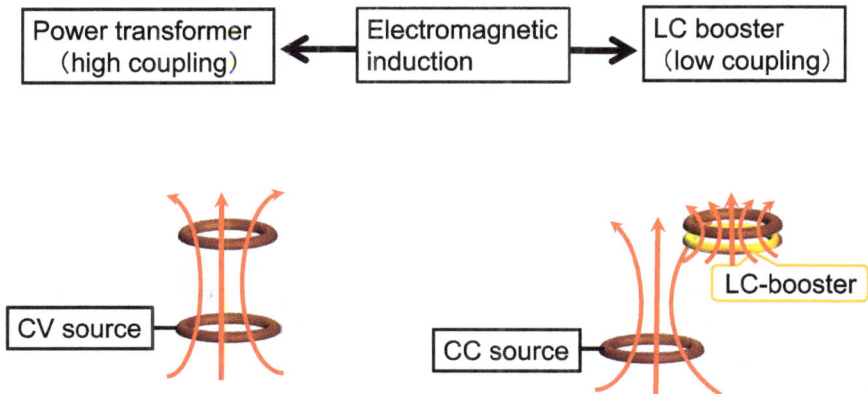

Figure 2.13   Comparison of the features of the high combination type of electromagnetic induction method with that of the low combination type

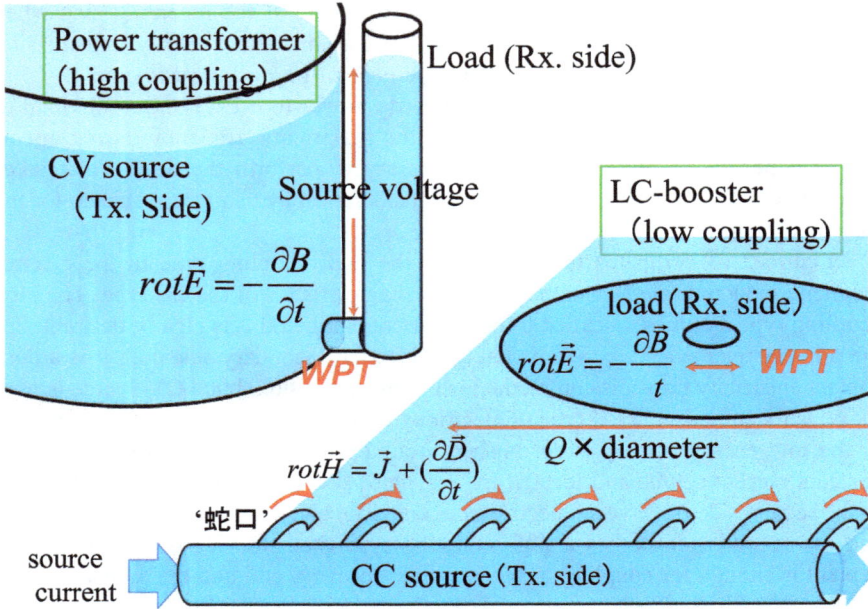

Figure 2.14   Energy flow of the high and low combination types of power transmission

pressure. When the position of the pipe slips off, it will be hardly transmitted water well.

The LC type is equivalent to the state in which a water supply pipe emptied the hole into is put on the ground, and the water that flowed out of the hole floods the ground (leak water). An outlet port of the size is assumed to become Q times by LC booster. Then, water flooded on the ground is efficiently drained out.

The position of the real outlet port is unimportant.

2.4   Medical applications

WPT is well-suited for medical applications. The electromagnetic induction method lends itself to this field, but the magnetic field resonance and electromagnetic induction methods of the LC type are under active, widespread study.

Regarding internal devices, one should avoid sending energy through a cable implanted in the skin from the standpoints of facilitating daily care and minimizing infections. Researchers and practitioners classify implantable medical devices in terms of the energy output, such as kinetic power (e.g., an artificial heart, sphincter, or esophagus) and electrical power (e.g., a cardiac pacemaker, a defibrillator, or an electrical stimulator).

The magnetic field resonance method and the electromagnetic induction method of the LC type are expected to attract attention in the fields of medical care and welfare in the future. It is in particular thought that the need of the wireless feeding technology also increases as the introduction such as robots by the battery drive.

References

[1] T. Matsuzaki and H. Matsuki: Properties of Transcutaneous Electrical Power Transmitting Coils for Functional Electrical Stimulation (FES), *J. Magn. Soc. Jpn.*, 18, 663–666 (1994).

[2] T. Takura, Y. Ota, K. Kato, *et al.*: Relationship between Efficiency and Figure-of-merit in Wireless Power Transfer through Electromagnetic induction, *J. Magn. Soc. Jpn.*, 35, 132–135 (2011).

[3] T. Misawa, T. Takaura, F. Sato, T. Sato, and H. Matsuki: Study of Material Selection of Litz Wire for Contactless Power Transmission, *J. Magn. Soc. Jpn.*, 37, 89–94 (2013).

[4] T. Takura, T. Misawa, F. Sato, T. Sato and H. Matsuki: Maximum Transmission Efficiency of LC-Booster Using Pick-up Coil with Capacitance, *J. Magn. Soc. Jpn.*, 37, 102–106 (2013).

[5] H. Yamaguchi, T. Takura, F. Sato, T. Sato and H. Matsuki: A Study of Primary Coil Interoperable with Heterogeneous Coil to Charge Contactless Electric Vehicles, *J. Magn. Soc. Jpn.*, 38, 33–36 (2014).

[6] T. Misawa, T. Takura, F. Sato, T. Sato and H. Matsuki: Parameter Design for High-efficiency Contactless Power Transmission under Low-impedance Load, *IEEE Trans. Magn.*, 49, 7, 4164–4166 (2013).

[7] Miura, H., Arai, S., Sato, F., Matsuki, H., and Sato, T: Synchronous Rectification for Contactless Power Supply Utilizing Mn-Zn Ferrite Core Coils, *J. Appl. Phys.*, 97, 7021–7023 (2005).

[8] ICNIRP: *Health Phys.*, 99, 818–836 (2010).

[9] ICNIRP: *Health Phys.*, 74, 494–522 (1998).

[10] ICNIRP: *Health Phys.*, 96, 504–514 (2009).

Chapter 3
Basic theory of resonance coupling WPT

Hiroshi Hirayama[1]

Wireless power transfer (WPT) technology has two origins of technical background. One is power electronics technology, in which WPT mechanism is considered a transformer. The other is radio frequency (RF) technology, in which WPT mechanism is considered a coupling of resonators.

The aim of this chapter is to unveil the essence of coupled-resonant WPT by introducing a unified model that enables us to understand power electronics-based WPT and RF-based WPT in the same manner.

In the first part of this chapter, a unified model of resonance coupling WPT is conducted. This model enables us to understand the mechanism of resonant WPT system from the viewpoint of electric/magnetic coupling and resonance.

In the later part, generalized WPT model is conducted. This model represents the whole WPT system, including an RF inverter and a rectifier. By using this model, various kinds of WPT system can be explained from the viewpoint of impedance matching, resonance, and coupling.

3.1 Classification of WPT systems

3.1.1 Classification of near- and far-field WPT

There are various ways of realizing WPT, such as far-field type (microwave WPT), magnetic induction type, and coupled-resonant type. From the viewpoint of "coupling" and "resonance," they can be categorized as shown in Figure 3.1.

For the far-field type, a load impedance of the receiver (Rx) side does not affect the state of the transmitter (Tx) side. Thus, coupling between Tx and Rx is weak. Antennas for the far-field type may have resonant mechanism (e.g., half-wavelength dipole antenna) or may not have (e.g., horn antenna).

Coupled-resonant and magnetic induction types are used in near-field region. In this type, load impedance of the Rx side effects the state of the Tx side. Thus, coupling between Tx and Rx is stronger. The coupled-resonant type has resonant mechanism, whereas the magnetic induction type does not have.

[1]Department of Electrical and Mechanical Engineering, Nagoya Institute of Technology, Japan

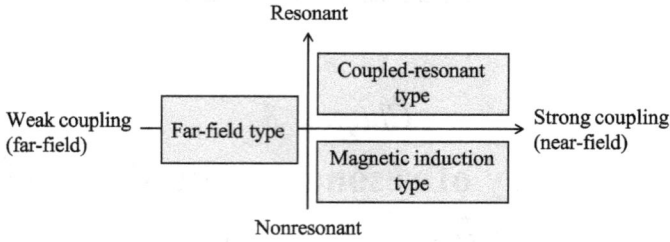

Figure 3.1   Classification of WPT systems by aspect of coupling and resonance

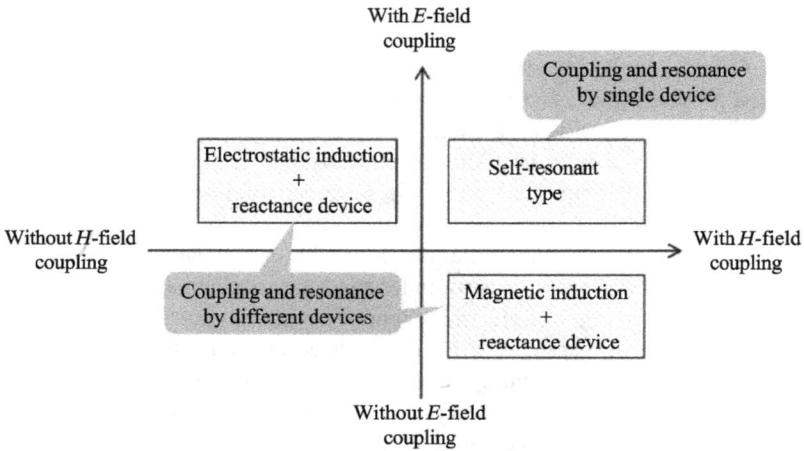

Figure 3.2   Classification of resonant WPT from the viewpoint of H- and E-field coupling

Transition between these types is ambiguous. For example, when two half-wavelength dipole antennas are settled in short wavelength distance, it becomes far-field type. However, when the distance between Tx and Rx becomes closer, it gradually becomes coupled-resonant type [1]. For magnetic induction type, by applying conjugate matching, it gradually becomes coupled-resonant type.

3.1.2   Classification of resonant WPT

Resonant WPT technologies are categorized as shown in Figure 3.2 from the viewpoint of electric- and magnetic field coupling. For understanding near-field WPT, it is better to consider coupling mechanism and resonant mechanism separately. When using only E- or H-field coupling, an external reactance device is necessary for resonance. On the other hand, self-resonant antenna has both E- and H-field coupling because resonance occurs at the frequency at which the amount of electric stored energy and magnetic stored energy becomes identical.

3.1.3 Relationship among WPT types

The relationship among nonresonant type, coupled-resonant type, and far-field type WPT is shown in Figure 3.3. In this figure, the horizontal axis represents transmission distance, whereas the vertical axis represents the role of magnetic and electric fields.

The most traditional WPT scheme is magnetic induction and electrostatic induction, shown as "Type 1" in Figure 3.3. In this scheme, the resonant capacitor may be used for power factor compensation due to a leakage flux. However, load impedance is determined by power demand of the load. As a result, when transmission distance increases, transmission efficiency decreases in accordance with a coupling coefficient.

One of the implementations of "resonant WPT" is magnetic induction with optimum complex impedance, shown as "Type 2" in Figure 3.3. The difference between Types 1 and 2 is that not only the imaginary part but also the real part of the load impedance is adjusted to achieve simultaneous conjugate matching to realize maximum power transfer efficiency for Type 2, while the real part of the load impedance is determined by power demand for Type 1. From the viewpoint of power transmission on the air, the principle of Type 2 is completely explained by magnetic induction. It is remarkable that WPT based on power electronics considers coupling coil as a part of the power conversion circuit [2]. On the other hand, in the WPT based on RF technology, RF power source and RF coupling device are designed independently, and then they are connected.

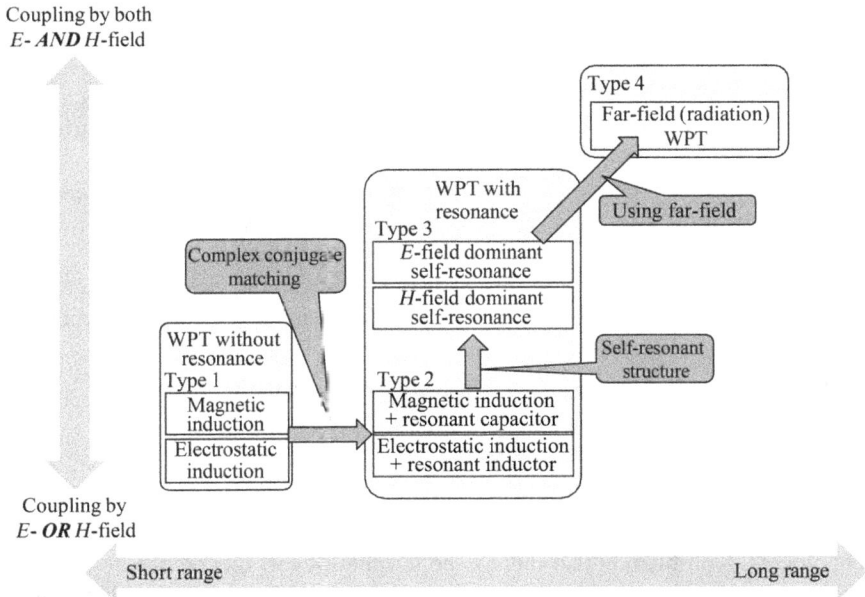

Coupling by both
E- AND H-field

Type 4
Far-field (radiation) WPT

WPT with resonance
Type 3
E-field dominant self-resonance
H-field dominant self-resonance

Using far-field

Complex conjugate matching

WPT without resonance
Type 1
Magnetic induction
Electrostatic induction

Type 2
Magnetic induction + resonant capacitor
Electrostatic induction + resonant inductor

Self-resonant structure

Coupling by
E- OR H-field

Short range   Long range

Figure 3.3   Relationship among various kinds of WPT scheme

		Coupling mechanism		Resonant mechanism	Impedance matching mechanism (feeding mechanism)	Schematic
		E-field	*H*-field			
Type 1	Electrostatic induction	Yes	No	Power factor compensation may be considered a resonant circuit	Not active following for load impedance	
	Magnetic induction	No	Yes			
Type 2	Coupled-resonant using electrostatic induction	Dominant	Negligible	Discrete reactance device is necessary for resonance	According to the load impedance or transmission distance, active following by circuit parameter in impedance matching circuit or transmission frequency is necessary to achieve simultaneous conjugate matching	
	Coupled-resonant using magnetic induction	Negligible	Dominant			
Type 3	Coupled-resonant with self-resonant coupler (*E*-field dominant)	Dominant	Small, but not negligible	Coupler acts as resonator		
	Coupled-resonant with self-resonant coupler (*H*-field dominant)	Small, but not negligible	Dominant			
Type 4	Far-field type	Coupling in far-field. Ratio of *E*-field to *H*-field is 377 Ω		Tx and Rx antennas resonate independently	Tx/Rx antennas are matched to the source/load independently	

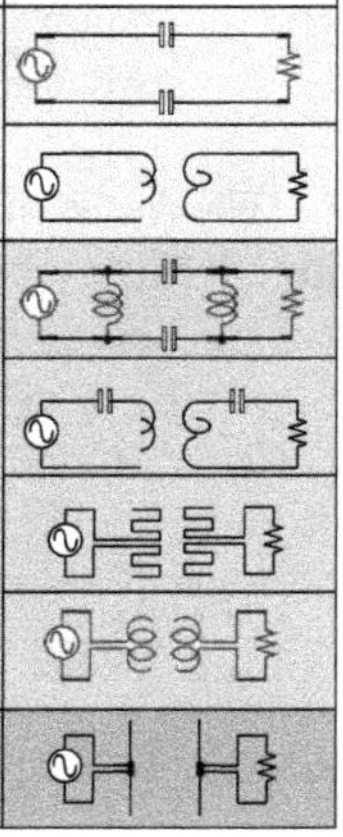

Figure 3.4 Coupling mechanism, resonant mechanism, and impedance matching mechanism for various kinds of WPT

Transition between Types 1 and 2 is ambiguous. For example, let us consider WPT for a moving vehicle using magnetic- [3] or electric field [4]. In these schemes, the receiving device on the vehicle may have resonant mechanism. However, transmitting devices under the road may not have resonant mechanism because that is just a conductor pair of transmission line.

Another implementation of "resonant WPT" is self-resonant antenna, shown as "Type 3" in Figure 3.3. As a result of the self-resonant, antenna stores electric and magnetic energy in opened space, not only magnetic field but also electric field concerns with power transfer.

WPT using far field extends transmission distance, shown as Type 4 in Figure 3.3. In this case, wave impedance becomes 377 Ω, which means power contributions of electric and magnetic fields become identical.

Figure 3.4 shows the relationship of Types 1–4 from the viewpoint of coupling mechanism, resonant mechanism, and feeding (impedance matching) mechanism.

3.2 Unified model of resonance coupling WPT

3.2.1 Concept of the "coupler"

A device to transduce power between circuit and air is the most important component in the WPT system. To integrate power electronics- and RF-based WPT technologies, a problem is that there is no terminology of this device.

For example, the term "coil" describes the shapes of the device, which cannot be applicable for electric field coupling-type WPT that uses an electrode. Furthermore, there are two meanings of the term "coil." One acts as an inductor, which is used in kHz band WPT system. Both ends are connected to the circuit. The other acts as a self-resonant resonator, which can be seen mainly in MHz band WPT system. Both ends are opened. In the circuit, this "coil" can be described as an inductor and a capacitor (LC) resonator, rather not an inductor.

The term "resonator" is also often confusing. In the kHz band WPT system, coupling coil and resonant capacitor consist of a "resonator." However, sometimes "resonator" designates only the coupling coil.

The term "antenna" is not adequate for all WPT systems because coupled-resonant WPT is used in the near-field region, not radiating the far-field region.

In this section, the term "coupler" is defined as a device that transduces electric power between circuit and air. In this definition, coupling coil, antenna, and electrode are involved in a coupler. For a resonator with a coupling coil and a resonant capacitor, only the coupling coil is a coupler. A coupler may be a self-resonant device or a nonresonant device.

3.2.2   Unified model based on resonance and coupling

Unified model of the coupled-resonant WPT [5] is shown in Figure 3.5 [6]. This model enables us to explain coupled-resonant WPT from the viewpoint of "resonance" and "coupling," which is uniformly applicable for power electronics-based WPT and RF-based WPT [7].

Power supply is connected to the transmitting (Tx) resonator. Load is connected to the receiving (Rx) resonator.

The resonator consists of a coupler and a reactance element. The coupler is defined in the previous section. A reactance element is a device that has reactive impedance, but not intended to interact between air and circuit. For example, a coupling coil is a coupler, but the coil in matching circuit is a reactance element. The coupler has capacitive and inductive reactance $X_{C_{cp}}$ and $X_{L_{cp}}$, respectively. The reactance element has capacitive and inductive reactance $X_{C_{ex}}$ and $X_{L_{ex}}$, respectively.

The phenomena "resonance" are explained as follows. The inductive reactance of the resonator consists of $X_{L_{ex}}$ and $X_{L_{cp}}$. The capacitive reactance of the resonator consists of $X_{C_{ex}}$ and $X_{C_{cp}}$. The resonance occurs at the frequency in which the

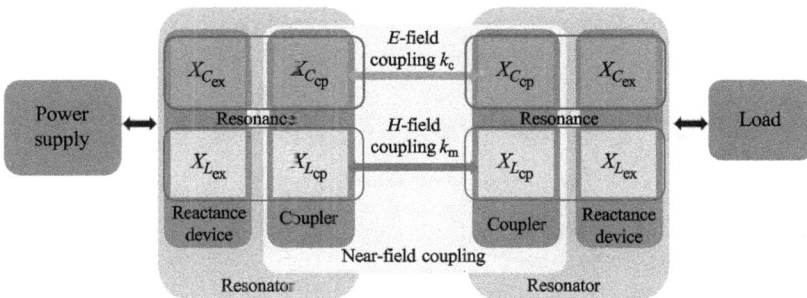

Figure 3.5   Unified model for coupled-resonant WPT

inductive and the capacitive reactance become the same. Definition of "resonant" is an amount of the magnetic stored energy, and the electric stored energy becomes the same in the resonator. The reactance device holds stored energy in a closed space, whereas the coupler holds it in an opened space.

The phenomena "coupling" is explained as follows. The capacitive reactance of the Tx resonator and the Rx resonator couple by electric field with an E-field coupling coefficient k_c. The inductive reactance of the Tx resonator and the Rx resonator couple by magnetic field with an H-field coupling coefficient k_m.

By using this model, coupled-resonant WPT can be understood from the phenomena "coupling" and "resonant" separated out.

Although the unified model shown in Figure 3.5 is a conceptual model, a numerical model and derivation of its parameters are discussed in Ref. [8].

3.2.3   Application for LC resonator

Circuit diagram using a coupling coil and a resonant capacitor is shown in Figure 3.6. By using the unified coupling and resonant model, this type of circuit topology is explained as shown in Figure 3.7. The coupler has only an inductive reactance without a capacitive reactance. Therefore, external reactance is necessary to achieve a resonance. As a result, electric field coupling is negligible between the couplers.

3.2.4   Application for electric field coupling WPT

A circuit diagram, using electric field coupling and complex-conjugate matching circuit, is shown in Figure 3.8. By using the unified coupling and resonant model,

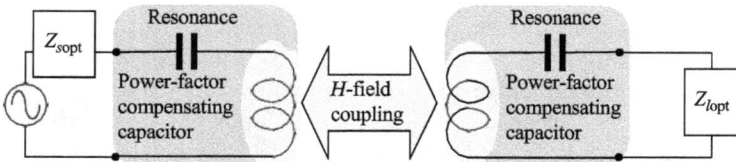

Figure 3.6   Schematic diagram of coupling coil and resonant capacitor WPT

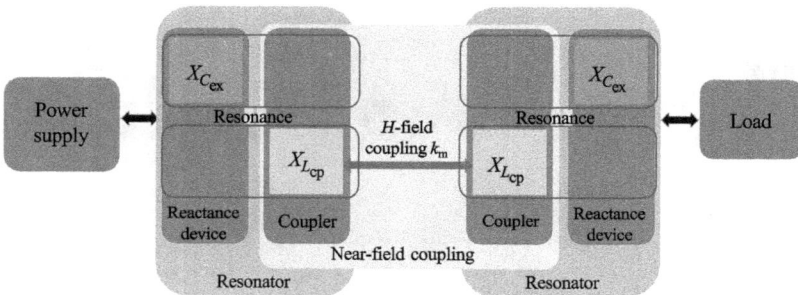

Figure 3.7   Unified model for coupling coil and resonant capacitor WPT

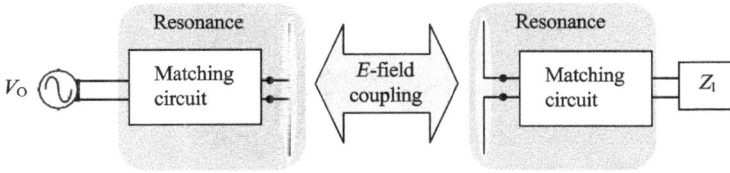

Figure 3.8   Schematic diagram of E-field coupling WPT

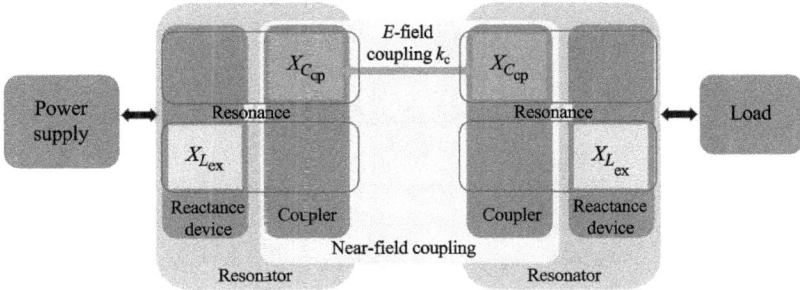

Figure 3.9   Unified model for E-field coupling WPT

this type of circuit topology is explained as shown in Figure 3.9. In this case, the coupler has only electric field coupling. The matching circuit compensates capacitive reactance of the coupler. As a result, the combination of coupler and matching circuit can be regarded as a resonator.

3.2.5   Application for a self-resonator

A self-resonant antenna is an alternative way to realize coupled-resonant WPT. A self-resonant antenna is decomposed into a magnetic loop due to current mainly flowing at the center of the antenna and an electric dipole due to electric charge mainly accumulated at the both ends of the antenna, shown in Figure 3.10.

By using the unified coupled-resonant WPT model, a self-resonant antenna is explained as shown in Figure 3.11. Since the self-resonant antenna has a capacitive reactance due to the charge and an inductive reactance due to the current, resonance is achieved without using an external reactance device. On the other hand, not only magnetic field coupling but also electric field coupling cannot be negligible because both magnetic energy and electric energy are stored in the opened space.

In the self-resonant system, transmission distance can be extended by using a repeater made of a single-wire conductor, which does not have a closed-circuit topology [7]. This is because the electric field concerns with power transfer. In the self-resonant scheme, transmission efficiency can be improved by adjusting electric and magnetic field coupling by modifying a geometry of the antenna [9].

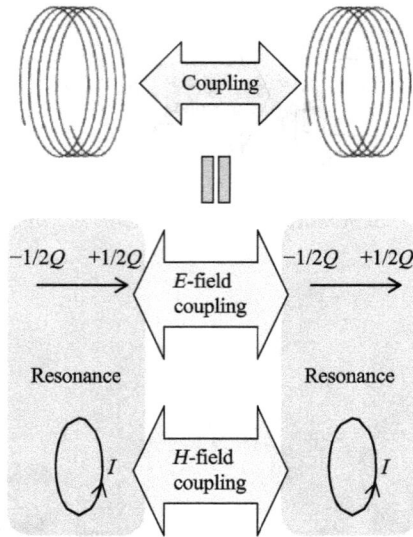

Figure 3.10   Schematic diagram of self-resonant WPT

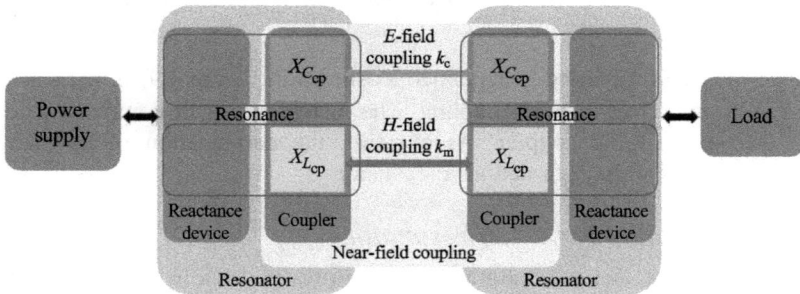

Figure 3.11   Unified model for self-resonant WPT

3.3 Generalized model of WPT

3.3.1 *Energy flow in WPT system*

The unified model in the previous section is applicable only for the WPT system using resonance. However, there are various kinds of WPT systems using near-field without resonance. In this section, the generalized model of WPT system is discussed. First of all, let us consider a power flow in direct current (DC) (i.e., 0 Hz) wired power transfer, shown in Figure 3.12. In this system, transferred power is $P=VI$; however, the physical entity of the transferred power is not the voltage and the current, but the electric field and magnetic field, which is known as a Poynting vector [10]. This fact is easily understood by the question, "In which region in space is the power transferred?" The answer is not in the wire, but around the wire. Even in three-phase wired power transfer, its physical mechanism of power transfer is explained by the Poynting vector [11]. Therefore, it is said that transverse electromagnetic (TEM) mode electromagnetic power is conveyed in the wire.

Next, let us consider how the power is conveyed in the WPT system changing its properties. In general, the block diagram of the WPT system becomes that as shown in Figure 3.13.

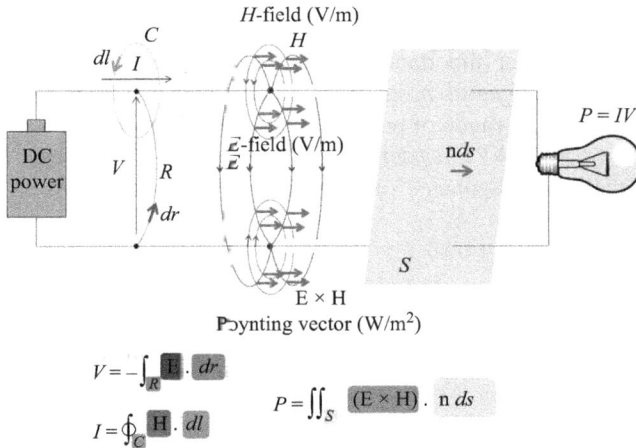

Figure 3.12 Power transfer in DC-wired system

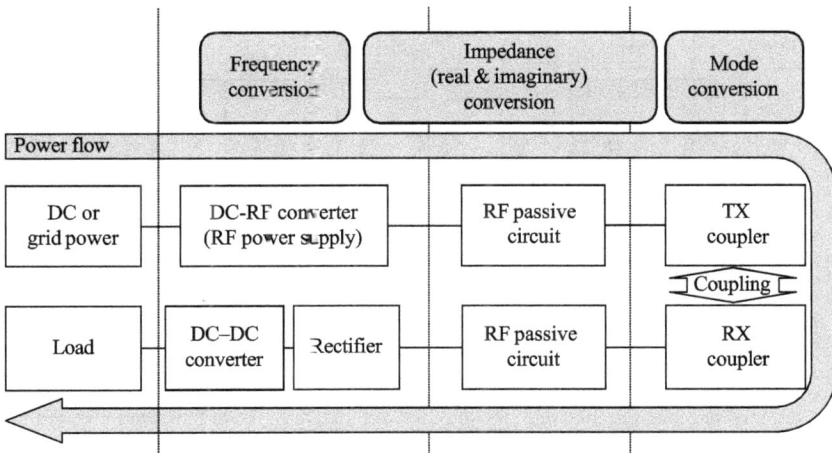

Figure 3.13   Power flow in WPT system: frequency, impedance (real and imaginary part), and propagation mode of the power is converted in the WPT system

At first, the RF inverter (or the RF power amplifier) changes the frequency of the power. Usually, the RF inverter also changes the voltage of the power. Considering that the power is *P=VI* and the impedance is *Z=V/I*, the RF inverter also changes the impedance of the power. The rectifier in Rx side also changes the frequency of the power.

In some systems, a capacitor is used for resonance or power factor compensation. In other systems, an LC network is used for complex conjugate matching. In general, an RF passive circuit is connected after the RF inverter. Considering the property of two-port passive network, it is said that the RF-passive circuit changes impedance (real and imaginary parts) of the power.

A coupler is the most important device in the WPT system. From the viewpoint of how the power changes its property by the coupler, it is said that the coupler changes the propagation mode of power between TEM mode and non-TEM mode.

In this way, the WPT system has functions of converting "frequency," "impedance (real and imaginary)," and "mode" of the electric power.

3.3.2   *Generalized model*

A generalized model for near-field WPT [5] is shown in Figure 3.14. This model describes the whole system of WPT and its parameters. Additionally, this model represents which part of the WPT system concerns the four functions of the WPT

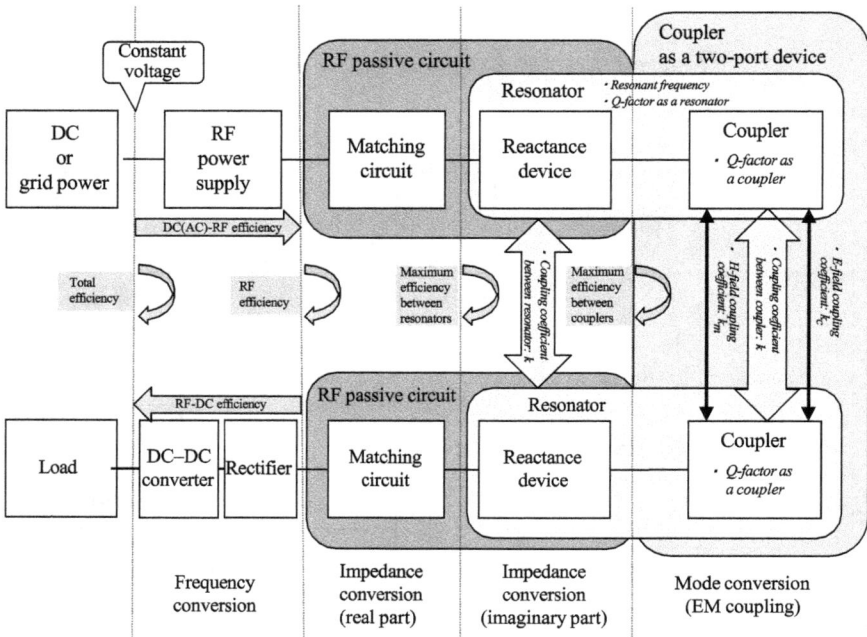

Figure 3.14   Generalized model for near-field WPT system

system, discussed in the previous section. By using this model, WPT is explained by aspects of "coupling," "resonance," and "impedance matching."

The "RF-passive circuit" consists of matching circuit and/or reactance device. The matching circuit has a function of converting real and imaginary part of the impedance. The reactance device has a function of converting imaginary part of the impedance.

In the case of WPT system using coupling coil and resonant capacitor shown in Figure 3.6, the coupler and the reactance device in Figure 3.14 correspond to the coupling coil and the resonant capacitor, respectively.

Coupling coefficient "between couplers" is defined between Tx coupler and Rx coupler, whereas coupling coefficient "between resonators" is defined between Tx resonator and Rx resonator. This model clearly distinguishes between "coupling coefficient between couplers (e.g., $k = M/L$)" and "coupling coefficient between resonators (e.g., $k = 2(\omega_h - \omega_l)/(\omega_h + \omega_l)$)." Also, "$Q$-factor of coupler (e.g., $Q = \omega L/R$)" and "Q-factor of resonator (e.g., calculated from 3-dB band width)" are readily discriminated.

3.3.3 Understanding of coupled-resonator WPT system through generalized model

When considering WPT system as a coupling of resonators, the generalized model shown in Figure 3.14 can be rewritten as shown in Figure 3.15. In this case, a

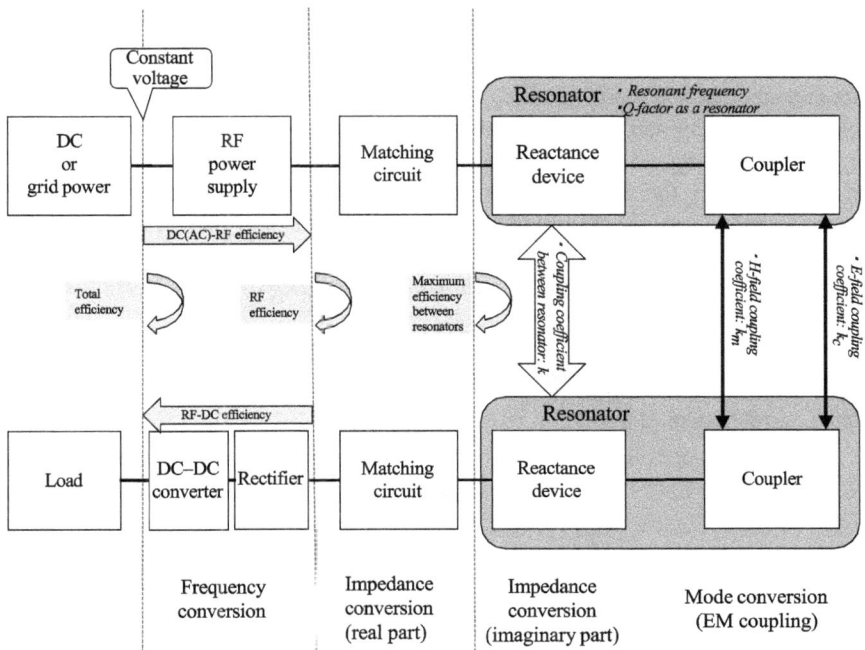

Figure 3.15   Generalized model for near-field WPT system: a case of coupling of resonators

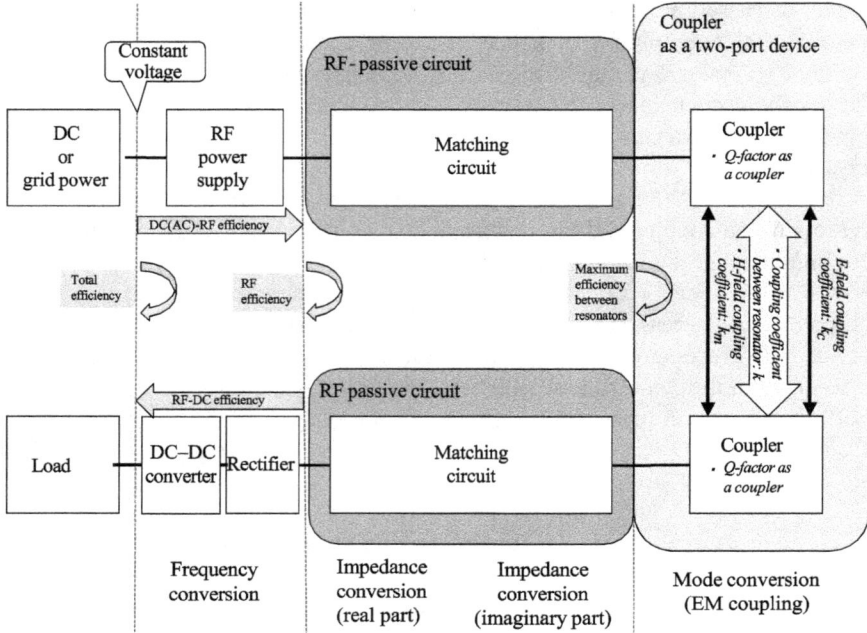

Figure 3.16   Generalized model for near-field WPT system: a case of coupler with a matching circuit

resonator acts as a role of not only electromagnetic (EM) coupling but also the impedance of imaginary part matching.

When coupling coil and resonant capacitor are used, both "reactance device" and "coupler" in the model consist of the "resonator." When a self-resonant antenna is used, only the "coupler" in the model consists of the "resonator."

3.3.4   Understanding of coupler-and-matching-circuit WPT system through generalized model

From the viewpoint of impedance matching, the generalized mode in Figure 3.14 becomes Figure 3.16. A matching circuit is used to realize the maximum available efficiency of the coupler [12].

When coupling coil and resonant capacitor are used, the resonant capacitor can be considered a part of the matching circuit. For this, the "resonant" phenomenon is not necessary.

Acknowledgment

This work was supported by the JSPS KAKENHI Grant-in-Aid for Scientific Research (C) Number 15K06061.

References

[1]   Q. Chen, K. Ozawa, Q. Yuan, and K. Sawaya, "Antenna characterization for wireless power-transmission system using near-field coupling," *IEEE Antennas and Propagation Magazine*, vol. 54, no. 4, pp. 108–116, 2012.

[2]   T. Hosotani and I. Awai, "A novel analysis of ZVS wireless power transfer system using coupled resonators," *Proceedings of IMWS*, pp. 235–238, May 2012.

[3]   S. Ahn, J. Pak, T. Song *et al.*, "Low frequency electromagnetic field reduction techniques for the on-line electric vehicle (OLEV)," *Proceedings of EMC*, pp. 625–630, July 2010.

[4]   T. Ohira, "Via-wheel power transfer to vehicles in motion," *Proceedings of WPTC*, pp. 242–246, May 2013.

[5]   H. Hirayama, "Unified coupling and resonant model for near-field wireless power transfer system," *Proceedings of ICCEM*, March 2017.

[6]   H. J. Kim, H. Hirayama, S. Kim, K. J. Han, R. Zhang, and J.-W. Choi, "Review of near-field wireless power and communication for biomedical applications," *IEEE Access*, no. 5, pp. 21264–21285.

[7]   N. Inagaki and S. Hori, "Classification and characterization of wireless power transfer systems of resonance method based on equivalent circuit derived from even- and odd mode reactance functions," *Proceedings of IWPT*, pp. 115–118, May 2011.

[8]   I. Awai, Y. Ikuta, Y. Sawahara, Y. Thang and T. Ishizaki, "Applications of a novel disk repeater," *Proceedings of WPTC*, pp. 114–117, May 2014.

[9]   Y. Sawahara, Y. Ikuta, Y. Zhang, T. Ishiazaki and I. Awai, "Proposal of a new disk-repeater system for contactless power transfer," *IEICE Transactions on Communications* vol. E98-B, no. 12, December 2015.

[10] J. H. Poynting, "On the transfer of energy in the electromagnetic field," *Philosophical Transactions of the Royal Society of London*, no. 175, pp. 343–361, January 1884.

[11] L. S. Czarnecki, "Could power properties of three-phase systems be described in terms of the Poynting vector?," *IEEE Transactions on Power Delivery*, vol. 21, no. 1, pp. 339–344, 2006.

[12] T. Ohira, "Maximum available efficiency formulation based on a black-box model of linear two-port power transfer systems," *IEICE ELEX*, vol. 11, no. 13, pp. 1–6, 2014.

Chapter 4

Multi-hop wireless power transmission

Yoshiaki Narusue[1] and Yoshihiro Kawahara[1]

Magnetic-resonant coupling is suitable for wireless power transmission in the mid-distance range since it offers a longer distance of transfer than that offered by transmission using electromagnetic induction. Further, its transfer efficiency for mid-range transmission is also observed to be higher than that of microwave-power transmission because the efficiency of an RF power source and a rectifier circuit decreases as the operating frequency increases. However, energy cannot be transferred with high efficiency over a distance that is not greater than several times the resonator radius even after applying magnetic-resonant coupling. Several applications of wireless power transmission require longer distances, which may be up to several times that of the transmitting-resonator radii; for example, a wireless power-transmission system to cover an entire room would be difficult to achieve if there is only one large transmitting resonator placed on the floor. Further, it would be possible to fabricate a resonator that is large enough to cover the floor. However, the larger the transmitting resonator, the larger the deviation between its size and that of a receiving device such as a smartphone or laptop PC. This further decreases the transfer efficiency.

A hint to solve this problem is the "relay effect," which is a unique characteristic of wireless power transmission using magnetic-resonant coupling. The relay effect is a phenomenon in which a relay resonator can serve as an intermediary for power transmission [1,2]. A relay resonator is used for relaying power, and its physical structure is similar to that of a transmitter or receiver.

Sending power from a transmitting resonator to a receiving resonator using relay resonators is called "multi-hop wireless power transmission." Using a single relay resonator, the transfer distance can be doubled; however, the number of relay resonators is not limited to one. Further, the transfer distance is tripled using two relay resonators, whereas this distance is quadrupled in the case of three relay resonators. Generally, the transfer distance is increased by $n + 1$ times if there are n relay resonators. Therefore, multiple relay resonators can be used to transmit electric power to a receiving resonator that is located far away in a manner that wireless power transmission cannot be achieved using a single transmitting resonator.

[1]Graduate School of Engineering, The University of Tokyo, Japan

Target applications that benefit from multi-hop wireless power transmission include indoor power supplies [3], unwiring of robot arms [4], desktop power-supply systems [5], and wireless charging of endoscopic capsules within the body [6]. Owing to the extended transfer distance, many applications that could not be realized using a pair of transmitting and receiving resonators are made achievable by expanding the transfer distance using multi-hop wireless power transmission. Further, as depicted in Section 4.2, it is possible to control the resonance condition of each relay on demand rather than statically extending the transfer distance between resonators, making it feasible to alter the position in order to supply power dynamically. Hence, multi-hop wireless power transmission is a promising candidate that can be used to develop applications in which the receiving resonator is observed to move.

However, such a transmission depicts many more design parameters than that of the wireless power transmission, which uses only a transmitter and receiver. In addition to the distance between the transmitting and receiving resonators and the impedance of a receiver load, we should also consider the distances between the transmitting and relay resonators, relay resonators, and a relay resonator and the receiving resonator. By appropriately designing all these parameters, it is possible to obtain a highly efficient multi-hop wireless-power-transmission system.

In Chapter 4, the design methods of multi-hop wireless power transmission have been illustrated. First, we provide an overview of the characteristics of such transmission; further, design methods are derived using the bandpass-filter (BPF) theorem and for realizing maximum efficiency in arbitrary hopping. Power efficiencies are evaluated in Section 4.6 with the design theories that are presented in this study.

4.1 Transfer-distance extension using the relay effect

Section 4.1 explains the relay effect and its characteristics. In order to maintain consistency throughout Chapter 4, the electromagnetic simulations of the wireless power transmission, which was used to provide illustrations, employ open-helical resonators having a self-resonant frequency of 13.56 MHz [7]. Altair Engineering Inc. FEKO [8] is used as an electromagnetic simulator. The parameters of an open-helical resonator are depicted in Table 4.1. According to the simulation results that were obtained using FEKO with these parameters, the self-resonant frequency of the resonator is observed to be approximately 13.56 MHz when the number of turns is 6.4. Hereafter, simulation is therefore performed using open-helical resonators

Table 4.1   Parameters of an open-helical resonator used to perform analysis and simulations

Resonant frequency	Diameter	Pitch	Diameter of copper wire
13.56 MHz	30 cm	5 mm	1 mm

with 6.4 turns. Here, it should be noted that this does NOT indicate that the results of Chapter 4 are valid only for wireless power transmission having an operating frequency of 13.56 MHz or containing only open-helical resonators.

The physical structure of a relay resonator used to perform multi-hop wireless power transmission is similar to that of a transmitting or receiving resonator. A transmitting or receiving resonator has a port to which a power amplifier or a load can be connected. However, the port in a relay resonator is connected to a short circuit. Thus, a relay resonator can be considered to be a receiving resonator that does not exhibit any load consuming electrical power. Because there is no load in the relay resonator, no electric power is consumed during the flow of an induced current, strengthening the magnetic flux in the vicinity of the relay resonator. The magnetic flux that is generated by the relay resonator is attributed to depict the relay effect.

Transfer distance can be extended not only in the vertical direction (Figure 4.1(a)) but also in the horizontal direction (Figure 4.1(b)). The vertical and horizontal arrangements of two-hop wireless power transmission in the presence of a single relay resonator are depicted in Figure 4.1. The relay effect is not only in a vertical direction, as depicted in Figure 4.1(a), but also in the horizontal hopping, as depicted in Figure 4.1(b). This multi-directionality is due to a characteristic of the wireless power transmission using magnetic-resonant coupling that indicates that highly efficient power transmission can be achieved even though the coupling coefficient is small. When the resonators are arranged in a horizontal direction, the maximum coupling coefficient is observed to be approximately 0.1. This coefficient is constant in the effective range of magnetic resonant coupling, which is obvious from the design that is depicted in Table 4.2. Additionally, a horizontal arrangement is assumed to be applied for an indoor power supply [3] in which the transfer range must be expanded in a two-dimensional plane, whereas the unwiring of a robot arm [4] assumes a vertical arrangement.

To confirm the extension of the transfer distance by the relay effect, electromagnetic simulations are performed using FEKO. Here, transfer efficiency is defined by $|S_{21}|^2$ at 13.56 MHz. Further, power efficiency is discussed in

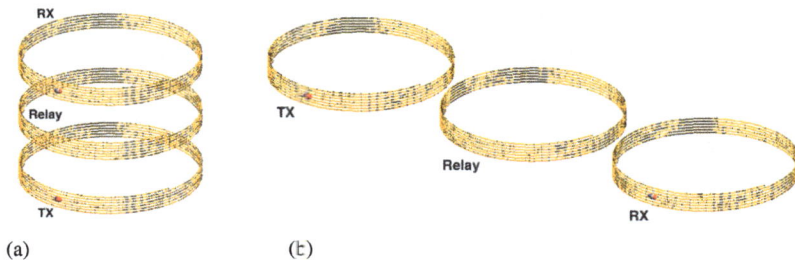

(a)   (b)

Figure 4.1   (a) Vertical and (b) horizontal arrangements of transmitting, relay, and receiving resonators that are used in multi-hop wireless power transmission

*Table 4.2   Coupling coefficients calculated with a design concept
based on the bandpass-filter theorem*

N	k_1	k_2	k_3	k_4
1	0.0446			
2	0.0315	0.0315		
3	0.0287	0.0185	0.0287	
4	0.0275	0.0153	0.0153	0.0275

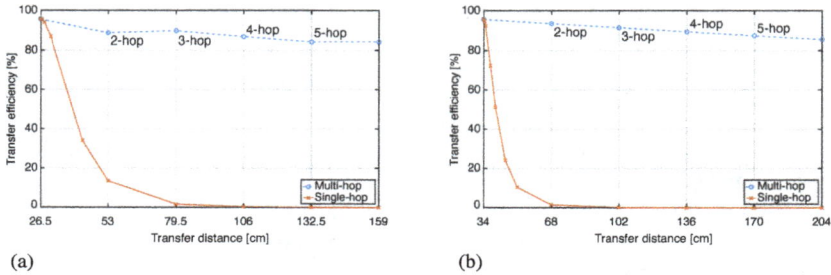

(a)   (b)

*Figure 4.2   Improvements in efficiencies using multi-hop wireless power
transmission. Transfer efficiency is plotted as a function of the transfer
distance. (a) Vertical arrangement and (b) horizontal arrangement.*

Section 4.6. The number of hops ranges from 1 to 6, indicating that the number of relay resonators ranges from 0 to 5. To enhance simplicity, the distances between the transmitting and relay resonators, relay resonators, and relay and receiving resonators are set to be equal. Here, the transfer distance (*d*) is defined as the distance from the center of the transmitting resonator to that of the receiving resonator.

Transfer efficiencies in the vertical arrangement, as obtained using electromagnetic simulations, are depicted in Figure 4.2(a). In the simulations, the distances between the resonators are set to 26.5 cm, which is the derived value based on the design methodology that is described in Section 4.5. From Figure 4.2(a), we observe that wireless power transmission is performed using multi-hop wireless power transmission with relay resonators having an efficiency exceeding 80% even at distances where the transfer efficiency is 1% or less using a single-hop system. Similarly, transfer efficiencies that are obtained by the simulations of the horizontal arrangement are depicted in Figure 4.2(b). The distances between the resonators are set to 34.0 cm, which was derived using the design method described in Section 4.5. The effect of improving the transfer efficiencies in the horizontal arrangement is confirmed, along with that of the vertical arrangement. Multi-hop wireless power transmission in which the relay resonators are linearly arranged is called "straight multi-hop wireless power transmission."

From Figure 4.2(b), we observe that the transfer efficiency of the horizontal arrangement decreases linearly as the number of hops increases. There are larger fluctuations in the transfer efficiency of the vertical-arrangement than that in the transfer efficiency of the horizontal arrangement. This is caused due to the influence of couplings between nonadjacent resonators; generally, the coupling strength attenuates more sharply in the horizontal direction rather than in the vertical direction. Therefore, the influence of couplings between nonadjacent resonators is observed to be smaller in the horizontal arrangement. It has additionally been reported that the transfer efficiency can be improved somewhat by considering the couplings between the nonadjacent resonators of the vertical-arrangement [9]. However, when there is not so much influence, it is common to ignore these couplings while designing a vertical-type multi-hop wireless power transmission system in order to simplify the design processes [10,11].

4.2   Multi-hop routing

By considering the power transmission in a two-dimensional plane, such as an indoor environment, straight multi-hop wireless power transmission is observed to be insufficient. To construct a wireless-power-transmission system that covers the floor and walls, as depicted in Figure 4.3, it is necessary to expand the transfer range two-dimensionally. The simplest way to expand the transmission range in a two-dimensional plane is to connect all the resonators to power supplies. However, the cost of manufacturing and installation of the hardware increases when many transmitting resonators are connected to the power supplies. Therefore, the use of relay resonators instead of transmitting resonators and power supplies has been investigated.

The simplest approach to expand the transmission range two-dimensionally using relay resonators may be to arrange these resonators in a two-dimensional plane. Electromagnetic simulations are performed in order to observe the behavior when the relay resonators are arranged in a two-dimensional plane around the transmitting resonator. The arrangement of resonators is depicted in Figure 4.4(a). All the resonators except for the transmitting resonator are observed to be relay

Figure 4.3   Illustration of indoor wireless power transmission using relay effects

(a)

(b)

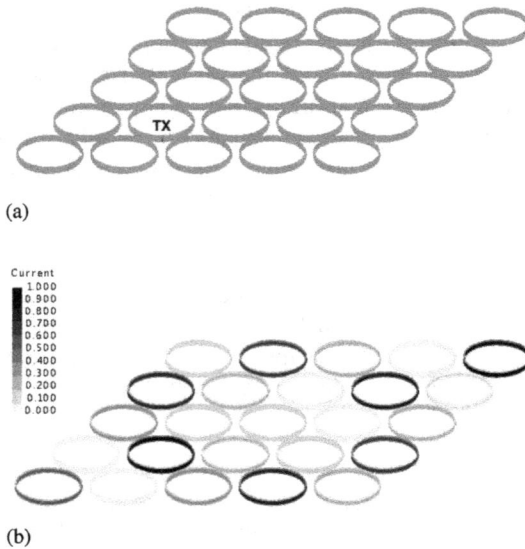

Figure 4.4   Simulated current distribution in a planar-resonator array.
(a) Simulation setup of a planar-resonator array. All resonators
except the transmitting resonator are observed to be on-state relays.
(b) Normalized current distribution in a planar-resonator array.

resonators, and they are arranged in a triangular-lattice arrangement. One side of this lattice is set to 34 cm, indicating that the distance between relay resonators is 34 cm. The simulation result of the normalized current distribution is depicted in Figure 4.4(b). As illustrated in Figure 4.4(b), the current distribution is not uniform at all. This indicates that the transfer efficiency strongly varies depending on the location and that almost no power can be transferred in particular around the relay resonator with a small current. Therefore, a highly efficient power supply cannot be achieved if the relay resonators are simply arranged on a two-dimensional plane. Note that a study has been conducted to investigate a method in order to achieve uniform current distribution in a two-dimensional planar-resonator array [12]; the result depicts that the impedance of each resonator should be adjusted based on the number of resonators and strength of couplings.

However, there is a method for enlarging the power-supply range in a two-dimensional plane using straight multi-hop wireless power transmission with ON/OFF switching of relay resonators. Highly efficient power transmission can be achieved by deploying relay resonators that are capable of being switched ON or OFF in a two-dimensional plane and by dynamically performing straight multi-hop wireless power transmission in a horizontal arrangement from the transmitting resonator to the receiving resonator using only ON-state relay resonators. Thus, ON-state relay resonators are connected from the vicinity of the transmitting resonator to that of the receiving resonator, and the other relay resonators are set to

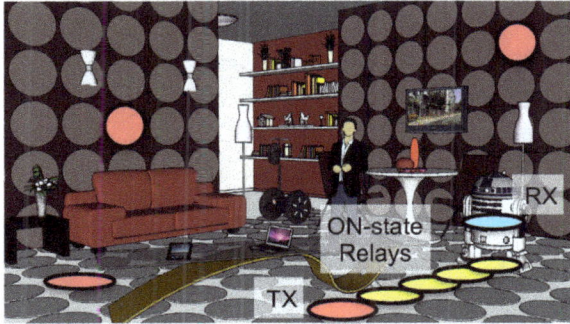

Figure 4.5   Expanding the transmission range in a two-dimensional plane using straight multi-hop wireless power transmission

the OFF-state, whereby a virtual path comprising ON-state relay resonators from the transmitting resonator to the receiving resonator is generated; this is called "multi-hop routing" in wireless power transmission. As an example, the indoor power-supply system that is depicted in Figure 4.5 can be explained. In the case of power transmission to a robot, it is possible to build a virtual path from the transmitting resonator to the robot by turning ON the relay resonators between the transmitting resonator and robot (as shown in Figure 4.5) and by turning OFF the other relays. Electrical power can be routed through the virtual path.

Normally, the port of a relay resonator is connected to a short circuit. However, it is possible to turn relays ON or OFF by inserting a switch into the port and controlling its impedance. By making the port of the relay resonator to be open-circuited, i.e., making it high-impedance, the relay resonator can be disabled. However, if the port of the relay resonator is short-circuited, it can be employed as a normal relay resonator.

ON/OFF switching of a relay resonator can be realized using an electric component **relay device**. In order to use in the MHz band, the impedance of the OFF state must be sufficiently large. It is difficult to provide low loss and sufficiently large OFF-state impedance using commercially available solid-state relay devices. However, sufficient performance can be expected using some mechanical relay devices, which depict a parasitic resistance of 50 mΩ or less in the short condition and a capacitance of only several pF in the open condition. If solid-state relay devices were implemented using recently developed GaN field-effect transistors, it would become possible to achieve sufficient performance efficiency using solid-state relay devices.

To investigate the feasibility of a virtual path, an electromagnetic simulation, as depicted in Figure 4.6(a), is conducted. The port impedance of the ON-state relay resonators connected from the transmitting resonator to the receiving resonator is short-circuited, and 1 pF is connected to the port of the other relay resonators to emulate the open-state impedance of a mechanical-relay device. The normalized current distribution is illustrated in Figure 4.6(b). As depicted in the figure, only the

(a)

(b)

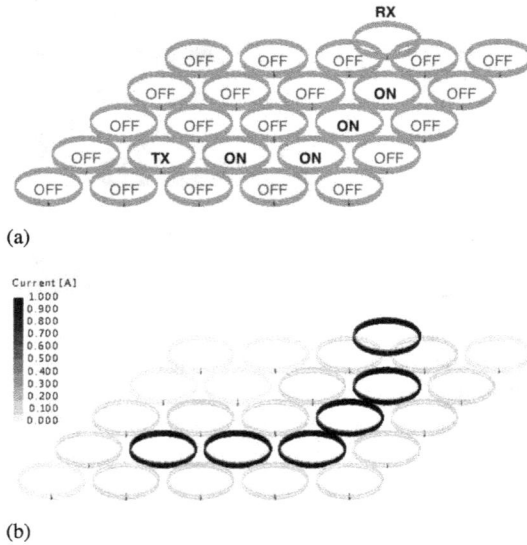

Figure 4.6   *Simulated current distribution of straight multi-hopping in a two-dimensional plane with the ON/OFF switching of the relay resonators. (a) Simulation setup to observe the feasibility of a virtual path by controlling the ON/OFF states of the relay resonators. (b) Normalized current distribution in a planar resonator array with ON/OFF switching of the relay resonators.*

ON-state relay resonators contribute to wireless power transmission, and almost no current is generated in the OFF-state relay resonator. This means that the OFF-state relay resonators depict almost no influence upon straight multi-hop wireless power transmission. Further, a virtual path comprising the ON-state resonators is formed. Thus, it is possible to build a virtual path freely and control the destination of power transmission by controlling the ON/OFF state of each relay resonator. The relay device required for ON/OFF switching of a relay resonator can be controlled using a microcontroller. The implementation of a system that automatically detects the position of the receiving resonator and realizes a virtual path from the transmitting resonator to the receiving resonator has been investigated using an RF position detector and a wireless-communication module [13].

Note that straight multi-hop wireless power transmission can physically depict a large-angle curve if its equivalent circuit can be represented using linearly coupled resonator arrays. Because the distance between nonadjacent resonators become shorter around the curve, it is preferable to make the angle of the curve gentle so that the distance will be as long as possible. In the case of a square lattice, the curve is 90°, whereas it is 120° in the case of a triangular lattice. Therefore, a triangular lattice-shaped resonator arrangement is more suitable than a square lattice-shaped arrangement while using multi-hop routing.

4.3   Equivalent circuit and transfer efficiency

A circuit that is equivalent to a series resonator employed for wireless power transmission using magnetic-resonant coupling is depicted in Figure 4.7(a), and an equivalent circuit of a relay resonator is depicted in Figure 4.7(b). Where L is the inductance of the resonator, C is the capacitance, and r is the parasitic resistance. The capacitance (C) is attributed to the physical structure in an open-type resonator or a lumped capacitor connected in series with a coil [7]. The self-resonant frequency of the resonator (f_0) is

$$f_0 = \frac{\omega_0}{2\pi} = \frac{1}{2\pi\sqrt{LC}},$$ (4.1)

where ω_0 is the angular frequency.

Because the port of a relay resonator is short-circuited, the equivalent circuit of the relay resonator is depicted in Figure 4.7(b). The equivalent circuit of straight multi-hop wireless power transmission is illustrated in Figure 4.8. Here, it is noted that the couplings between nonadjacent resonators are ignored in the equivalent circuit. Each M_i ($i = 1, 2, \ldots, N$) is the mutual inductance between resonator $i - 1$ and resonator i. R_s is the source impedance of a power source, and R_{load} is the load resistance.

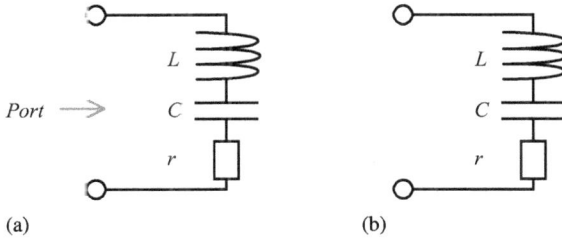

Figure 4.7   *Equivalent circuits of the (a) TX/RX resonators and (b) a relay resonator*

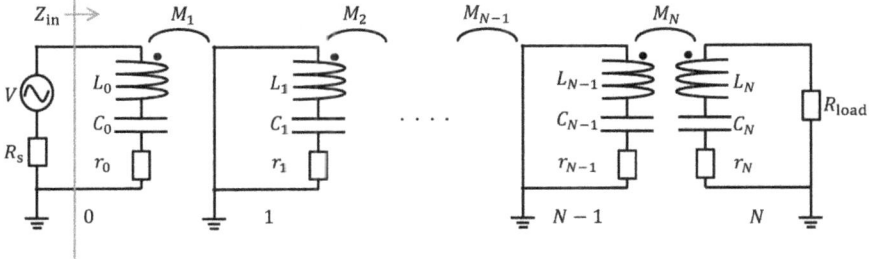

Figure 4.8   *Equivalent circuit of straight multi-hop wireless power transmission*

It is understood that the magnetic coupling of mutual inductance (M) is equivalent to that of the K-inverter of $K = \omega M$. As depicted in Figure 4.9, the K-inverter converts the impedance of Z into

$$Z' = \frac{K^2}{Z}.$$ (4.2)

In wireless power transmission using magnetic-resonant coupling, a K-inverter has been frequently utilized to perform system design and analysis. The design methods in Chapter 4 are further derived based on a K-inverter. Using a K-inverter, the equivalent circuit that is depicted in Figure 4.8 is converted into the illustration that is depicted in Figure 4.10.

Further, we note that only the case with a properly designed straight multi-hop wireless power transmission was analyzed and discussed. However, in practice, high transfer efficiency cannot be maintained unless the load impedance and coupling coefficients between the resonators are properly adjusted. In this study, straight multi-hop wireless power transmission having a horizontal arrangement in which the distances between the resonators are closer to 34 cm will be analyzed as an example of low-efficiency transmission. The distances between the resonators are set to 31 cm. The transfer efficiencies obtained by electromagnetic simulations are illustrated in Figure 4.11. While the distances between the resonators are too

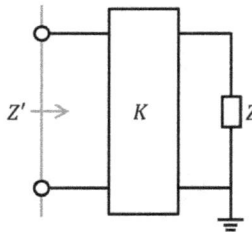

Figure 4.9   Impedance conversion using a K-inverter

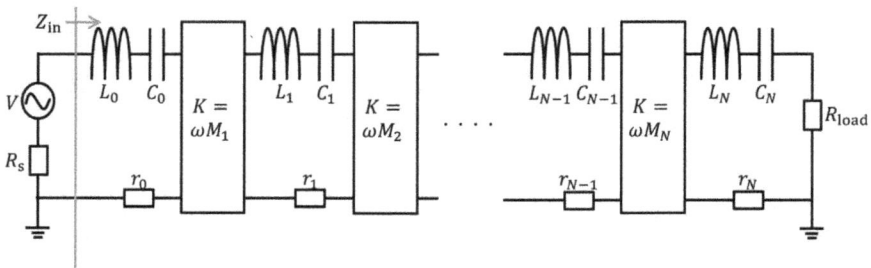

Figure 4.10   Representation using K-inverters in an equivalent circuit for straight multi-hop wireless power transmission

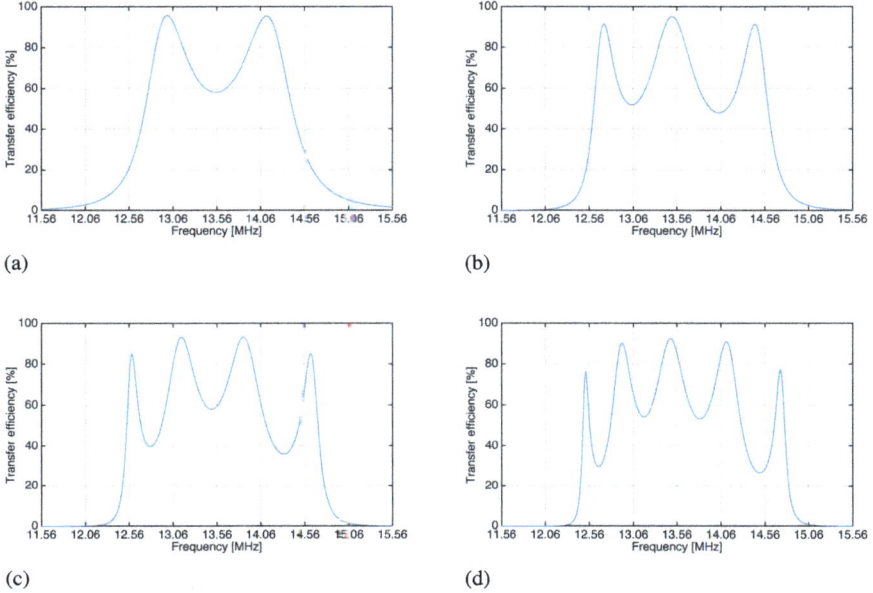

Figure 4.11   Transfer efficiency of over coupled multi-hop wireless power transmission: (a) 1-hop, (b) 2-hop, (c) 3-hop, and (d) 4-hop

short, there are multiple peaks in the frequency characteristics of the transfer efficiency, thereby proving that the transfer efficiency at the self-resonant frequency of the resonators is not necessarily high. Generally, if the coupling coefficients between the resonators are too large, there are as many peak frequencies as the number of resonators.

This peak frequency varies depending on the number of resonators and coupling coefficients. However, we have to consider the regulations as well; therefore, we have to fix the operating frequency used for power transmission. It is necessary to design a wireless power transmission system such that a high transfer efficiency can always be achieved using a fixed frequency even though the number of resonators changes. In straight multi-hop wireless-power-transmission systems, it is common to design all the resonators using the same self-resonant frequency (f_0), and the resonant frequency is also used as the operating frequency.

The discussion so far has used the transfer efficiency defined as $|S_{21}|^2$. While employing a switching power supply known to have an efficiency greater than 50-Ω power sources, the transfer efficiency $|S_{21}|^2$ cannot be fixed any longer. It is common to evaluate the wireless power transmission with a switching power source using power efficiency (η) rather than the transfer efficiency. However, impedance matching is also important for switching between power supplies. There is optimal input impedance for a class-E or class-DE inverter, which is one of the most highly efficient switching power supplies [14]. Using class-D inverters, which are also commonly used in wireless power transmission, the load impedance is observed to significantly affect the efficiency of the power source in the MHz band. Hence, it is desirable that the input impedance that is observed by the power source should not change even though the number of hops may change. In a system, such as an indoor power supply, the movement of the receiving resonator is attributed to a change in the number of hops. Thus, it is necessary to maintain high transfer efficiency with a constant input impedance regardless of the number of hop changes owing to the movement of the receiver.

Instead of simply using power efficiency η, the design method based on transfer efficiency $|S_{21}|^2$ by considering impedance matching is applicable not only to a 50-Ω system but also to a system design with a switching power supply. Therefore, we derive a design theorem based on the transfer efficiency $|S_{21}|^2$ and evaluate the design theories using both the transfer $|S_{21}|^2$ and power efficiencies η in Chapter 4.

4.4   Design based on the BPF theorem

This study presents the first proposed design methodology to perform such a transmission. It was illustrated that the BPF theorem can further improve the transfer efficiency of straight multi-hop wireless power transmission with indirect feeding and flatten the frequency characteristics of the transfer efficiency near the self-resonant frequency of the resonators [15]. However, [16] proposed a system-design concept based on the BPF theorem for straight multi-hop wireless power transmission with direct feeding. The design theorem presented in [16] is derived as an example owing to its ease of derivation.

The BPF-design theorem using K-inverters has been investigated for a long time. The equivalent circuit of a BPF using K-inverters is depicted in Figure 4.12(a). The conditions for this equivalent circuit to operate as a Butterworth BPF are as follows [17]:

$$\omega_c = \frac{1}{\sqrt{L_i'C_i'}}, \tag{4.3}$$

$$K_{0,1} = \sqrt{\frac{(\omega_2 - \omega_1)L_1'R_A}{g_0 g_1}}, \tag{4.4}$$

$$K_{i,i+1} = (\omega_2 - \omega_1)\sqrt{\frac{L_i'L_{i+1}'}{g_i g_{i+1}}}, \tag{4.5}$$

$$K_{n,n+1} = \sqrt{\frac{(\omega_2 - \omega_1)L_n'R_B}{g_n g_{n+1}}}, \tag{4.6}$$

where ω_c is the center frequency, ω_1 and ω_2 are cutoff frequencies, $(\omega_1 < \omega_c < \omega_2)$, R_A is the source impedance of the power source, R_B is the load

(a)

(b)

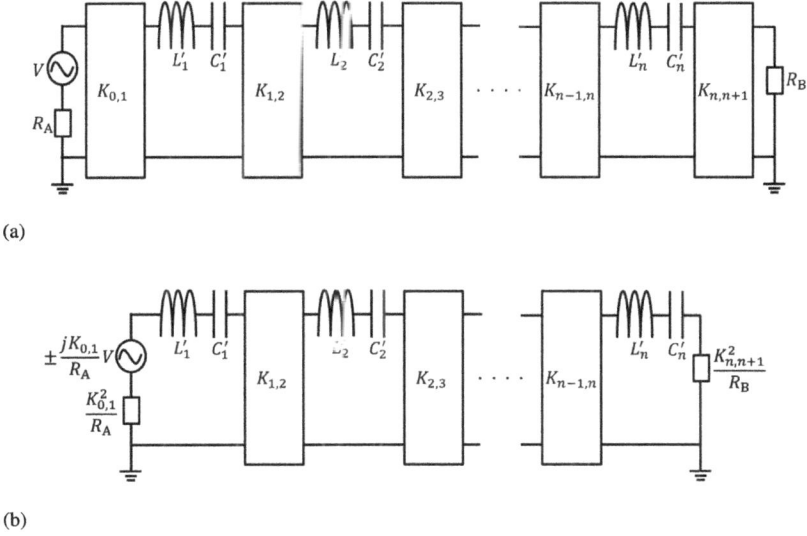

Figure 4.12 (a) *Transformation of the equivalent circuit of a bandpass filter using K-inverters. (b) Transformation that is similar to that of an equivalent circuit of straight multi-hop wireless power transmission.*

resistance value, and

$$g_0 = 1, \tag{4.7}$$

$$g_i = 2 \sin \left(\frac{2i - 1}{2n} \pi \right), \tag{4.8}$$

$$g_{n+1} = 1. \tag{4.9}$$

This equivalent circuit is converted into a circuit of a straight multi-hop wireless power transmission that is depicted in Figure 4.10. The *ABCD* parameter of the *K*-inverter is

$$\begin{bmatrix} A & B \\ C & D \end{bmatrix} = \begin{bmatrix} 0 & \mp jK \\ \pm \dfrac{1}{jK} & 0 \end{bmatrix}. \tag{4.10}$$

Using (4.10), the equivalent circuit of the BPF can be converted to the equivalent circuit illustrated in Figure 4.12(b). The source impedance and load resistance are converted using the *K*-inverter into $K_{0,1}^2/R_A$ and $K_{n,n+1}^2/R_B$, respectively.

By comparing Figures 4.10 and 4.12(b), the conditions for equalizing the equivalent circuit of the BPF and that of a multi-hop wireless power transmission are derived. Because the BPF theorem does not consider the parasitic loss of each element, the parasitic resistances of the resonators in Figure 4.10 are ignored in this study.

First, comparing the source impedance and load resistance gives

$$R_s = \frac{K_{0,1}^2}{R_A},$$
(4.11)

$$R_{load} = \frac{K_{n,n+1}^2}{R_B}.$$
(4.12)

Here, by substituting (4.4) and (4.6),

$$R_s = \frac{(\omega_2 - \omega_1)L_1'}{g_0 g_1}$$
(4.13)

$$= \frac{(\omega_2 - \omega_1)L_1'}{2 \sin\left(\frac{\pi}{2n}\right)},$$
(4.14)

$$R_{load} = \frac{(\omega_2 - \omega_1)L_n'}{2 \sin\left(\frac{\pi}{2n}\right)},$$
(4.15)

are obtained. Further, by comparing each inductance and capacitance, we obtain

$$N = n - 1,$$
(4.16)

$$L_i = L_{i+1}',$$
(4.17)

$$C_i = C_{i+1}'.$$
(4.18)

From (4.3), we observe that each resonator i satisfies

$$\omega_0 = \frac{1}{\sqrt{L_i C_i}} = \frac{1}{\sqrt{L_i' C_i'}} = \omega_c.$$
(4.19)

Therefore, the BPF has the same central frequency as the self-resonant frequency. Finally, each K-inverter is compared. The K-inverter in the BPF depicts no frequency dependence, whereas $K_{i,i+1}$ in the wireless-power-transmission system is proportional to the frequency. Therefore, the equivalent circuits of the BPF and multi-hop wireless power transmission do not completely match. However, because only the characteristic around the central frequency, f_c, is important for multi-hop wireless power transmission, the frequency to be considered in the design is approximated to be $f \approx f_c = f_0$. Thus, we compare the K-inverters in Figures 4.10 and 4.12(b) to obtain the following condition:

$$K_{i,i+1} = \omega_c M_i.$$
(4.20)

Substituting (4.5) into (4.20),

$$M_i = \left(\frac{\omega_2 - \omega_1}{\omega_c}\right)\sqrt{\frac{L_i''L_{i+1}''}{g_i g_{i+1}}} \tag{4.21}$$

$$= \left(\frac{\omega_2 - \omega_1}{\omega_c}\right)\sqrt{\frac{L_i'L_{i+1}'}{4\sin\left(\frac{2i-1}{2n}\pi\right)\sin\left(\frac{2i+1}{2n}\pi\right)}} \tag{4.22}$$

is obtained. Thus, we can derive the coupling coefficient (k_i) as

$$k_i = \left(\frac{\omega_2 - \omega_1}{\omega_c}\right)\frac{1}{2\sqrt{\sin\left(\frac{2i-1}{2n}\pi\right)\sin\left(\frac{2i+1}{2n}\pi\right)}}. \tag{4.23}$$

By combining (4.14)–(4.23) and deleting the parameters of the BPF, the design formula of the straight N-hop wireless-power-transmission system can be derived as follows:

$$\omega_0 = \frac{1}{\sqrt{L_i C_i}}, \tag{4.24}$$

$$Q_1 := \frac{\omega_0 L_0}{R_s}, \tag{4.25}$$

$$= \frac{\omega_0 L_N}{R_{\text{load}}}, \tag{4.26}$$

$$k_i = \frac{\sin\left(\frac{\pi}{2N+2}\right)}{Q_1\sqrt{\sin\left(\frac{2i-1}{2N+2}\pi\right)\sin\left(\frac{2i+1}{2N+2}\pi\right)}}. \tag{4.27}$$

Here, the loaded quality factor (Q_1) was introduced. Note that the pass bandwidth (f_{bw} (Hz)) can be calculated as follows:

$$f_{\text{bw}} = \frac{\omega_2 - \omega_1}{2\pi} = \frac{2f_0\sin\left(\frac{\pi}{2N+2}\right)}{Q}. \tag{4.28}$$

To confirm the effectiveness of this design, electromagnetic simulations are performed using FEKO. An open-helical resonator with a diameter of 30 cm is used for the resonator, and the source impedance and load resistance are 50 Ω. The resonator is arranged in a horizontal layout. The number of hops ranges from 1 to 4. To calculate the coupling coefficients using (4.27), the inductance L and capacitance C of the resonator are calculated by simulation results and are observed to be $L = 13.17$ (μH) and $C = 10.47$ (pF). The coupling coefficients that were calculated using these values are presented in Table 4.2.

The simulated transfer efficiencies are illustrated in Figure 4.13. It can be confirmed that the frequency characteristics of the transfer efficiency contain the

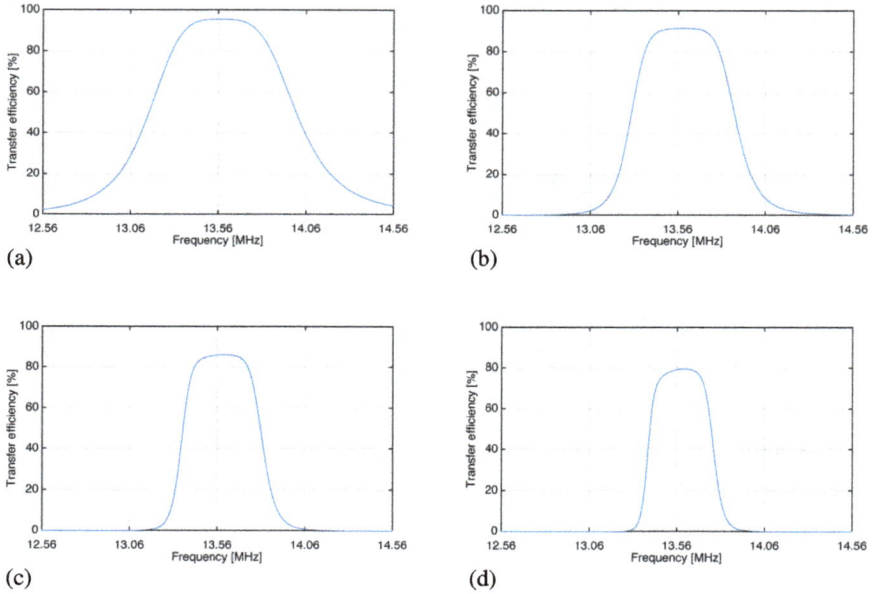

Figure 4.13   Simulated transfer efficiency $|S_{21}|^2$ of multi-hop wireless power transmission with a design concept based on the bandpass-filter theorem: (a) 1-hop, (b) 2-hop, (c) 3-hop, and (d) 4-hop

Figure 4.14   Simulation plane of magnetic-field intensity with a transferred power of 100 W in multi-hop wireless power transmission with a design concept that is based on the bandpass-filter theorem

characteristics of BPF and that the transfer efficiency is maximized at the resonators' self-resonance frequency, $f_0 = 13.56$ MHz.

Subsequently, the magnetic-field intensity is investigated by conducting electromagnetic simulations. The simulation plane is depicted in Figure 4.14. The transferred power is set to 100 W. The leftmost resonator is the transmitter, whereas the rightmost is the receiver. The magnetic-field-strength distribution is illustrated

(a)   (b)

(c)   (d)

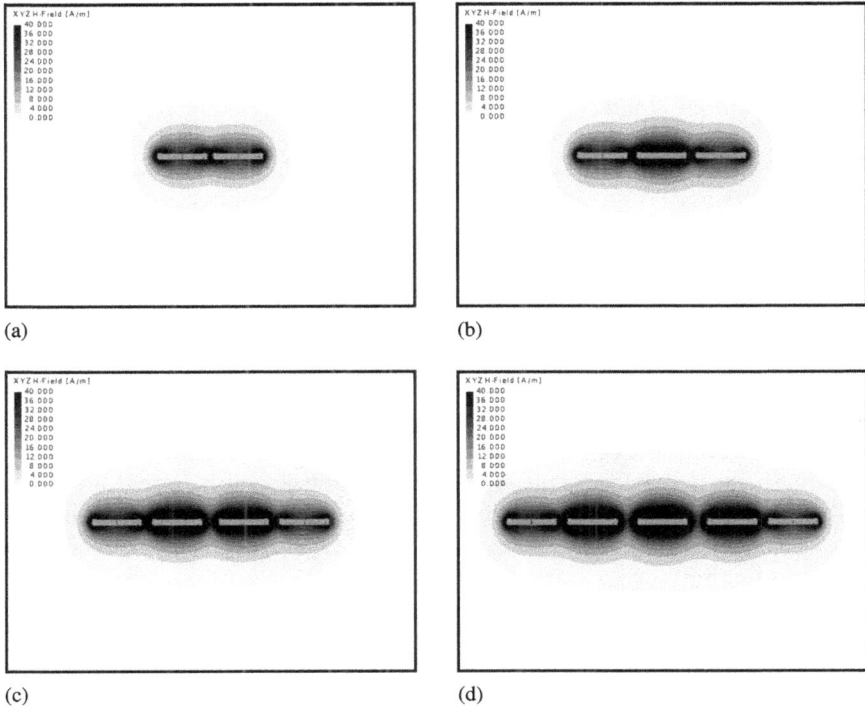

Figure 4.15   Magnetic-field intensity with a transferred power of 100 W in multi-hop wireless power transmission with a design concept based on the bandpass-filter theorem: (a) 1-hop, (b) 2-hop, (c) 3-hop, and (d) 4-hop

in Figure 4.15. In the design method based on the BPF theorem, the magnetic-field intensity is observed to increase as the number of hops increases, and the magnetic-field strength increases considerably near the central resonator.

4.5   Design theorem for arbitrary-hop power transmission

Considering a system in which the number of hops varies dynamically with the movement of the receiving resonator, as in the indoor power-supply system that is depicted in Figure 4.3, the design method based on the BPF theorem is not considered to be appropriate. According to the BPF theorem, the coupling coefficients between the resonators depend on the number of hops N, as indicated by (4.27) and Table 4.2; hence, it is necessary to adjust the coupling coefficients while the receiver is moving. To maintain an efficient power supply while the number of hops changes, the distances between the resonators must be readjusted each time the number of hops changes, which is an unrealistic expectation in indoor power-supply systems. To overcome this problem, a design theorem for arbitrary-hop wireless power transmission, which does not require readjustment of the coupling coefficients even if the number of hops changes, has been proposed [3]. The design theorem for arbitrary-hop wireless power transmission is derived below.

While deriving the design theorem for arbitrary-hop wireless power transmission, each parasitic resistance is ignored. Note that the power efficiency (η) by considering the parasitic resistances will be discussed in Section 4.6. While ignoring each parasitic resistance, all the input power to the transmitting resonator is consumed by the load of the receiving resonator because the reactance-circuit networks cannot consume any power because the power source and receiver are connected via reactance-circuit networks, as depicted in Figure 4.8. Therefore, the condition for $|S_{21}|^2 = 1$ is impedance matching, i.e., $Z_{\mathrm{in}} = R_{\mathrm{s}}$.

First, the input impedance (Z_{in}) is analyzed. In Section 4.5, the input impedance of the N-hop wireless-power-transmission system is denoted as $Z_{\mathrm{in}}(N)$. In this design theorem, the self-resonant frequency (f_0) of the resonators is used as the operating frequency. Because the self-resonant frequency of the resonators is used for the power supply, the equivalent circuit using the K-inverter depicted in Figure 4.10 can be simplified to that illustrated in Figure 4.16. Here, $Z_{\mathrm{in}}(N)$ can be derived as follows:

$$Z_{\mathrm{in}}(N) = \cfrac{(\omega_0 M_1)^2}{\cfrac{(\omega_0 M_2)^2}{\cfrac{\vdots}{\cfrac{(\omega_0 M_N)^2}{R_{\mathrm{load}}}}}} \tag{4.29}$$

$$= \begin{cases} \cfrac{\prod\limits_{i:\mathrm{odd}} (\omega_0 M_i)^2}{\prod\limits_{i:\mathrm{even}} (\omega_0 M_i)^2} \cfrac{1}{R_{\mathrm{load}}} & N : \text{odd} \\[20pt] \cfrac{\prod\limits_{i:\mathrm{odd}} (\omega_0 M_i)^2}{\prod\limits_{i:\mathrm{even}} (\omega_0 M_i)^2} R_{\mathrm{load}} & N : \text{even} \end{cases} \tag{4.30}$$

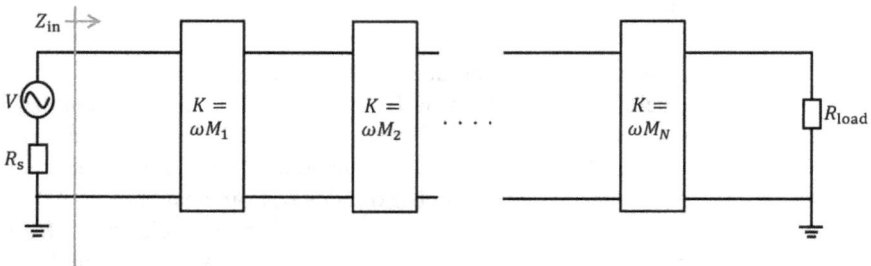

Figure 4.16   Simplified equivalent circuit operating at the self-resonance frequency of the resonators by ignoring the parasitic resistances

$$= \begin{cases} \dfrac{\prod\limits_{i:\text{odd}} M_i^2}{\prod\limits_{i:\text{even}} M_i^2 R_{\text{load}}} \omega_0^2 & N : \text{odd} \\[2em] \dfrac{\prod\limits_{i:\text{odd}} M_i^2}{\prod\limits_{i:\text{even}} M_i^2} R_{\text{load}} & N : \text{even} \end{cases}. \tag{4.31}$$

To derive design formulae for an arbitrary number of hops, the system configurations for N and $N+1$ hops are compared based on Figure 4.17. To maintain maximum efficiency without readjusting the coupling coefficients even when the number of hops changes, it is necessary to derive M_N satisfying $Z_{\text{in}}(N+1) = R_s$ based on the assumption that $Z_{\text{in}}(N) = R_s$.

First, according to (4.31), the design formula by which impedance matching can be achieved in a one-hop system is $(\omega_0 M_1)^2/R_{\text{load}} = R_s$. This provides the mutual inductance M_{LAST} between the transmitting and receiving resonators in a

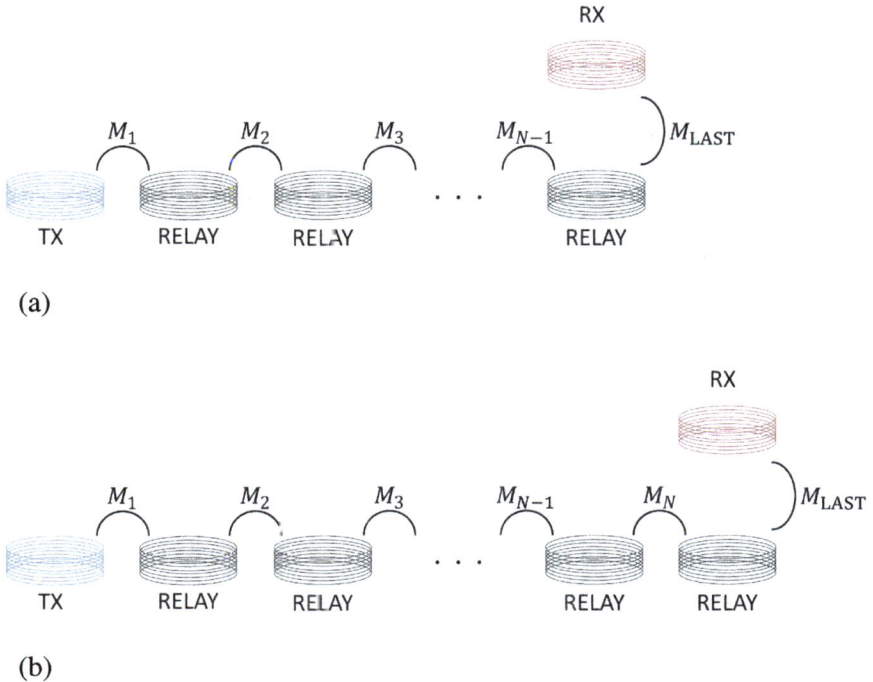

(a)

(b)

Figure 4.17   Comparison between the configurations of (a) N-hop and (b) (N+1)-hop wireless power transmission

one-hop system or between the last relay and receiving resonators as

$$\frac{(\omega_0 M_{\text{LAST}})^2}{R_{\text{load}}} = R_{\text{s}}. \tag{4.32}$$

Further, by considering a system of $N+1$ hops, the mutual inductance (M_N) between the $(N-1)$th and Nth resonators can be derived. Using (4.31) (when N is an odd number),

$$Z_{\text{in}}(N+1) = \frac{\displaystyle\prod_{i=1,3,\dots,N} M_i^2}{\left(\displaystyle\prod_{i=2,4,\dots,N-1} M_i^2\right) M_{\text{LAST}}^2} R_{\text{load}} \tag{4.33}$$

$$= \left(\frac{\left(\displaystyle\prod_{i=1,3,\dots,N-2} M_i^2\right) M_{\text{LAST}}^2 \, \omega_0^2}{\displaystyle\prod_{i=2,4,\dots,N-1} M_i^2 \quad R_{\text{load}}}\right) \frac{M_N^2 R_{\text{load}}^2}{\omega_0^2 M_{\text{LAST}}^4} \tag{4.34}$$

$$= Z_{\text{in}}(N) \left(\frac{\omega_0 M_N}{R_{\text{s}}}\right)^2 \tag{4.35}$$

holds true. Equation (4.32) was used for the transformation from (4.34) to (4.35). Additionally, when N is an even number, we obtain

$$Z_{\text{in}}(N+1) = \frac{\left(\displaystyle\prod_{i=1,3,\dots,N-1} M_i^2\right) M_{\text{LAST}}^2 \, \omega_0^2}{\displaystyle\prod_{i=2,4,\dots,N} M_i^2 \quad R_{\text{load}}} \tag{4.36}$$

$$= \left(\frac{\displaystyle\prod_{i=1,3,\dots,N-1} M_i^2}{\left(\displaystyle\prod_{i=2,4,\dots,N-2} M_i^2\right) M_{\text{LAST}}^2} R_{\text{load}}\right) \frac{\omega_0^2 M_{\text{LAST}}^4}{M_N^2 R_{\text{load}}^2} \tag{4.37}$$

$$= Z_{\text{in}}(N) \left(\frac{\omega_0 M_N}{R_{\text{s}}}\right)^{-2}. \tag{4.38}$$

If $Z_{\text{in}}(N) = R_{\text{s}}$ is satisfied in the N-hop system,

$$Z_{\text{in}}(N+1) = \left(\frac{\omega_0 M_N}{R_{\text{s}}}\right)^{\pm 2} \quad Z_{\text{in}}(N) = \left(\frac{\omega_0 M_N}{R_{\text{s}}}\right)^{\pm 2} R_{\text{s}} \tag{4.39}$$

are obtained. Therefore, when $\omega_0 M_N / R_{\text{s}} = 1$, impedance matching can also be achieved using a configuration of $N+1$ hops. Because (4.39) does not depend upon N, the mutual inductances between the transmitting and first relay resonators and between the adjacent relay resonators should be constant, indicating that the relay resonators are equally spaced because of the identical resonators that were used. The design formula for the mutual inductance (M_{RELAY}) can be given as

$$\frac{(\omega_0 M_{\text{RELAY}})^2}{R_{\text{s}}} = R_{\text{s}}. \tag{4.40}$$

Thus, the design formulas for arbitrary-hop wireless power transmission are as follows:

$$\text{mutual inductance in the last hop}: \quad \frac{(\omega_0 M_{\text{LAST}})^2}{R_{\text{load}}} = R_{\text{s}}, \tag{4.41}$$

$$\text{the other mutual inductances}: \quad \frac{(\omega_0 M_{\text{RELAY}})^2}{R_{\text{s}}} = R_{\text{s}}. \tag{4.42}$$

When the height of a receiving resonator is determined in advance using an application scenario, the mutual impedance at that particular height is initially measured, and R_{s}, R_{load}, or both should be designed to satisfy (4.41). To adjust R_{s} and R_{load}, impedance-conversion circuits between the transmitting and first relay resonators and between the receiving resonator and load, respectively, are observed to be effective. In the case of R_{load}, it is also possible to perform load conversion using a DC–DC converter or a switching regulator [18,19].

Electromagnetic simulations are conducted using FEKO to confirm the effectiveness of this design. For comparison with the BPF-theorem-based design concept, the resonator arrangement is horizontal, and the receiving resonator is also arranged on the same plane. An open-helical resonator with a diameter of 30 cm is used, and the source impedance and load resistance are observed to be 50 Ω. According to the simulation results, the distance between the resonators that satisfies (4.41–4.42) is 34.0 cm. Simulation results of the transfer efficiency for the numbers of hops ranging from 1 to 4 are depicted in Figure 4.18. It can be observed that the transfer efficiency is maximized at the resonators' self-resonant frequency, $f_0 = 13.56$ (MHz), after performing any number of hops.

Subsequently, the magnetic-field intensity was surveyed using electromagnetic simulations. The transferred power is fixed to be 100 W. The magnetic-field-strength distribution is illustrated in Figure 4.19. It can be observed that this distribution is substantially uniform in the design method depicting an arbitrary-hop wireless-power transmission.

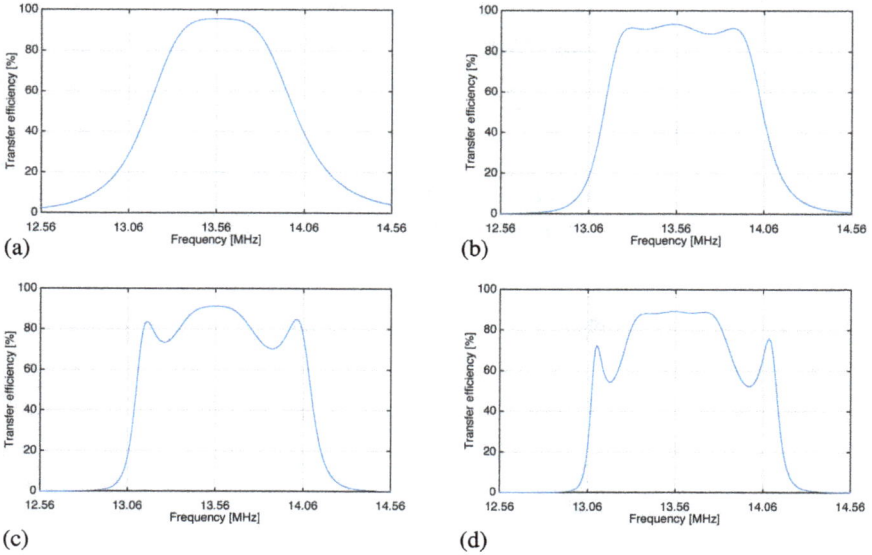

Figure 4.18   Simulated transfer efficiency, $|S_{21}|^2$, of multi-hop wireless power transmission using the design theorem for arbitrary hops: (a) 1-hop, (b) 2-hop, (c) 3-hop, and (d) 4-hop

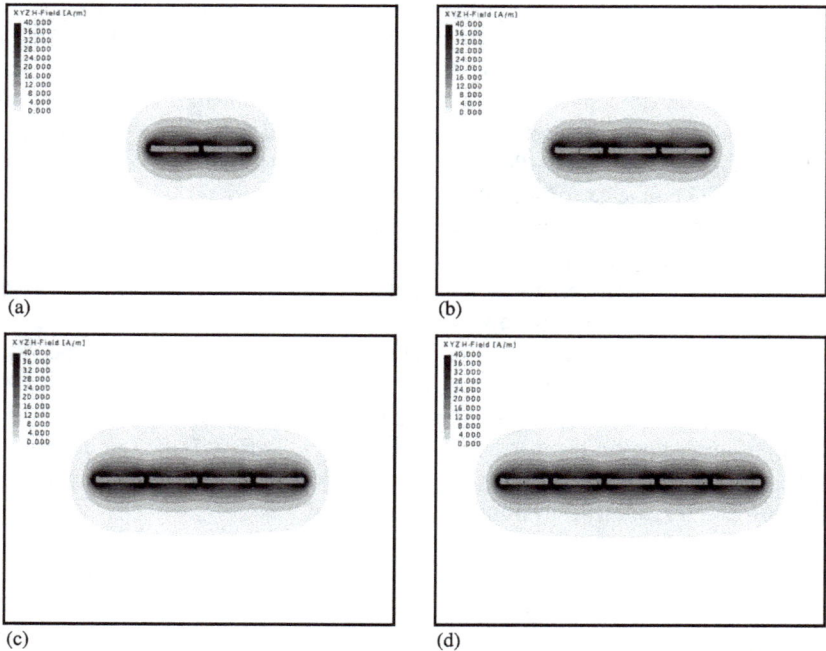

Figure 4.19   Magnetic-field intensity with a transferred power of 100 W in multi-hop wireless power transmission using the design theorem for arbitrary hops: (a) 1-hop, (b) 2-hop, (c) 3-hop, and (d) 4-hop

4.6 Power-efficiency estimation

The design concept using BPF theorem and arbitrary hops was derived based on the transfer efficiency $|S_{21}|^2$. However, evaluation using power efficiency (η) has become more common recently. Therefore, Section 4.6 estimates the power efficiency of multi-hop wireless-power-transmission systems based on design concepts using the BPF theorem and theorem for arbitrary hops.

Let us consider the power efficiency at the self-resonance frequency of the resonators. The equivalent circuit depicted in Figure 4.20 has been introduced for explanatory purposes. Power efficiency (η) can be defined as

$$\eta = \frac{R_{\text{load}} |I_{\text{load}}|^2}{\text{Re}(Z_{\text{in}}) |I_{\text{in}}|^2}, \tag{4.43}$$

where I_{in} is the input current to the transmitting resonator, and I_{load} is the load current. Using the input impedance (Z_i) of the ith K-inverter defined in Figure 4.20, η can be calculated as

$$\eta = \frac{R_{\text{load}}}{R_{\text{load}} + r_N} \prod_{1 \le i \le N} \left(\frac{\text{Re}(Z_i)}{\text{Re}(Z_i) + r_{i-1}} \right). \tag{4.44}$$

Therefore, power efficiency η can be determined if we know each $\text{Re}(Z_i)$.

To ensure simplicity, we assume that the unloaded quality factors of all resonators are equal and that their value is denoted as Q_u. The impedance (Z_i) is estimated by neglecting the parasitic resistance. Here, the derivation process of Z_i is omitted owing to its complexity. In the case of the design theorem using the BPF theorem, the impedance (Z_i) can be estimated as

$$Z_i \approx \frac{\omega_0 L_{i-1} \sin\left(\frac{\pi}{2N+2}\right)}{Q_1 \sin\left(\frac{2i-1}{2N+2}\pi\right)}, \tag{4.45}$$

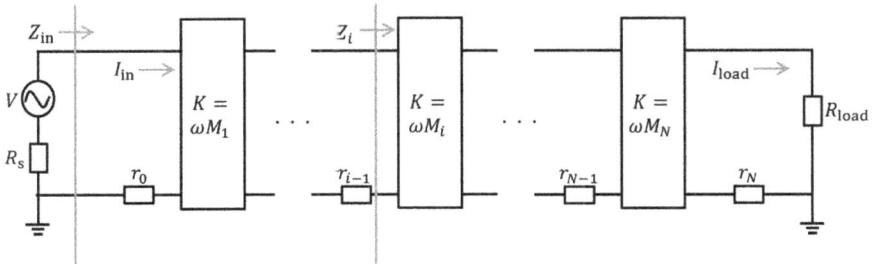

Figure 4.20 Definition of Z_i at the resonators' self-resonant frequency (f_0)

where Q_1 is defined in Section 4.4 as

$$Q_1 = \frac{\omega_0 L_0}{R_\mathrm{s}} = \frac{\omega_0 L_N}{R_\mathrm{load}}. \tag{4.46}$$

Additionally, in the case of a design method with arbitrary hops,

$$Z_i \approx R_\mathrm{s} \tag{4.47}$$

is obtained.

By substituting (4.45) into (4.44), the power efficiency (η_BPF) of a multi-hop wireless-power-transmission system based on the BPF theorem is derived as

$$\eta_\mathrm{BPF} \approx \frac{R_\mathrm{load}}{R_\mathrm{load} + r_N} \prod_{1 \le i \le N} \left(\frac{\frac{\omega_0 L_{i-1}\,\sin\left(\frac{\pi}{2N+2}\right)}{Q_1\,\sin\left(\frac{2i-1}{2N+2}\pi\right)}}{\frac{\omega_0 L_{i-1}\,\sin\left(\frac{\pi}{2N+2}\right)}{Q_1\,\sin\left(\frac{2i-1}{2N+2}\pi\right)} + r_{i-1}} \right) \tag{4.48}$$

$$= \frac{1}{1 + \frac{r_N Q_1}{\omega_0 L_N}} \prod_{1 \le i \le N} \left(\frac{1}{1 + \frac{r_{i-1} Q_1 \sin\left(\frac{2i-1}{2N+2}\pi\right)}{\omega_0 L_{i-1}\,\sin\left(\frac{\pi}{2N+2}\right)}} \right) \tag{4.49}$$

$$= \prod_{1 \le i \le N+1} \left(\frac{1}{1 + \frac{Q_1 \sin\left(\frac{2i-1}{2N+2}\pi\right)}{Q_u \sin\left(\frac{\pi}{2N+2}\right)}} \right). \tag{4.50}$$

In particular, in a situation when $Q_1 / \left(Q_u \sin\left(\frac{\pi}{2N+2}\right) \right) \ll 1$ is true, η_BPF can be approximated as

$$\eta_\mathrm{BPF} \approx \prod_{1 \le i \le N+1} \left(1 - \frac{Q_1 \sin\left(\frac{2i-1}{2N+2}\pi\right)}{Q_u \sin\left(\frac{\pi}{2N+2}\right)} \right). \tag{4.51}$$

Similarly, in the case of a design theorem for arbitrary hops, the power efficiency η_AH can be obtained as follows:

$$\eta_\mathrm{AH} \approx \frac{R_\mathrm{load}}{R_\mathrm{load} + r_N} \prod_{1 \le i \le N} \left(\frac{R_\mathrm{s}}{R_\mathrm{s} + r_{i-1}} \right) \tag{4.52}$$

$$= \frac{1}{1 + \frac{r_N}{R_\mathrm{load}}} \prod_{1 \le i \le N} \left(\frac{1}{1 + \frac{r_{i-1}}{R_\mathrm{s}}} \right) \tag{4.53}$$

$$\approx \left(1 - \frac{r_N}{R_\mathrm{load}} \right) \prod_{1 \le i \le N} \left(1 - \frac{r_{i-1}}{R_\mathrm{s}} \right). \tag{4.54}$$

Table 4.3   Comparison of the simulated and estimated power efficiencies (η)

Hop count	One-hop	Two-hop	Three-hop	Four-hop
BPF (simulated) (%)	95.7	91.6	86.1	79.3
BPF (estimated by (4.50)) (%)	95.7	91.7	86.3	79.9
BPF (estimated by (4.51)) (%)	95.6	91.4	85.7	78.9
Arbitrary hops (simulated) (%)	95.7	93.7	91.6	89.6
Arbitrary hops (estimated by (4.55)) (%)	95.6	93.5	91.5	89.5
Arbitrary hops (estimated by (4.56)) (%)	95.6	93.4	91.2	89.0

In particular, when $\left(\sum_{1 \le i \le N} r_{i-1} \right) / R_s \ll 1$ is true and $\left(\sum_{1 \le i \le N} r_{i-1} \right)^2 / R_s^2$ is negligible, η_{BPF} can be approximated as

$$\eta_{\mathrm{AH}} \approx \left(1 - \frac{r_N}{R_{\mathrm{load}}} \right) \left(1 - \frac{\sum_{1 \le i \le N} r_{i-1}}{R_s} \right). \tag{4.55}$$

According to (4.55), it can be observed that the power efficiency (η_{AH}) decreases linearly as the number of relay resonators increases.

The power efficiencies were calculated using the formulae that were estimated and were compared with those obtained by electromagnetic simulations. Each parameter is set to be the same as depicted in the simulations of Sections 4.4 and 4.5. The calculated and simulated power efficiencies are listed in Table 4.3. Using both the BPF theorem and theorem for arbitrary hops, the numerical values obtained by the electromagnetic simulations and estimation formulae are observed to be in good agreement. From the viewpoint of power efficiency, the design theorem for arbitrary hops is observed to be better than that obtained on the basis of the BPF theorem.

4.7   Output characteristics with a switching power supply

Here, we discuss the output characteristics of multi-hop wireless power transmission at the self-resonant frequency of the resonators. As the load impedance may change owing to the system conditions in a practical situation, it is important to consider the behavior of the output voltage and/or current according to the change in the load impedance Z.

The output characteristics of a one-hop system employing a switching power source and series resonators can be approximated to those of a current source. The amplitude and phase of the current flowing in the receiving resonator exhibited a small dependence on the load impedance in the one-hop system. However, the output characteristics of multi-hop wireless power transmission depend on whether the number of hops is odd or even. When the number of hops is odd, the output characteristics are identical to those of the one-hop case. Ideally, the amplitude and phase of the output current depend on the input and couplings and do not depend on the load impedance; this is called the constant-current (CC) mode. When the number of hops is even, the output characteristics can be regarded as identical to those of the voltage source. The amplitude and phase of the voltage incident on the load do not depend on the load impedance; this is called the constant-voltage (CV) mode.

For simplicity, the analysis assumes that the switching power source has extremely low output impedance, which is approximated as zero. In addition, the quality factors of the resonators are assumed to be sufficiently high such that the parasitic resistances of the resonators are negligible.

The equivalent circuit at self-resonant frequency under these assumptions is shown in Figure 4.21. Please ensure that Z rather than R_{load} denotes the load impedance.

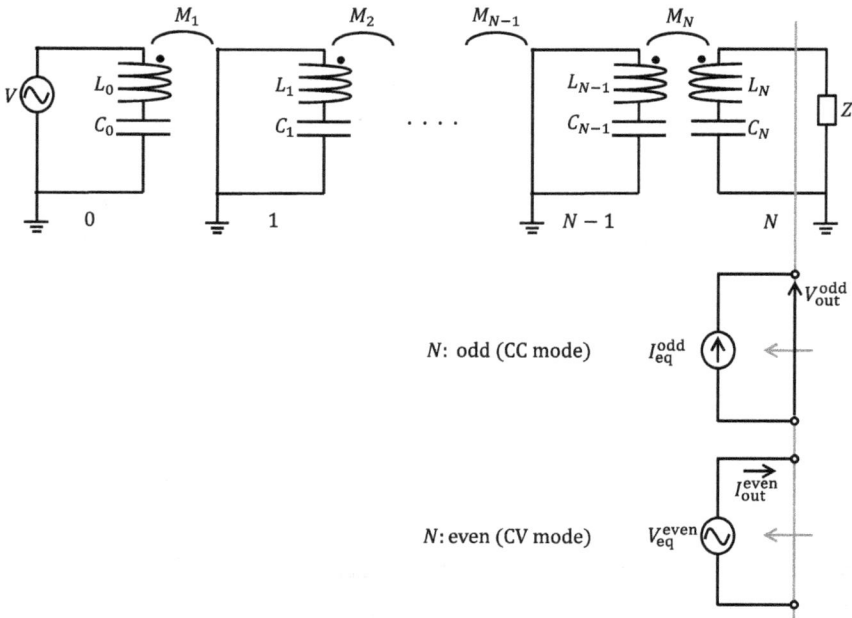

Figure 4.21   Equivalent circuit seen from a load at the self-resonant frequency of resonators under the assumptions that an output impedance of input RF source and parasitic resistances of resonators are negligible

In the odd cases (CC mode), the complex amplitude $I_{\text{eq}}^{\text{odd}}$ of an equivalent current source is formulated as

$$I_{\text{eq}}^{\text{odd}} = \frac{\prod\limits_{i:\text{even}} (j\omega_0 M_i)}{\prod\limits_{i:\text{odd}} (-j\omega_0 M_i)}\, V = j^N \frac{\prod\limits_{i:\text{even}} M_i}{\omega_0 \prod\limits_{i:\text{odd}} M_i}\, V. \tag{4.56}$$

The output voltages $V_{\text{out}}^{\text{odd}}$ and $ZI_{\text{eq}}^{\text{odd}}$ are as follows:

$$V_{\text{out}}^{\text{odd}} = j^N \frac{\prod\limits_{i:\text{even}} M_i}{\omega_0 \prod\limits_{i:\text{odd}} M_i}\, ZV. \tag{4.57}$$

In the even cases (CV mode), the complex amplitude $V_{\text{eq}}^{\text{even}}$ of the equivalent voltage source is formulated as

$$V_{\text{eq}}^{\text{even}} = \frac{\prod\limits_{i:\text{even}} (j\omega_0 M_i)}{\prod\limits_{i:\text{odd}} (-j\omega_0 M_i)}\, V = j^N \frac{\prod\limits_{i:\text{even}} M_i}{\prod\limits_{i:\text{odd}} M_i}\, V. \tag{4.58}$$

The output current $I_{\text{out}}^{\text{even}}$, which is $V_{\text{eq}}^{\text{even}}/Z$, is represented as

$$I_{\text{out}}^{\text{even}} = j^N \frac{\prod\limits_{i:\text{even}} M_i}{\prod\limits_{i:\text{odd}} M_i}\, \frac{V}{Z}. \tag{4.59}$$

Specifically, in a system designed for arbitrary-hop power transmission, $I_{\text{eq}}^{\text{odd}}$ and $V_{\text{eq}}^{\text{even}}$ can be represented by (4.41) and (4.42) as

$$I_{\text{eq}}^{\text{odd}} = j^N \frac{V}{R_{\text{load}}}, \tag{4.60}$$

$$V_{\text{eq}}^{\text{even}} = j^N \frac{R_{\text{load}}}{R_{\text{s}}}\, V, \tag{4.61}$$

where R_{load} and R_{s} are the design parameters that determine the mutual couplings and do not correspond to the actual values of the load and output impedances of the power source, respectively. Please ensure that neither $I_{\text{eq}}^{\text{odd}}$ nor $V_{\text{eq}}^{\text{even}}$ is dependent on load impedance Z.

For example, when a receiver-side switching regulator is used to generate stable DC voltage, the CV mode is more suitable than the CC mode [19]. A matching circuit, such as an LCC in a transmitter or receiver, is a potential candidate to fix the output characteristics in either the CC or CV mode. An appropriate matching circuit can convert the CC mode into the CV mode and vice versa. By dynamically validating and invalidating the matching circuit based on the number of hops, the output characteristics can be fixed in either the CC or CV mode, regardless of the number of hops.

Chapter 4 introduced multi-hop wireless power transmission and its characteristics as well as the derivation of design theories for straight multi-hop wireless power transmission. In such a transmission, it is possible to extend the transfer distance using relay resonators; however, design parameters increase along with the number of resonators. To overcome this problem, design concepts were designed using the two most basic methods of BPF theorem and a design theorem for arbitrary hops. At the end of Chapter 4, it has to be noted that there are other design theories than those that were presented in this study. Hence, it would be helpful to select or develop the most suitable concept for each application scenario in order to maximize the efficiency.

References

[1]   Zhang F, Hackworth SA, Fu W, *et al.* Relay effect of wireless power transfer using strongly coupled magnetic resonances. *IEEE Transactions on Magnetics.* 2011;47(5):1478–1481.

[2]   Imura T. Equivalent circuit for repeater antenna for wireless power transfer via magnetic resonant coupling considering signed coupling. In: Proc. IEEE Conf. Ind. Electron. Appl.; 2011. p. 1501–1506.

[3]   Narusue Y, Kawahara Y, and Asami T. Impedance matching method for any-hop straight wireless power transmission using magnetic resonance. In: Proc. IEEE Radio Wireless Symp.; 2013. p. 193–195.

[4]   Hui SYR, Zhong W, and Lee CK. A critical review of recent progress in mid-range wireless power transfer. *IEEE Transactions on Power Electronics.* 2014;29(9):4500–4511.

[5]   Mori K, Lim H, Iguchi S, *et al.* Positioning-free resonant wireless power transmission sheet with staggered repeater coil array (SRCA). IEEE Antennas and Wireless Propagation Letters. 2012 January; p. 1710–1713.

[6]   Sun T, Xie X, Li G, *et al.* A two-hop wireless power transfer system with an efficiency-enhanced power receiver for motion-free capsule endoscopy inspection. *IEEE Transactions on Biomedical Engineering.* 2012;59 (11):3247–3254.

[7]   Imura T, Okabe H, Uchida T, *et al.* Study on open and short end helical antennas with capacitor in series of wireless power transfer using magnetic resonant couplings. In: Proc. 35th Annual Conf. of IEEE Ind. Electron.; 2009. p. 3848–3853.

[8]   FEKO. Altair Engineering, Inc. Available from: https://altair.com/feko. [Accessed 27 Feb 2024]

[9]   Zhong W, Lee CK, and Hui SR. General analysis on the use of Tesla's resonators in domino forms for wireless power transfer. *IEEE Transactions on Industrial Electronics.* 2013;60(1):261–270.

[10]   Kim J, Son HC, Kim KH, *et al.* Efficiency analysis of magnetic resonance wireless power transfer with intermediate resonant coil. *IEEE Antennas and Wireless Propagation Letters.* 2011;10:389–392.

[11]   Ahn D and Hong S. A study on magnetic field repeaters in wireless power transfer. *IEEE Transactions on Industrial Electronics.* 2013;60(1):360–371.

[12]   Narusue Y, Kawahara Y, and Asami T. Hercules: resonant magnetic coupling modules for 2D wireless energy sharing. *IPSJ Journal.* 2015; 56(1):250–259 (In Japanese).

[13]   Hashizume A, Narusue Y, Kawahara Y, *et al.* Receiver localization for a wireless power transfer system with a 2D relay resonator array. In: Proc. IEEE International Conf. Computational Electromagnetics. IEEE; 2017. p. 127–129.

[14]   Kazimierczuk MK. *RF power amplifiers.* John Wiley & Sons; 2008.

[15]   Awai I, Komori T, and Ishizaki T. Design and experiment of multi-stage resonator-coupled WPT system. In: Proc. IEEE MTT-S Int. Microw. Workshop Series Innov. Wireless Power Transmission: Technol., Syst., Appl.; 2011. p. 123–126.

[16]   Wei W, Miyasaka T, Kawahara Y, *et al.* Maximizing wireless power transmission efficiency with linear deployment resonator array and band pass filter theory. In: Adjunct Proc. Pervasive; 2012.

[17]   George LM, Leo Y, and Jones E. *Microwave filters, impedance-matching networks, and coupling structures.* McGraw-Hill Book Company; 1980.

[18]   Moriwaki Y, Imura T, and Hori Y. Basic study on reduction of reflected power using DC/DC converters in wireless power transfer system via magnetic resonant coupling. In: Proc. IEEE Int. Telecommunications Energy Conf.; 2011. p. 1–5.

[19]   Narusue Y, Kawahara Y, and Asami T. Maximizing the efficiency of wireless power transfer with a receiver-side switching voltage regulator. *Wireless Power Transfer.* 2017;4(1):42–54.

Chapter 5

Circuit theory on wireless couplers

Takashi Ohira[1]

5.1 Introduction

A passive device that delivers electric power to a remote point across a space is called a *wireless coupler*. A well-known example is a magnetic-field coupler consisting of twin loops, solenoidal, or spiral coils placed at a distance from each other, as shown in Figure 5.1(a). Another example is an electric-field coupler consisting of two pairs of metal solid or mesh plates normally facing each other across a space, as shown in Figure 5.1(b). In addition to these basic schemes, any passive system having two radio-frequency (RF) ports, even involving distributed-constant

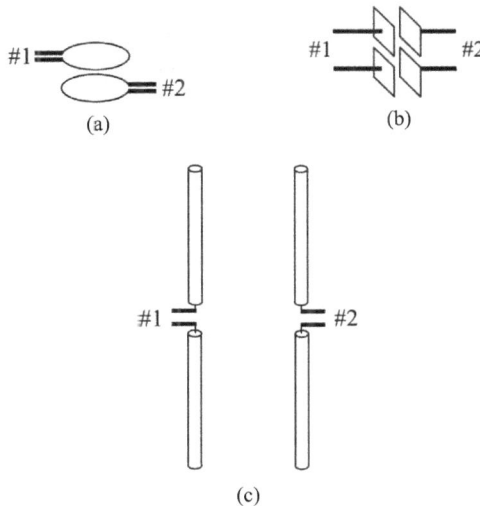

Figure 5.1   Three kinds of wireless couplers: (a) inductive, (b) capacitive, and (c) radiative

[1]Research Center for Future Vehicle City, Toyohashi University of Technology, Japan

elements, such as a pair of antennas located near each other as shown in Figure 5.1 (c), may work as a wireless coupler as well.

Wireless couplers are usually made from passive materials such as conductive metals or dielectric insulators. One can thus know their *S* matrix (two-port *S* parameters) by using an electromagnetic-field simulator. Especially for structures in a size much smaller than the wavelength, such as the ones shown as (a) and (b) in Figure 5.1, one can make their circuit equivalents consisting of lumped-constant elements.

Once the coupler of interest is characterized with an equivalent circuit or *S* matrix, we can evaluate how well it works or not for power transfer. In that evaluation, the key index we focus on is called *kQ*. It originated in the product of *coupling coefficient k* and *quality factor Q* for inductive couplers. Later, *kQ* was found to work for any kind of coupler as a *single figure of merit* that cannot always be separated into *k* and *Q* [1,2].

This chapter first introduces *kQ* formulas of simple inductive and capacitive couplers by way of their equivalent circuits. We then derive the optimum load impedance formula that brings the coupler into play at its best efficiency. The formulas are finally generalized to a unified theory that can apply to any kind of coupler represented by a two-port reciprocal black box.

5.2   Inductive coupler

5.2.1   Equivalent circuit

We start from the inductive coupler, as shown in Figure 5.1(a). For simplicity, the coupler is assumed to have a symmetrical shape between input and output. It is thus equivalent to the one-to-one transformer, as shown in Figure 5.2, where two identical coils exchange RF energy with each other via magnetic interlinkage flux.

The symmetrical transformer is characterized by its element parameters, that is, self-inductance *L* and mutual inductance *M*. Each coil has parasitic resistance in series that represents the power loss due to RF dissipation inside and radiation to the outside. These element parameters are summarized into a two-by-two impedance matrix as

$$\begin{bmatrix} z_{11} & z_{12} \\ z_{12} & z_{22} \end{bmatrix} = R \begin{bmatrix} 1 & 0 \\ 0 & 1 \end{bmatrix} + j\omega \begin{bmatrix} L & M \\ M & L \end{bmatrix} \tag{5.1}$$

where *j* stands for imaginary unit $\sqrt{1}$. Note that ω stands for angular frequency of excitation, which is not to be confused with the one from the circuit's resonance.

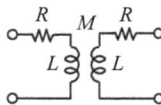

Figure 5.2   Inductive coupler equivalent circuit

We should keep this matrix in mind as a foundation to learn the circuit theory on inductive couplers during this section.

5.2.2 Coupling coefficient

To use the coupler in WPT systems, the primary coil is excited with RF current. The current generates magnetic flux around it. Part of the flux links to the secondary coil, and the rest of it leaks out. The primary-to-secondary interlinkage flux ratio is called *coupling coefficient*, denoted as *k*. In terms of the element parameters, *k* is expressed as

$$k = \frac{M}{L} \tag{5.2}$$

where *L* and *M* designate self and mutual inductance, respectively, as shown in Figure 5.2. Since *M* cannot exceed *L* in nature, *k* ranges from zero to unity.

5.2.3 Quality factor

Imagine a stand-alone coil that has inductance *L* and resistance *R* in series. This is exactly what we find in the right or left half part of the circuit, as shown in Figure 5.2. We can formulate it by focusing on the diagonal component of the impedance matrix. Actually from (5.1), the coil's complex impedance is expressed as

$$z_{11} = z_{22} = R + j\omega L. \tag{5.3}$$

From this equation, we extract its imaginary-to-real part ratio

$$Q = \frac{\omega L}{R} \tag{5.4}$$

which is usually known as a coil's *quality factor* or *Q factor*. Again, note that ω is not the circuit's resonant frequency but the incoming wave frequency.

Not only for inductors, but *Q* works as a figure of merit also for other passive and active one-port devices such as resonators and oscillators [3]. What we further learn in this section is that *Q* works even for *two-port* systems such as inductive couplers. This comes true not from sole *Q* but in conjunction with coupling coefficient *k*.

5.2.4 Coupling quality factor

Just multiplying *k* and *Q* from (5.2) and (5.4), it is quite straightforward to deduce

$$kQ = \frac{\omega M}{R}. \tag{5.5}$$

This is called *coupling Q factor* or *kQ product*. The physical role of *kQ* in general two-port systems is described later, but one thing we can notice here is that *kQ* for the inductive coupler is invariant to self-inductance *L*. This is because *L*, stemming from k and one from Q, cancels each other out. This implies that external addition of any reactor or resonator to the coupler can never contribute to its kQ enhancement.

The above kQ formula is valid only for a symmetrical scheme, where the two coils have an identical property. In case they have different L and/or R, we should be able to modify the formula after learning the versatile black box model outlined later in Section 5.4.

5.2.5 Optimum impedance

The mission of wireless couplers is to deliver RF power from a source to a load. To effectively carry out this mission, we should avoid unsolicited wave reflection at the ports. In other words, power transfer efficiency depends not only on the coupler itself but also on the load to be connected at the output port.

Particularly, the inductive coupler shown in Figure 5.2 works best when the secondary coil is loaded with impedance:

$$Z_{opt} = \omega M - j\omega L \tag{5.6}$$

where we assume a good coupler that satisfies $kQ >> 1$ or $R << \omega M$. A mathematical proof of this formula is described in [6], but here we look at it from a physical aspect. The right-hand side of (5.6) finds that its first term (ωM) implies the optimum load resistance, and its last term ($-j\omega L$) compensates for the coil reactance. It may look somewhat unusual but is true that mutual inductance M contributes here not to imaginary but to the real part of Z_{opt}.

In developing practical WPT systems, we cannot always choose an appropriate load. It is sometimes provided a priori, whose impedance may be different from the optimum. For such situations, we should be able to satisfy (5.6) by inserting an output-impedance-matching circuit between the secondary coil and the load. This is the same way as that employed to avoid unwanted wave reflection in other RF devices and systems.

As well as the load impedance, another issue of concern is the input impedance Z_{in} of the coupler. Due to possible multiple reflections between its input and output ports, Z_{in} changes with the load impedance. When the load or the abovementioned matching circuit meets (5.6), the coupler exhibits input impedance:

$$Z_{in} = \omega M + j\omega L. \tag{5.7}$$

This is indispensable information when we design an RF power source or an input-impedance-matching circuit to be used between the source and the primary coil.

5.2.6 Maximum efficiency

The prime concern in wireless coupler engineering is power transfer efficiency. It is generally defined as

$$\eta = \frac{P_{out}}{P_{in}} \tag{5.8}$$

where P_{in} and P_{out}, respectively. denote power input to the primary coil and output from the secondary. As described in the previous section, η depends not only on the coupler itself but also on the load to be connected at the output port. Only when the load meets (5.6), can η achieve its peak, which is denoted as η_{max}.

Particularly for the inductive coupler shown in Figure 5.2, η_{max} is explicitly expressed in terms of the equivalent element parameters, that is,

$$\eta_{max} = \frac{\omega M - R}{\omega M + R}. \tag{5.9}$$

This is the formula of *maximum power transfer efficiency* for inductive couplers. Note again that we assume $R << \omega M$ for simplicity. This assumption usually applies in pragmatic high-efficiency couplers. For those interested in rigorous formulas, consult the general theory described in Section 5.4.

5.3 Capacitive coupler

We move on to the capacitive coupler, as shown in Figure 5.1(b). Electric-field coupling was initially supposed to be applicable only to low-power systems such as wireless charging of mobile phone batteries. However, later, it was demonstrated to make a battery-less electric vehicle run with kilowatt-class wireless power via electric field [5]. Capacitive coupling is now considered to have become a strong option for a wide variety of wireless power transfer applications. This section formulates capacitive coupling as the contrary alternative to an inductive one. The duality theorem between wound coils and parallel plates may provide useful assistance in translating the following formulation.

5.3.1 Equivalent circuit

For simplicity again, the coupler is assumed to have a symmetrical shape. It consists of two parallel pairs of metal plates normally facing each other, which is equivalent to the twin capacitors, as shown in Figure 5.3.

The circuit features two main capacitors C_2 between the facing plates. As parasitic effects, it includes stray capacitors C_1 and shunt resistors R. The resistors represent the power loss due to RF conductive current leakage between the terminals on both sides. These element parameters are summarized into a two-by-two

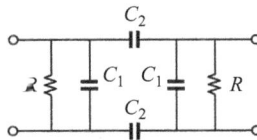

Figure 5.3   Capacitive coupler equivalent circuit

admittance matrix as

$$
\begin{bmatrix} y_{11} & y_{12} \\ y_{12} & y_{22} \end{bmatrix} = \left(\frac{1}{R} + j\omega C_1 \right) \begin{bmatrix} 1 & 0 \\ 0 & 1 \end{bmatrix} + \frac{1}{2} j\omega C_2 \begin{bmatrix} 1 & -1 \\ -1 & 1 \end{bmatrix}. \tag{5.10}
$$

We should keep this matrix in mind as a foundation to learn the circuit theory on capacitive couplers during this section.

5.3.2 Coupling coefficient

To use the coupler in WPT systems, an RF voltage is applied to the input port. The voltage creates electric flux around the plates. Part of the flux reaches the output port, and the rest of it leaks out. The primary-to-secondary flux ratio is called *coupling coefficient*, denoted as k. In terms of the element parameters, k is expressed as

$$
k = \frac{C_2}{2C_1 + C_2}. \tag{5.11}
$$

If high k is needed, we must increase the C_2-to-C_1 ratio. However, k is limited to unity because both C_2 and C_1 are always positive in nature.

5.3.3 Quality factor

In the same way as we derived (5.3), we focus on the diagonal component of the admittance matrix in (5.10). It is expressed as

$$
y_{11} = y_{22} = \frac{1}{R} + j\omega \left(C_1 + \frac{1}{2} C_2 \right) \tag{5.12}
$$

from which we extract its imaginary-to-real part ratio to yield

$$
Q = \omega \left(C_1 + \frac{1}{2} C_2 \right) R. \tag{5.13}
$$

This is the Q factor formula for the capacitive coupler. Be sure once again not to mix up incoming wave frequency ω with the circuit's resonance. It is worth comparing this formula with the inductive coupler. We find Q increases with R in (5.13), which is opposite to that seen in (5.4).

5.3.4 Coupling quality factor

From (5.10) and (5.12), it is quite straightforward again to deduce

$$
kQ = \frac{1}{2} \omega C_2 R. \tag{5.14}
$$

This is the kQ formula of capacitive couplers. Looking at the right-hand side, we should increase C_2 and R for couplers to achieve high performance. We also find here that kQ is invariant to stray capacity C_1. In other words, we can append any external reactor or compensator (assumed lossless) to the port while keeping kQ unchanged from its original value.

5.3.5 Optimum admittance

The capacitive coupler shown in Figure 5.3 works best when it is loaded with admittance:

$$Y_{\text{opt}} = \frac{1}{2}\omega C_2 - j\omega C_1 \tag{5.15}$$

where we assume a good coupler satisfies $kQ \gg 1$. The first term of the right-hand side implies the optimum load conductance, and the last term compensates for the reactance of C_1. It may look unusual again but is true that C_1 contributes not to imaginary but to the real part of Y_{opt}.

As well as the load admittance, another issue of concern is the input admittance Y_{in} of the coupler. When the load meets (5.15), the coupler exhibits input admittance:

$$Y_{\text{in}} = \frac{1}{2}\omega C_2 + j\omega C_1. \tag{5.16}$$

This is indispensable information when we design an RF power source or an input-matching circuit to be used between the source and the coupler.

5.3.6 Maximum efficiency

It is exactly common to (5.8) that the power transfer efficiency *is defined as*

$$\eta = \frac{P_{\text{out}}}{P_{\text{in}}}. \tag{5.17}$$

Only when the load meets (5.15), η achieves its peak η_{max}. For the capacitive coupler, it is expressed as

$$\eta_{\text{max}} = \frac{\omega C_2 R - 2}{\omega C_2 R + 2}. \tag{5.18}$$

Note that it is valid for good couplers that satisfy $\omega C_2 R \gg 2$. The right-hand side of this equation advises us again that C_2 and R should be increased in designing high-performance couplers. When either C_2 or R goes to infinite, η_{max} converges into unity, that is, 100%.

5.4 Generalized formulas

The inductive and capacitive couplers we learned in the previous sections have ideal schemes, so that they are indeed useful to understand the basis of wireless power transfer. However, they are too simple to represent pragmatic systems that often have complicated structures. For example, a pair of half-wave dipole antennas

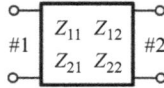

Figure 5.4   Two-port black box representing wireless couplers

shown in Figure 5.1(c) is one that works as a wireless coupler [7]. It is quite difficult to establish an accurate equivalent circuit of the antenna due to its strong wave radiation effects. There are also many other examples suffering from serious parasitic effects, for which we cannot even find an analytical way to estimate k or Q. To address these problems, this section extends the formulas to general ones that we can apply to any scheme of wireless coupler.

5.4.1   Two-port black box

We introduce a two-port black box as shown in Figure 5.4 and exploit it as a versatile model to formulate kQ, optimum impedance, and maximum efficiency. The adjective word *black* means that we do not have to know what is inside the box. All we need is its two-by-two impedance matrix, which can be just observed at the input and output ports.

5.4.2   Impedance matrix

A wireless coupler, in any scheme, is mathematically equivalent as a whole to a linear two-port circuit. It is fully characterized by the impedance matrix consisting of four complex-number components. They can be each decomposed into their real (resistive) and imaginary (reactive) parts, and thus, the matrix is entirely expressed as

$$\begin{bmatrix} Z_{11} & Z_{12} \\ Z_{21} & Z_{22} \end{bmatrix} = \begin{bmatrix} R_{11} & R_{12} \\ R_{21} & R_{22} \end{bmatrix} + j \begin{bmatrix} X_{11} & X_{12} \\ X_{21} & X_{22} \end{bmatrix} \tag{5.19}$$

where j stands for imaginary unit $\sqrt{-1}$. Subscripts 1 and 2 designate the port address. The matrix components can be calculated by an electromagnetic-field simulator if we know the coupler's material and structure. Even if not, they are at least measurable by employing a vector network analyzer when we obtain a fabricated sample.

The eight components on the right-hand side of (5.19) may exhibit any real number. However, they actually have constraints of

$$\begin{aligned} R_{11} &\geq 0, \quad R_{22} \geq 0 \\ R_{11}R_{22} &\geq R_{12}R_{21} \end{aligned} \tag{5.20}$$

because the coupler system generally involves no active devices such as transistors. Also, thanks to reciprocity between ports #1 and #2, the non-diagonal components are always commutative as

$$\begin{aligned} R_{12} &= R_{21} \\ X_{12} &= X_{21} \end{aligned} \tag{5.21}$$

Note that this keeps true regardless of whether the system is symmetrical or not with respect to the two ports.

One last item we should address before starting formulation is a set of two quadratic quantities

$$\Delta = R_{11}R_{22} - R_{12}R_{21}$$
$$\Sigma = R_{11}R_{22} + X_{12}X_{21}$$

(5.22)

introduced for mathematical elegance. Be careful that Σ is a hybrid of resistance and reactance, while Δ is a pure resistance matrix determinant. They are both nonnegative real numbers, which enable kQ and related formulas to come in quite comfortable or easy-to-remember fashions.

5.4.3 Generalized kQ

Since there is no information on what is inside the black box, we cannot find its equivalent circuit in a simple way. Might one try to synthesize it in some clever way, it would result in such convolution that both of the k and Q definitions become theoretically ambiguous. Still, even in that situation, there exists a sound way to evaluate the box from the aspect of power transfer potential [4,5]. We introduce a single figure:

$$kQ = \sqrt{\frac{\Sigma}{\Delta} - 1}$$

(5.23)

where Σ and Δ are the quadratics defined in (5.22). This is what we call *generalized kQ* formula. It covers all kinds of wireless couplers, including those shown in Figure 5.1(a), (b), and even (c). The *single* figure means that it works regardless of whether k and Q can be separable or not.

A mathematical proof of (5.23) is complicated and takes time to review [1]. Readers are best advised to first experience the formula by applying it to the simple example, as shown in Figure 5.2. By picking up the necessary components from (5.1), and putting them into (5.22), we get the quadratics reducing to

$$\Delta = R^2$$
$$\Sigma = R^2 + \omega^2 M^2$$

(5.24)

Then substituting them into (5.23), we finally reach

$$kQ = \sqrt{\frac{R^2 + \omega^2 M^2}{R^2} - 1} = \frac{\omega M}{R}.$$

(5.25)

This is not anything but identical to (5.5). We thus recognize that (5.5) is a special case of the general formula shown in (5.23).

5.4.4   Optimum load and input impedance

As well as kQ, the impedance-domain quantities are extended to be applicable to an arbitrary type of wireless coupler. The optimum load impedance of the black box shown in Figure 5.4 is

$$Z_{opt} = R_{opt} + jX_{opt}$$
$$\left(\begin{array}{l} R_{opt} = \dfrac{\sqrt{\Delta\Sigma}}{R_{11}} \\[2mm] X_{opt} = \dfrac{R_{21}X_{21}}{R_{11}} - X_{22} \end{array} \right. \tag{5.26}$$

where R_{ij} and X_{ij} are the circuit parameters seen in (519)–(5.22).

Readers can see how (5.26) works for the same example as used in the last section. Remembering (5.1) for the matrix components and (5.24) for the quadratics, we find (5.26) reduces to

$$R_{opt} = \sqrt{R^2 + \omega^2 M^2}$$
$$X_{opt} = -j\omega L \tag{5.27}$$

Under the good coupler assumption, that is, $R << \omega M$, the above square root just approximates ωM. This results in

$$Z_{opt} = \omega M - j\omega L \tag{5.28}$$

which is identical to (5.6).

Another issue we should extend is the input impedance. When the black box is terminated with the optimum load specified by (5.26) at port #2, the input impedance at port #1 exhibits

$$Z_{in} = R_{in} + jX_{in}$$
$$\left(\begin{array}{l} R_{in} = \dfrac{\sqrt{\Delta\Sigma}}{R_{22}} \\[2mm] X_{in} = X_{11} - \dfrac{R_{12}X_{12}}{R_{22}} \end{array} \right. \tag{5.29}$$

Note that reactance X_{in} has opposite polarity to that of X_{opt}, while resistance is always positive. This implies the complex-conjugate technique for impedance matching in usual RF devices. For students, a good exercise is to illustrate that (5.29) reduces to (5.7) particularly for the simple inductive coupler shown in Figure 5.2.

5.4.5   Maximum efficiency

The final issue we should extend is the maximum power transfer efficiency. We refer again to the black box, as shown in Figure 5.4. The definition of power transfer efficiency between ports #1 and #2 is the same as that in (5.8) and (5.17). When the box is terminated with the optimum load specified by (5.26) at port #2, the power transfer efficiency achieves its peak:

$$\eta_{max} = \frac{\sqrt{\Sigma} - \sqrt{\Delta}}{\sqrt{\Sigma} + \sqrt{\Delta}} \qquad (5.30)$$

This is the general formula of η_{max} valid for arbitrary wireless couplers. For example, one can apply it to the simple inductive coupler, as shown in Figure 5.2. Adopting Δ and Σ from (5.24), the above general formula reduces to

$$\eta_{max} = \frac{\sqrt{R^2 + \omega^2 M^2} - R}{\sqrt{R^2 + \omega^2 M^2} + R} \qquad (5.31)$$

which approximates (5.9) by assuming $R \ll \omega M$.

Indeed (5.30) is convenient for numerical computation since it makes η_{max} directly countable from Δ and Σ. However, from the aspect of physics, it looks rather complicated due to the four square roots involved. We therefore renovate it into

$$\eta_{max} = \frac{\rho - 1}{\rho + 1} \qquad (5.32)$$

where ρ is a dimensionless parameter heuristically introduced for the sake of mathematical elegance. This formula smoothly comes into one's attention and can be kept in mind with comfort. To make (5.32) identical to (5.30), we define ρ as

$$\rho = \sqrt{\frac{\Sigma}{\Delta}} = \sqrt{1 + (kQ)^2} \qquad (5.33)$$

where the final right-hand side is yielded with the help of the generalized kQ formula that appeared in (5.23). From (5.32) and (5.33), we find that kQ always dominates η_{max} by way of ρ. As kQ ranges from zero to infinite, η_{max} monotonously increases from zero to unity. To know how kQ and η_{max} relate more in detail, geometry effectively helps. Draw a chart as shown in Figure 5.5 by following this five-step instruction:

1. An arc stretches its string ST, of which the middle point is marked M.
2. Create a movable point Q on the rightward extension of ST.

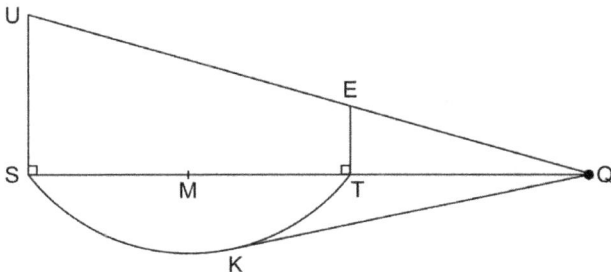

Figure 5.5   Graphical analogy to how kQ dominates η_{max}

3. From Q, draw a tangent to the arc. Their contact point is marked K.
4. Stand two perpendicular lines US and ET upright at both ends of ST.
5. Points U, S, M, and T are fixed to keep US = SM = MT = 1.

After completing the drawing, imagine that point Q moves from T to rightward infinite far. Length KQ accordingly increases from zero to infinite. Length ET also increases from zero but is limited to unity. One may now hit upon an idea that these two lengths represent the behaviors of kQ and η_{max}, respectively. That is exactly true and proved by geometry as follows.

Start from defining kQ as the length of tangential line piece KQ. Next, regarding the arc as part of a circle, the power point theorem finds SQ × TQ = $(kQ)^2$. Since the left-hand side can be rewritten as (MQ + 1) × (MQ − 1), we obtain

$$MQ = \sqrt{1 + (kQ)^2}.$$

Referring to (5.33), we notice MQ = ρ, and therefore, SQ = $\rho + 1$ and TQ = $\rho - 1$. Then, look at the two perpendicular lines US and ET. They measure the height of two similar right triangles USQ and ETQ, resulting in US:ET = SQ:TQ. Remembering US = 1, we finally conclude that

$$ET = \frac{TQ}{SQ} = \frac{\rho - 1}{\rho + 1}.$$

This is identical to η_{max} seen in (5.32).

In summary, a flow chart involving every parameter that has appeared in this section is shown in Figure 5.6. Each box in this chart denotes a parameter's name or

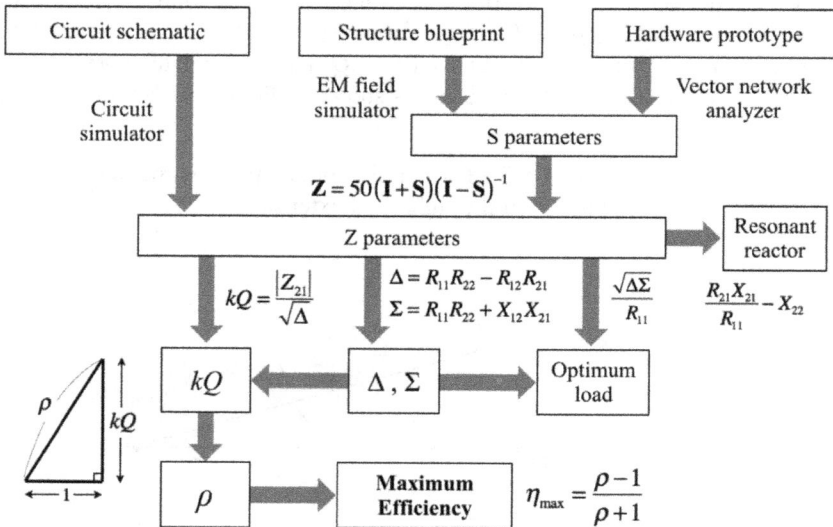

Figure 5.6   Flow chart to estimate η_{max} of a wireless coupler

its mathematical symbol. The arrows from each box imply the numerical operations needed to proceed to the next box. To start, one may choose the circuit schematic, structure blueprint, or hardware prototype. For any of these, the chart leads to the final destination η_{max}.

5.5   Conclusion

We have opened up a lucid theoretical vista to wireless couplers. Simple inductive and capacitive couplers are characterized with lumped-constant equivalent circuits. Coupling coefficient and Q factor are explicitly expressed in terms of the circuit-element parameters. This formulation leads us to an understanding of the fundamental behavior of wireless couplers.

For practical applications, a two-port reciprocal black box is introduced as a versatile model for a wireless coupler of any kind. Generalized k_Q and related formulas are derived in regard to the impedance matrix components of the black box. How k_Q relates to the maximum power transfer efficiency is also visualized in a graphical chart. The theory we learn here provides an essential piece of knowledge for RF engineering and will help the design and development of prospective wireless power transfer systems.

Acknowledgment

This work is supported in part by the Government of Japan (Contract: MEXT KAKENHI 17K06384).

Appendix A

A.1   Measurement of kQ in practice

Since the concept of kQ has not yet been widely deployed for WPT engineers, this appendix gives a practical example of kQ measurement. The best way to get familiar with something new, in general, is the experience of a simple experiment. The key instrument in this experiment is a two-port vector network analyzer. After careful calibration of the analyzer, we can accurately observe the S parameters of a device under test. The setup is so simple that we just connect the analyzer to the device with two coaxial cables, as shown in Figure A1(a).

For the device under test, we employ a pair of flat spiral coils. They are made of copper-Litz wires wound up to 35 mm in diameter. We place them face to face on a grid-section sheet to fix them exactly at the specified distance to each other, as shown in Figure A1(b). The coil pair works as a set of magnetic-flux emitter and acceptor. Each of them has a coaxial jack to connect the cable, as shown in Figure A1(b).

We should be careful to keep the coils away from foreign metal obstacles. Otherwise, an unexpected eddy current may take place, which degrades the magnetic coupling performance. We even recommend the use of a styrofoam spacer under the coupler to avoid possible effects of the floor material.

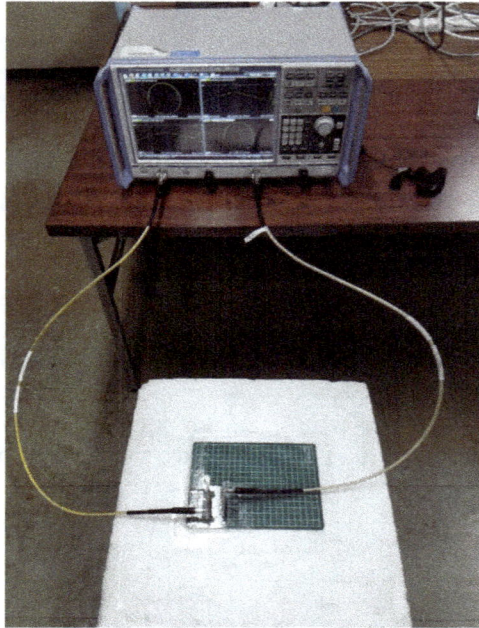

(a)

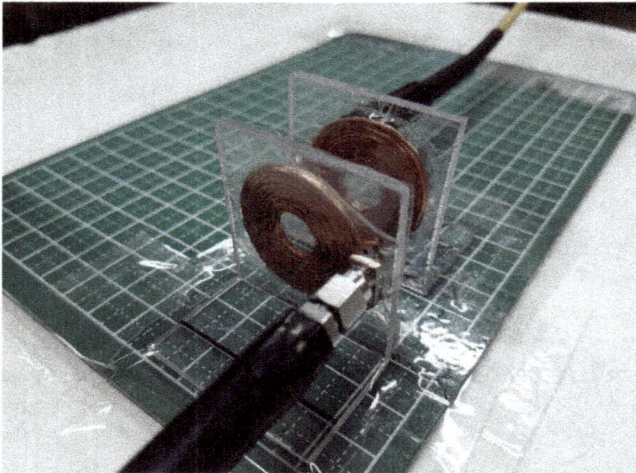

(b)

Figure A1   Wireless coupler evaluation with respect to kQ: (a) experimental setup
for two-port S parameter measurement; (b) closer look at the coupler

The *S* parameters observed at five typical frequencies from 10 kHz to 100 MHz are shown in Table A1. From these four *S* parameters, compose a two-by-two matrix

Table A1 Measured S parameters as for 5 mm in coil-to-coil distance

Frequency	S_{11}		S_{12}		S_{21}		S_{22}	
	Real	Imag.	Real	Imag.	Real	Imag.	Real	Imag.
10 kHz	−0.996	0.031	−0.000428	−0.0135	−0.000373	−0.0134	−0.995	0.0295
100 kHz	−0.946	0.28	−0.0363	−0.118	−0.0358	−0.118	−0.944	0.284
1 MHz	0.182	0.845	−0.398	0.0830	−0399	0.0824	0.198	0.841
10 MHz	0.978	0.154	−0.0152	0.0838	−0.0157	0.0840	0.979	0.146
100 MHz	0.827	−0.312	0.0336	0.0238	0.0336	0.0237	0.886	−0.345

$$\mathbf{S} = \begin{bmatrix} S_{11} & S_{12} \\ S_{21} & S_{22} \end{bmatrix}$$

and put it into

$$\mathbf{Z} = 50(\mathbf{I} + \mathbf{S})(\mathbf{I} - \mathbf{S})^{-1}$$

where 50 comes from the reference impedance of the network analyzer used in the S parameter measurement. $\mathbf{I}$ stands for identity matrix. Superscript "-1" designates matrix inversion. For convenience, let us refresh our memories with their two-by-two matrix formulas:

$$\mathbf{I} = \begin{bmatrix} 1 & 0 \\ 0 & 1 \end{bmatrix}$$

$$\begin{bmatrix} a & b \\ c & d \end{bmatrix}^{-1} = \frac{1}{ad - bc}\begin{bmatrix} d & -b \\ -c & a \end{bmatrix}$$

As a numerical example, extract the four S parameters at 1 MHz from Table A1, and put them into the above translation formula. We then get

$$\mathbf{Z} = 50\left(\begin{bmatrix} 1 & 0 \\ 0 & 1 \end{bmatrix} + \begin{bmatrix} 0.182 + j0.845 & -0.398 + j0.0830 \\ -0.399 + j0.0824 & 0.198 + j0.841 \end{bmatrix}\right) \cdot$$
$$\left(\begin{bmatrix} 1 & 0 \\ 0 & 1 \end{bmatrix} - \begin{bmatrix} 0.182 + j0.845 & -0.398 + j0.0830 \\ -0.399 + j0.0824 & 0.198 + j0.841 \end{bmatrix}\right)^{-1}$$
$$= \begin{bmatrix} 3.55 + j70.2 & -1.48 - j30.9 \\ -1.43 - j30.9 & 3.64 + j71.4 \end{bmatrix}$$

This gives the four Z parameters by referring to the definition of impedance matrix:

$$\mathbf{Z} = \begin{bmatrix} Z_{11} & Z_{12} \\ Z_{21} & Z_{22} \end{bmatrix}$$

The Z parameters obtained at the five frequencies are shown in Table A2.

Table A2   Translated Z parameters (Ω) from Table A1

Frequency	Z_{11}		Z_{12}		Z_{21}		Z_{22}	
	Real	Imag.	Real	Imag.	Real	Imag.	Real	Imag.
10 kHz	0.0827	0.778	−0.000446	−0.339	0.000846	−0.337	0.100	0.741
100 kHz	0.155	7.27	−0.0317	−3.18	−0.0204	−3.18	0.178	7.38
1 MHz	3.55	70.2	−1.48	−30.9	−1.43	−30.9	3.64	71.4
10 MHz	51.6	933	−39.7	−539	−37.2	−541	57.8	981
100 MHz	83.7	−248	−4.59	−31.4	−4.60	−31.3	32.5	−262

From these Z parameters, we next calculate two quadratic quantities:

$$\Delta = R_{11}R_{22} - R_{12}R_{21}$$
$$= 3.55 \times 3.64 - (-1.48) \times (-1.43)$$
$$= 10.8$$

$$\Sigma = R_{11}R_{22} + X_{12}X_{21}$$
$$= 3.55 \times 3.64 + (-30.9) \times (-30.9)$$
$$= 966$$

where their definitions come from (5.22). Substituting them into (5.23), we finally obtain

$$kQ = \sqrt{\frac{\Sigma}{\Delta} - 1}$$
$$= \sqrt{\frac{966}{10.8} - 1}$$
$$= 9.40$$

The results obtained at the five frequencies are summarized in Table A3.

Continuously sweeping the frequency, we measure S parameters and translate them into kQ. The results are projected in Figure A2. We find that kQ increases when the two coils come near to each other. In the frequency responses, we also find two peaks of kQ: one at around 100 kHz and another at around 20 MHz.

The kQ measurement procedure described here is just based on the conventional S parameters observed at the two ports of the system under test. There is no need to create equivalent circuit models of the system to extract its k or Q separately. This means that we can apply the above formulas not only to such a simple coil pair as presented in this appendix but also to any other coupling scheme, no matter how complicated their structures are. Preferably, we can build an automatic kQ measurement system by installing the program into a software-controlled vector network analyzer.

Acknowledgment

The author would like to thank Shinji Abe for his technical work in the experiment and numerical computation carried out at Toyohashi University of Technology.

Table A3   Translated kQ from Table A2

Frequency	Δ	Σ	kQ
10 kHz	0.00830	0.123	3.71
100 kHz	0.0270	10.1	19.3
1 MHz	10.8	966	9.40
10 MHz	1,510	294,000	13.9
100 MHz	2,700	3,710	0.610

A.2   Geometrical demonstration from kQ to η*max*

The maximum power transfer efficiency of wireless couplers, denoted as η_{max}, is exclusively dominated by the coupler's kQ, that is, coupling coefficient k multiplied by quality factor Q. This rule is mathematically expressed as

$$\eta_{\text{max}} = \frac{\sqrt{1 + k^2 Q^2} - 1}{\sqrt{1 + k^2 Q^2} + 1}. \tag{A1}$$

This equation may look complicated for engineers to imagine how η_{max} behaves as a function of kQ. To project this relation onto graphics, we introduce such an angle θ as to meet

$$kQ = \tan 2\theta. \tag{A2}$$

Employing this angle, (A1) becomes amazingly elegant as

$$\eta_{\text{max}} = \tan^2 \theta. \tag{A3}$$

That is to say, kQ and η_{max} are correlated by way of θ. In this sense, $\tan \theta$ should be called efficiency tangent by analogy with so-called loss tangent. Square on $\tan \theta$ implies a quantity measurable on a power scale rather than an amplitude scale.

Based on (A2) and (A3), we can visualize the relation (A1), as shown in Figure A3. This geometrical demonstration can be performed exactly by the following procedure. Create a right triangle with its unity-length base and height kQ.

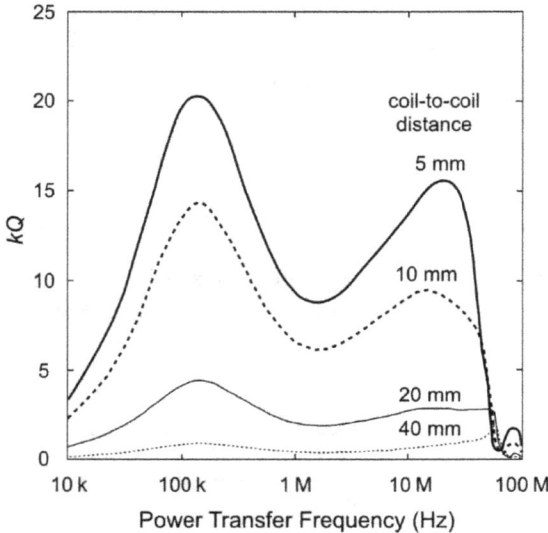

Figure A2   Exhibited kQ in continuous frequency sweep

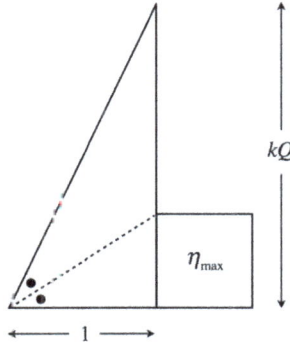

Figure A3 *Right triangle and square to demonstrate how kQ and η_{max} relate to each other*

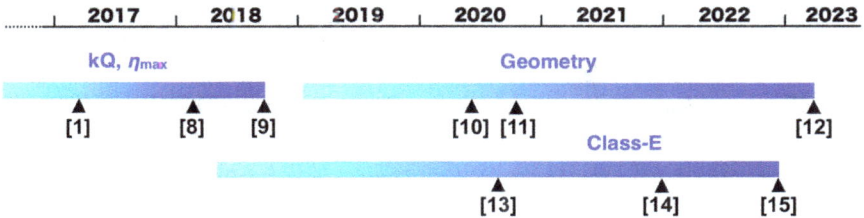

Figure A4 *Chronological chart of recent progress in WPT theory*

Draw a broken line that precisely bisects the left angle. Then, a small triangle appears inside, which stands $\tan\theta$ tall. Create a square with its side common to the small triangle in height. The square's area finally signifies η_{max}. Looking at these graphics, we realize a basic law of wireless power transfer: η_{max} ranges from zero to unity when kQ ranges from zero to infinity.

A.3 Progress in WPT theory

Wireless power transfer systems are usually designed by using electromagnetic-field simulators and analog circuit simulators. Although the simulators are so smart and useful, a basic question always remains: What is the underlying theory? If we can establish the basic theory, it is quite instructive for both educational and engineering purposes. The recent progress in WPT theory is shown in Figure A4. The progress covers not only the wireless couplers [1,8,9] but also geometrical approach to WPT system design [10–12], as well as class-E operation of power conversion such as inverters and rectifiers [13–15]. Students and engineers are strongly recommended to visit the nine references.

References

[1]   T. Ohira, "The kQ product as viewed by an analog circuit engineer," *IEEE Circuits and Systems Magazine*, vol. 17, no. 1, pp. 27–32, 2017.

[2]   Q.-T. Duong and M. Okada, "*kQ*-product formula for multiple transmitter inductive power transfer system," *IEICE Electronics Express*, vol. 14, no. 3, pp. 1–8, 2017.

[3]   T. Ohira, "Enigma on resonant quality," *IEEE Microwave Magazine*, vol. 18, no. 2, p. 119, 2017.

[4]   T. Ohira, "How to estimate the coupling Q factor from two-port S-parameters (invited)," *IEEE AP-S International Conference Computational Electromagnetics*, pp.120–121, Kumamoto, March 2017.

[5]   T. Ohira, "A battery-less electric roadway vehicle runs for the first time in the world (invited)," *IEEE MTT-S International Conference Microwave Intelligent Mobility*, MO4-1, Nagoya, March 2017.

[6]   T. Ohira, "Enigma on two-port resonance," *IEEE Microwave Magazine*, vol. 18, no. 4, p. 149, 2017.

[7]   T. Ohira and N. Sakai, "Dipole antenna pair revisited from kQ product and Poincaré distance for wireless power transfer (invited)," *IEEE Conference Antenna Measurements Applications*, Tsukuba, Dec. 2017.

[8]   T. Ohira, "Power transfer efficiency formulation for reciprocal and non-reciprocal linear passive two-port systems," *IEICE Electronics Express*, vol. 15, no. 3, pp. 1–6, 2018.

[9]   T. Ohira, "Power transfer theory on linear passive two-port systems," *IEICE Transaction on Electronics*, vol. E101-C, no. 10, pp. 719–726, Oct. 2018.

[10] T. Ohira, "Beltrami-Klein disk model as viewed for use in impedance trajectory projection," *IEICE Communications Express*, vol. 9, no. 7, pp. 256–261, 2020.

[11] T. Ohira, "A radio engineer's voyage to double-century-old plane geometry," *IEEE Microwave Magazine*, vol. 21, no. 11, pp. 60–67, 2020.

[12] T. Ohira, "Plane geometry inspires wave engineering starters," *IEEE Microwave Magazine*, vol. 24, no. 3, pp. 93–98, 2023.

[13] T. Ohira, "Load impedance perturbation formulas for class-E power amplifiers," *IEICE Communications Express*, vol. 9, no. 10, pp. 482–488, 2020.

[14] T. Ohira, "Linear algebra elucidates class-E power amplifiers," *IEEE Microwave Magazine*, vol. 23, no. 1, pp. 83–105, 2022.

[15] T. Ohira, "Linear algebra elucidates class-E diode rectifiers," *IEEE Microwave Magazine*, vol. 23, no. 12, pp. 113–122, 2022.

Chapter 6

Inverter/rectifier technologies

Hiroo Sekiya[1]

This chapter introduces a design theory for optimal magnetic-coupling wireless power transfer (WPT) systems from a circuit-theory viewpoint. WPT systems are generally divided into three parts: DC/AC inverter, coupling component, and AC/DC rectifier. To achieve high-efficiency WPT systems, it is crucial to minimize power losses in each part. This chapter also introduces high-efficiency DC–AC inverters and AC–DC rectifiers that achieve high power-conversion efficiency at high frequencies, particularly due to satisfying soft-switching conditions. However, optimizing each part individually is not sufficient. As certain parts affect one another, it is essential to design the WPT system optimally as an integrated system of all three parts. This concept is demonstrated through a design example of the class-E^2 WPT system.

6.1 Introduction

WPT systems have garnered significant attention from both industry and academia. They are expected to be applied in various applications, such as battery chargers for electric vehicles, smartphones, and consumer equipment [1–5]. Moreover, WPT systems have the potential to bring about a dramatic societal impact, much like communication technology, owing to the profound influence of "wireless" technology. Among the various coupling methods in WPT systems, this chapter specifically focuses on magnetic-field coupling, which is crucial for achieving high power-delivery efficiency.

In general, a WPT system is divided into three main parts: the DC–AC inverter, coupling, and DC–AC rectifier. In magnetic-field coupling WPT systems, these parts involve the use of two coils, often represented as a transformer with a low coupling coefficient in the electric-circuit model. The transmitter comprises the DC–AC inverter [6–30] and the primary coil, whereas the receiver includes the secondary coil and the AC–DC rectifier [6,31–40]. The DC–AC inverter is responsible for generating AC current flowing through the primary coil in the transmitter. In the receiver, the received AC current needs to be converted into DC voltage, which is achieved by the

[1]Graduate School of Engineering, Chiba University, Japan

rectifier. To achieve high power-delivery efficiency in the WPT system, it is crucial to minimize power losses at each part. However, optimizing each part individually is not sufficient since the interactions among the parts make it vital to design the WPT system as an integrated system [4,5,10,18].

This chapter delves into the design theory and techniques of WPT systems with magnetic-field coupling. The WPT system is modeled using the electric circuit, and optimal designs are discussed from a circuit-theory viewpoint. The necessity and effectiveness of magnetic resonance are expressed using circuit theory. Additionally, the chapter introduces circuit topologies and fundamental operations of rectifiers and inverters that achieve high power-conversion efficiency at high frequencies. Finally, a design example of the optimal WPT system is presented in this chapter.

6.2   WPT system construction

Figure 6.1 illustrates a typical construction of magnetic-field coupling WPT systems. In this chapter, we focus on magnetic-field coupling WPT systems with DC

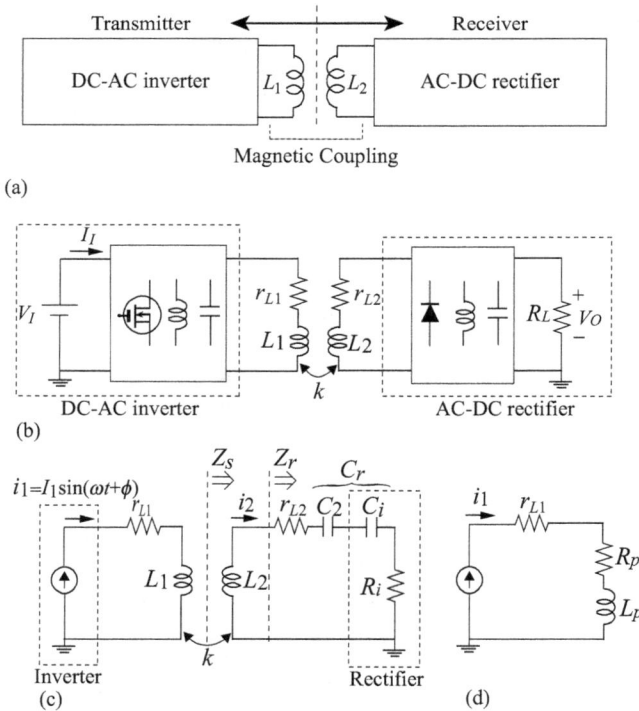

Figure 6.1   Construction of magnetic-field coupling WPT systems: (a) block diagram, (b) circuit model, (c) circuit model with an AC current source, and (d) circuit model without a transformer

input and DC output. Generally, a WPT system is divided into three main parts: the DC–AC inverter, coupling, and AC–DC rectifier, as shown in Figure 6.1(a). In this figure, L_1 and L_2 represent the primary coil and the secondary coil, respectively. The magnetic-field coupling is achieved through the interaction of these two coils, effectively forming a transformer-like configuration, thus categorizing the system as a magnetic-field WPT system and considered one of the isolated resonant converters [6].

The transmitter section includes the DC–AC inverter and the primary coil. Conversely, the receiver section consists of the secondary coil and the AC–DC rectifier.

The DC–AC inverter usually consists of input DC-voltage source, MOSFETs as switching devices, and passive components *L* and *C*. Conversely, the diode is often used in the AC–DC rectifier. The output voltage/power can be obtained from the load resistance R_L. In WPT systems, the coupling coefficient between two coils is very small, which is different from resonant converters. Therefore, the WPT system is expressed as shown in Figure 6.1(b).

In this chapter, the power-delivery efficiency is defined as

$$\eta = \frac{P_O}{P_I} = \frac{V_o^2}{R_L V_I I_I},$$

(6.1)

where V_o is the output voltage of the rectifier and V_I and I_I are the input voltage and current of the inverter, respectively. The power delivery efficiency is also expressed as

$$\eta = \eta_{inv}\eta_{coi}\eta_{rec},$$

(6.2)

where η_{inv}, η_{coi}, and η_{rec} are power-conversion efficiencies of inverter, coupling coils, and rectifier, respectively. Power-conversion efficiency at each part is not independent of other-part parameters. Each part should be optimized with taking into account the other parts. Namely, it is important to design optimal inverter, coupling coils, and rectifier simultaneously, which means that the WPT system should be design as an integrated system of three parts.

6.3   General theory of optimal WPT system designs

6.3.1   *Coupling coils*

In magnetic-field coupling, the coupling part is modeled as a transformer with low coupling coefficient. A large-amplitude sinusoidal AC current flows through the coupling coils. Because a large-amplitude AC current generates high core loss, the air-core coils are often used in WPT systems at high frequencies, in particular. In air-core coils, however, high conduction losses occur because large number of winding turns is necessary for keeping sufficient self-inductance. Therefore, equivalent series resistance (ESR) of the primary and secondary coils, which are expressed as r_{L1} and r_{L2}, respectively, are considered in the electrical model as shown in Figure 6.1(b).

6.3.1.1   Self-inductance of air-core solenoid

In this chapter, it is considered to adopt air-core solenoid for coupled coils as shown in Figure 6.2, where N is the turn number per layer, N_l is the number of layers, d_i is the diameter of the bare wire, d_o is the diameter of the wire including the insulation, p is the distance between centers of wires, d_C is the diameter of the coil, and d_{tr} is the interval between two coils. The self-inductance of an air-core solenoid can be obtained from

$$L = \frac{K_N \mu_0 \pi r^2 N_t^2}{h},$$
(6.3)

where $N_t = NN_l$ is the total turn number of the coil, $\mu_0 = 4\pi \times 10^{-7}$ is the permeability of vacuum, and $h = N(d_o + p) - p$ is the length of the solenoid as shown in Figure 6.2(b). Additionally, K_N is the coefficient for finite length of solenoid, called Nagaoka coefficient [41], which is expressed by

$$K_N = \frac{4}{3\pi\sqrt{1 - k_N^2}}\left(\frac{1 - k_N^2}{k_N^2}K(k_N) - \frac{1 - 2k_N^2}{k_N^2}E(k_N) - k_N\right),$$
(6.4)

where k_N is

$$k_N = \frac{1}{\sqrt{\left(\frac{h}{d_C}\right)^2 + 1}},$$
(6.5)

and $K(k_N)$ and $E(k_N)$ are the first kind and second kind of the complete elliptic integral for k_N, respectively. Namely, we have

$$K(k_N) = \int_0^{\frac{\pi}{2}} \frac{1}{\sqrt{1 - k_N^2 \sin^2\theta}} d\theta,$$
(6.6)

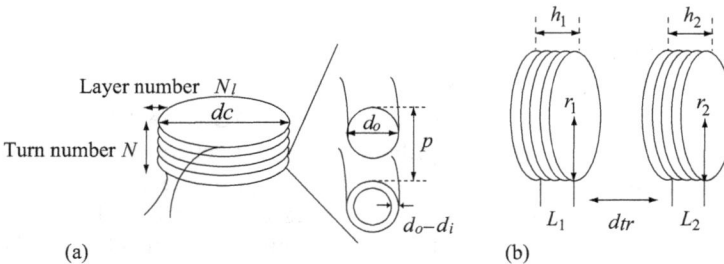

(a)   (b)

Figure 6.2   Modeling of air-core solenoid: (a) coil parameters and (b) transformer parameter

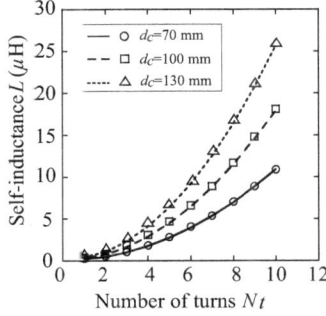

Figure 6.3   Theoretical (line) and measurement (plots) values of self-inductances as a function of turn number for $d_i = 1.0$ mm, $N_l = 1$, $h = 13$ mm and fixed coil diameters

and

$$E(k_N) = \int_0^{\frac{\pi}{2}} \sqrt{1 - k_N^2 \sin^2 6} d\theta. \tag{6.7}$$

Figure 6.3 shows the theoretical and measurement values of self-inductances as a function of turn number for $d_i = 1.0$ mm, $N_l = 1$, $h=13$ mm and fixed coil diameters.

6.3.1.2   Mutual inductance

Two coils are connected through the magnetic field for WPT. The mutual inductance can be obtained theoretically from the Neumann formula [42,43], which is expressed as

$$M = \frac{\mu}{4\pi} \oint \oint \frac{dl_1 dl_2}{R_{12}}, \tag{6.8}$$

where dl_1 and dl_2 are the minute length of the closed loops of the coils and R_{12} is the distance between dl_1 and dl_2. When the coils have a circular shape, we have

$$M = \sum_{i=0}^{N_1-1} \sum_{j=0}^{N_2-1} \frac{\mu r_1 r_2}{4\pi} \oint_0 \oint_c \frac{\cos(\phi_1 - \phi_2) d\phi_1 d\phi_2}{\sqrt{r_1^2 + r_2^2 - 2r_1 r_2 \cos(\phi_1 - \phi_2) + (D + ip_1 + jp_2)^2}}, \tag{6.9}$$

where the number of subscript expresses the primary and the secondary, and $r_k = d_{Ck}/2$ is the radius of coil as shown in Figure 6.2(b). From the mutual inductance, the coupling coefficient k can be derived from

$$k = M/\sqrt{L_1 L_2}. \tag{6.10}$$

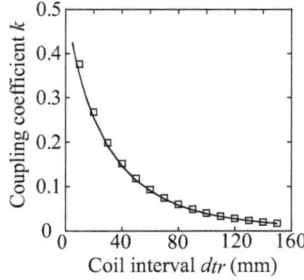

Figure 6.4   Theoretical (line) and measurement (plots) values of coupling coefficient as a function of coil interval d_{tr} for $h_1 = h_2 = 13$ mm, $d_{C1} = d_{C2} = 119$ mm, $d_{i1} = 1.0$ mm, $d_{i2} = 0.9$ mm, $N_1 = 10$, $N_2 = 12$, and $N_{l1} = N_{l2} = 1$

Figure 6.4 shows the theoretical and the measured coupling coefficients as a function of coil interval for $h_1 = h_2 = 13$ mm, $d_{C1} = d_{C2} = 119$ mm, $d_1 = 1.0$ mm, $d_2 = 0.9$ mm, $N_1 = 10$, $N_2 = 12$, and $N_{l_1} = N_{l_2} = 1$. It is seen from Figure 6.4 that the coupling coefficient k decreases as the coil interval decreases.

6.3.1.3   Equivalent series resistance

Generally, there are two power-loss factors in a coil, which are the core loss and the winding loss. Because it is assumed that the air-core solenoids are adopted to the coupling part of WPT systems in this chapter, it is necessary to consider only the winding loss as the power-loss factor of coils. When the AC current flows through the coil, the skin and proximity effects occur. These effects can be expressed by the equivalent series resistance, namely,

$$r_L = F_R r_{dc}, \tag{6.11}$$

where r_{dc} is the DC resistance of the coil and F_R is the ratio of AC-to-DC resistances. The DC resistance is expressed as

$$r_{dc} = \frac{\rho_w l_w}{A_w} = \frac{4\rho_w N_T N_l}{\pi d_i^2} = \frac{4\rho_w d_C N_l}{\pi d_i^2}, \tag{6.12}$$

where $\rho_w = 1.72 \times 10^{-8}\ \Omega\,m$ is the resistivity of the copper, A_w is the cross-sectional area of the wire. Additionally, F_R is obtained from Dowell's equation for round winding wire [44,45],

$$F_R = A\left[\frac{\sinh(2A) + \sin(2A)}{\cosh(2A) - \cos(2A)} + \frac{2(N_l^2 - 1)}{3}\frac{\sinh(A) - \sin(A)}{\cosh(A) + \cos(A)}\right], \tag{6.13}$$

where A is expressed by using skin depth δ_w as

$$A = \left(\frac{\pi}{4}\right)^{\frac{3}{4}}\frac{d_i}{\delta_w}\sqrt{\frac{d_i}{p}}. \tag{6.14}$$

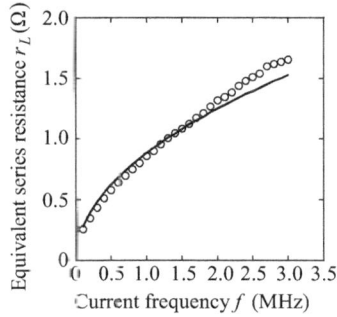

Figure 6.5   *Theoretical (line) and measurement (plots) values of equivalent series*
resistance as a function of f for $d_C = 119$ mm, $d_i = 1$ mm, $N_t = 10$,
$N_l = 1$, and $p = 1.5$ mm

The skin depth is obtained from

$$\delta_w = \sqrt{\frac{\rho_w}{\pi u_0 f}}, \tag{6.15}$$

where f is the fundamental frequency of AC current. Figure 6.5 shows the ESRs of coils as a function of AC-current frequency for $d_C = 119$ mm, $d_i = 1$ mm, $N_t = 10$, $N_l = 1$, and $p = 1.5$ mm.

Because the coupling part occupies large area of WPT systems, the coil designs are restricted by the permissible physical size of the coil. It is an important point that the coupling part can be expressed by three electric-circuit parameters, namely, self-inductance, coupling coefficient, and ESR, from physical parameters, such as wire diameter, coil interval, and so on, by using the above analytical expressions. It is possible to consider the electric-circuit parameter restrictions and the physical-size restrictions of coils simultaneously by using the previous theoretical relationships.

6.3.2   Optimal design of coupling part

Figure 6.1(b)–(d) shows a circuit model of WPT systems and its transformation for analysis. The rectifier is transformed into the equivalent capacitance C_i and resistance R_i connected in series as shown in Figure 6.1(c), where C_2 is the secondary resonant capacitance, which is the adjustable parameter. Because the purpose of the inverter is to generate a sinusoidal AC current, the inverter can be modeled as the AC-current source as shown in Figure 6.1(c). The voltage across R_i becomes AC voltage. Therefore, it can be stated that the DC–DC WPT system is expressed as the AC–AC one through the proper circuit transformations. This means that there is a common design theory of WPT systems among DC–DC, AC–AC, AC–DC, and DC–AC WPT systems.

6.3.2.1 Impedance transformation of rectifier and transformer

The impedances of secondary side without and with secondary inductance, which are defined as in Figure 6.1(c), are expressed as

$$Z_r = R_i + r_{L2} + \frac{1}{j\omega C_r}, \tag{6.16}$$

and

$$Z_s = R_i + r_{L2} + j\left(\omega L_2 - \frac{1}{j\omega C_r}\right), \tag{6.17}$$

respectively, where C_r is the equivalent capacitance of C_2 and C_i connected in series, namely,

$$C_r = \frac{C_2 C_i}{C_2 + C_i}. \tag{6.18}$$

Additionally, the secondary side can gather into the primary side. As a result, the WPT system can be expressed without transformer as shown in Figure 6.1(d). The equivalent resistance and inductance seen from primary side is expressed as

$$R_p = \frac{k^2 \omega^2 L_1 L_2 (R_i + r_{L2})}{(R_i + r_{L2})^2 + \left(\omega L_2 - \frac{1}{\omega C_r}\right)^2}, \tag{6.19}$$

and

$$L_p = \frac{k^2 L_1 \left[(R_i + r_{L2})^2 - \frac{L_2}{C_r}\frac{1}{\omega C_r}^2\right]}{(R_i + r_{L2})^2 + \left(\omega L_2 - \frac{1}{\omega C_r}\right)^2} + L_1(1 - k^2), \tag{6.20}$$

respectively.

6.3.3 Efficiency of coupling part

In the coupling part, the power dissipations occur at the ESRs of the primary and secondary coils. Because R_i is the equivalent resistance of the rectifier as shown in Figure 6.1(c), the power dissipation at R_i is used as a input power of the rectifier. It is seen from Figure 6.1(c) that the power efficiency at the secondary coil is

$$\eta_2 = \frac{R_i}{R_i + r_{L2}} = \frac{1}{1 + \frac{r_{L2}}{R_i}}. \tag{6.21}$$

Similarly, the power efficiency at the primary coil is

$$\eta_1 = \frac{R_p}{R_p + r_{L1}} = \frac{1}{1 + \frac{r_{L1}}{R_p}}. \tag{6.22}$$

From (6.21) and (6.22), power conversion efficiency at the coupling part is

$$\eta_{coi} = \eta_1\eta_2 = \frac{1}{1 + \frac{r_{L1}}{R_p}}\frac{1}{1 + \frac{r_{L2}}{R_i}}. \tag{6.23}$$

It is seen from (6.23) that r_L and r_{L2} should be small for efficiency enhancement. It is also seen from (6.23) that R_p and R_i should be large. R_p is, however, a function of R_i and R_p decreases as R_i increases. Therefore, there is an optimal value of R_i for efficiency maximization for given r_{L1} and r_{L2}.

6.3.3.1 Optimal coupling-part design and secondary resonance

R_i is independent of L_2 and C_2. Therefore, we have

$$\omega L_2 - \frac{1}{\omega C_r} = 0, \tag{6.24}$$

for obtaining maximum value of R_p. This means that the secondary resonance needs to be achieved by adjusting C_2 with taking into account the rectifier equivalent capacitance. It is well known that the magnetic resonance coupling is effective for efficiency enhancement as well as transmission distance extension. This knowledge can be explained by the maximization of R_p from the circuit theory.

By substituting (6.24) into (6.19) and (6.20), we obtain the values of R_p and L_p with secondary resonance as

$$R_{popt} = \frac{k^2\omega^2 L_1 L_2}{R_i + r_{L2}} \tag{6.25}$$

and

$$L_{popt} = L_1. \tag{6.26}$$

It is seen from (6.25) and (6.26) that the power factor becomes 1 by making secondary resonance. Substituting (6.25) into (6.23), the optimal power conversion efficiency at the coupling part is

$$\eta_{coiopt} = \frac{1}{1 + \frac{r_{L1}(R_i + r_{L2})}{k^2\omega^2 L_1 L_2}}\frac{1}{1 + \frac{r_{L2}}{R_i}}. \tag{6.27}$$

By differentiating (6.27) with respect to R_i, we obtain the condition for achieving the maximum η_{coiopt} as

$$\frac{d\eta_{coimax}}{dR_i} = \frac{k^2\omega^2 L_1 L_2(r_{L1}r_{L2}^2 - r_{L1}R_i^2 + k^2\omega^2 L_1 L_2 r_{L2})}{[r_{L1}(R_i + r_{L2})^2 + k^2\omega^2 L_1 L_2(R_i + r_{L2})]^2} = 0. \tag{6.28}$$

Namely, the optimal value of equivalent resistance of the rectifier is

$$R_{iopt} = \sqrt{\frac{r_{L1}r_{L2}^2 + k^2\omega^2 L_1 L_2 r_{L2}}{r_{L1}}}. \tag{6.29}$$

6.3.4 Design strategies of rectifier and inverter

It is seen from the above that the optimal WPT system design is started from the coil designs. The rectifier is designed for power-delivery efficiency enhancement. Namely, it is necessary to design a rectifier, whose equivalent resistance satisfies (6.29). Additionally, the inverter is designed for achieving the rated output power.

It is also important assumptions in this design theory that the power losses at inverter and rectifier are very small. Obviously, it is necessary to apply high-efficiency inverter and rectifier to the WPT system for achieving high power-delivery efficiency. From the next section, high-efficiency rectifiers and inverters are introduced.

6.4 High-efficiency rectifier

An AC–DC rectifier is usually adopted for the receiver part of WPT systems. The rectifier often includes diodes as switching devices. The diodes turn ON and OFF according to waveforms of diode current and voltage. This chapter considers the series resonant topology at the secondary side as shown in Figure 6.1(c). Therefore, several types of current driven rectifiers are introduced in this section.

6.4.1 Class-D rectifier

Figure 6.6(a) shows the circuit topology of the class-D rectifier [31,32], which consists of two diodes D_1 and D_2 and low-pass filter $C_f - R_L$. Figure 6.6(b) shows example waveforms of the class-D rectifier, where $\theta=\omega t=2\pi ft$ represents the angular time. When the rectifier-input current i_2 flows through the positive direction, the diode D_2 is in the ON state and D_1 is in the OFF state. Because the output voltage is constant due to a low-pass filter, the voltage across a diode is zero in the ON state and V_o in the OFF state. As a result, the rectangular shape voltage appears at the diodes, which is the class-D operation.

Because there is no phase difference between the input current and diode voltage, the equivalent capacitance of the class-D rectifier becomes zero. The equivalent resistance of the class-D rectifier is expressed as

$$R_i = \frac{\pi^2}{2} R_L. \tag{6.30}$$

This means that the equivalent resistance R_i is determined by resistance R_L. Therefore, R_i is not an adjustable parameter.

6.4.2 Effects of diode parasitic capacitance

Figure 6.6 includes no parasitic capacitance of the diodes. In real components, however, rectifier diodes must have parasitic capacitance. Figure 6.7 shows the circuit topology and example waveforms of the class-D rectifier with diode

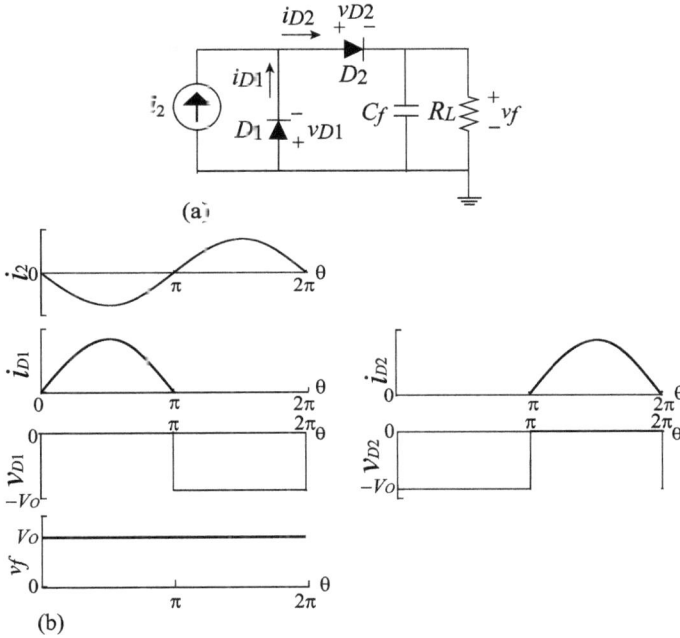

Figure 6.6 Class-D rectifier: (a) circuit topology and (b) example waveforms

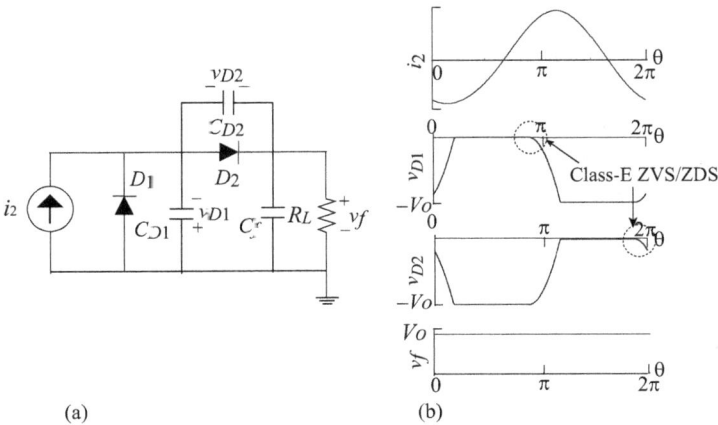

Figure 6.7 Class-DE rectifier: (a) circuit topology and (b) example waveforms

parasitic capacitance. When a diode turns OFF, the voltage across the diode never jumps to $-V_o$ but increases continuously as shown in Figure 6.7(b). An opposite diode turns ON when the diode voltage reaches zero. Because the sum of two diode voltages is always $-V_o$, there are durations that both diodes are in the OFF states, which is called "dead time." Additionally, not only the voltage but also the slope of the voltage dv/dt is zero at the diode turn-OFF instant. These turn-OFF switching conditions are called the class-E zero-voltage switching and zero-derivative switching (ZVS/ZDS) conditions. Namely, the rectifier as shown in Figure 6.7(a) is called the class-DE rectifier [6,33,34]. Because of the class-E switching, high efficiency can be achieved at high frequencies, in particular.

Because there must be parasitic capacitance at diodes, diode rectifiers must satisfy the class-E ZVS/ZDS conditions. Additionally, it is possible to adjust the capacitance when additional capacitance is connected to the diode in parallel, which is called shunt capacitance. By adjusting shunt capacitances, the equivalent resistance of the rectifier R_i becomes a controllable parameter. Additionally, the equivalent capacitance of the diode C_i is not zero and should be considered in the WPT design. For example, R_i and C_i of the class-DE rectifier as shown in Figure 6.7(a) are [34]

$$R_i = \frac{[1 - \cos(2\pi D_d)]^2}{2\pi^2} R_L \tag{6.31}$$

and

$$C_i = \frac{2\pi(C_{D1} + C_{D2})}{\sin(4\pi D_d) + 2\pi(1 - D_d)}, \tag{6.32}$$

respectively, where D_d is the dead time of the diodes, which satisfies

$$\omega(C_{D1} + C_{D2})R_L = \frac{2\pi[1 + \cos(2\pi D_d)]}{1 - \cos(2\pi D_d)}. \tag{6.33}$$

6.4.3   Class-E rectifier

Figure 6.8 shows a circuit topology of the class-E rectifier [6,35–37] and its example waveforms. The class-E rectifier consists of the diode D with shunt capacitance C_D and an $L_f - C_f - R_L$ low-pass filter.

When the output current is larger than the input current, the diode is in the ON state and the current flows through the diode. When the output current is the same as the input current, the diode turns OFF. Due to the existence of shunt capacitance, the class-E ZVS/ZDS conditions are satisfied at the turn-OFF instant. Therefore, the class-E rectifier is also suitable for high-frequency applications. In the OFF state, the difference between the input current and the output one flows through the shunt capacitance. Therefore, pulse-shape voltage appears across the diode as shown in Figure 6.8(b). The peak voltage of the diode is much higher than the output voltage through the peak diode voltages of the class-D, and DE rectifiers are

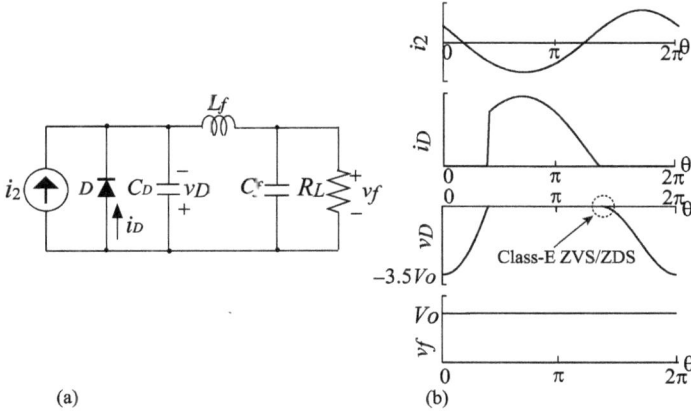

Figure 6.8   Class-E rectifier: (a) circuit topology and (b) example waveforms

the same as the output voltage, which is one of the drawbacks of the class-E rectifier. The peak value depends on the duty ratio of the diode. The average in the diode voltage becomes the output voltage through the low-pass filter.

The equivalent resistance of the class-E rectifier is expressed as a function of the load resistance R_L and the phase shift between input current and diode voltage ϕ_d, which are

$$R_i = 2R_L \sin^2 \phi_d. \tag{6.34}$$

The phase shift ϕ_d determines the duty ratio of the diode as

$$\tan \phi_d = \frac{1 - \cos(2\pi D)}{2\pi(1 - D_d) + \sin(2\pi D_d)}. \tag{6.35}$$

Additionally, the relationship between $\omega C_D R_L$ and D_d is expressed as [6]

$$\omega C_D R_L = \frac{1}{2\pi} \left\{ 1 - \cos(2\pi D_a) - 2\pi^2(1 - D_d)^2 + \frac{[2\pi(1 - D_d) + \sin(2\pi D_d)]^2}{1 - \cos(2\pi D_d)} \right\}. \tag{6.36}$$

It is seen from (6.34) – (6.36) that the shunt capacitance of the diode C_D is fixed uniquely when the load resistance and input resistance are given. Namely, it is possible to obtain the required input resistance R_i by changing the shunt capacitance.

Similar to the class-DE rectifier, the input capacitance of the class-E rectifier C_i also appears due to shunt capacitance, which is expressed as

$$C_i = 4\pi C_D / [4\pi(1 - D_d) + 4\sin(2\pi D_d) - \sin(4\pi D_d)\cos(2\phi_d)$$
$$-2\sin(2\phi_d)\sin^2(2\pi D_d) - 8\pi(1 - D_d)\sin \phi_d \sin(2\pi D_d - \phi_d)]. \tag{6.37}$$

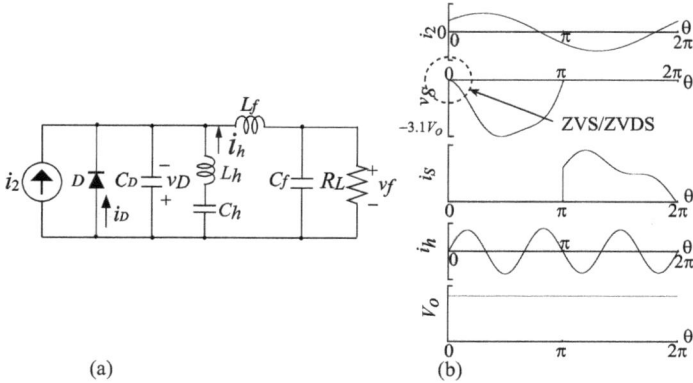

Figure 6.9   Class-E/F rectifier: (a) circuit topology and (b) example waveforms of the class E/F$_3$ rectifier

6.4.4   Class-E/F rectifier

Figure 6.9 shows a circuit topology of the class-E/F rectifier [38–40] and example waveforms of the class-E/F$_3$ rectifier. Compared with the class-E rectifier, the class-E/F rectifier has an additional resonant filter $L_h - C_h$, which is tuned by harmonic frequency. Only the DC current flows through the low-pass filter $L_f - C_f$. Therefore, the harmonic resonant current, which flows through the additional resonant filter, can go to the diode or its shunt capacitance. By generating a harmonic current with proper amplitude and phase shift, it is possible to reduce the peak value of the diode voltage and current as shown in Figure 6.9. As a result, the output power capability is enhanced. In the class-E/F rectifier, the equivalent resistance can also be adjusted by changing the shunt capacitance value.

6.5   High-efficiency inverters

In the transmitter part of WPT systems, a DC–AC inverter is often adopted. The inverter has switching devices such as MOSFET, which turns ON and OFF according to a driving signal. The fundamental frequency component of the switch voltage is extracted through an $L - C - R$ series resonant filter and AC output can be obtained. Because the timing of the switching is determined externally by the driving signal, there are cases that the switch voltage is not zero at turn-ON instant, which generates switching losses. Therefore, it is required that the inverter satisfies soft-switching conditions for achieving high power conversion efficiency. This chapter introduces several classes of series resonant inverters with considerations of soft-switching conditions.

6.5.1   Class-D inverter

Figure 6.10(a) shows a circuit topology of the class-D inverter [6–9,11–14], which is composed of input voltage source V_I, two switching devices S_1 and S_2,

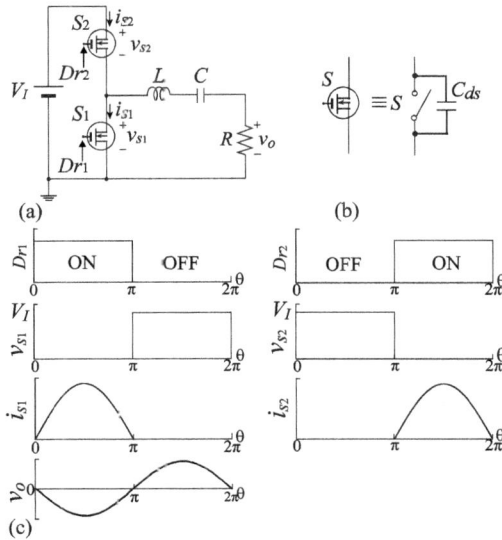

Figure 6.10 Class-D inverter: (a) circuit topology, (b) MOSFET modeling with parasitic capacitance, and (c) example waveforms

series resonant filter L–C, whose resonant frequency is tuned to the same as the operating frequency, and load resistance R. Figure 6.10(c) shows example waveforms of the class-D inverter. The MOSFETs S_1 and S_2 turn ON and OFF alternatively. When the duty ratios of both the switching devices are $D = 0.5$, the square-waveform voltage v_{S1}, whose amplitude is the same as the input voltage V_I, appears across S_1. Through the resonant filter, the fundamental frequency component of v_{S1} is extracted and the sinusoidal output waveform appears at the load resistance as shown in Figure 6.10(c). In the class-D inverter operation, the switch voltage and current never appear simultaneously because of ideal switching operations. Therefore, the power loss at the switching device is regarded as zero.

Though they do not appear on the circuit topology in Figure 6.10(a), however, the MOSFET devices have a parasitic capacitance between drain and source as shown in Figure 6.10(b), where C_{ds} is the drain-to-source capacitance of the MOSFET. When a MOSFET is in the OFF state, the drain-to-source voltage of the MOSFET is the same as the input voltage V_I. Because the class-D inverter has two switches, the switching losses of the class-D inverter are expressed as

$$p_{Cds} = 2 \times \frac{1}{2} C_{ds} v_{S}^2 f = C_{ds} V_I^2 f. \tag{6.38}$$

Note that the switching loss occurs at every turn-ON switching. Usually, the value of C_{ds} is small and the switching losses can be ignored at low frequencies. At high frequencies, however, the switching losses cannot be neglected because of the

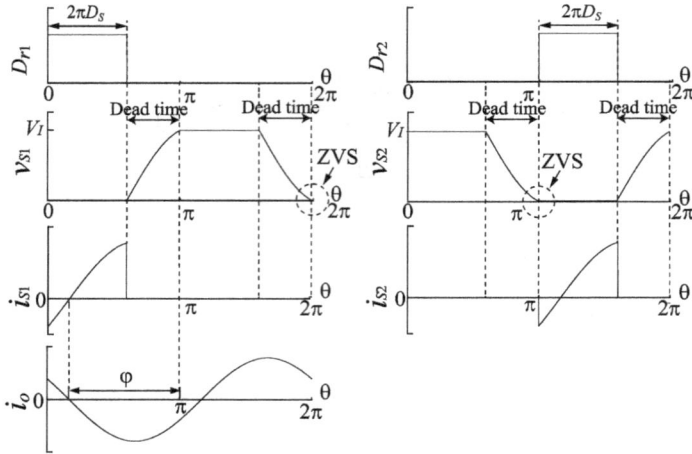

Figure 6.11   Example waveforms of the class-D ZVS inverter

increase in the operating frequency f. When the operating frequency is in the megahertz order, the power conversion efficiency degradation due to switching losses is the main problem of the class-D inverter.

For avoiding the switching losses of the class-D inverter, ZVS is often adopted [12–14]. Figure 6.11 shows the example waveforms of the class-D ZVS inverter. The ZVS condition can be achieved by making a dead time, which is the duration that two switches are zero simultaneously. During the dead time, the charge of the parasitic capacitance of a MOSFET moves to that of another MOSFET. By body-diode turning ON, the class-D inverter satisfies the ZVS condition. For achieving ZVS, it is necessary that the resonant filter is tuned inductively. Therefore, the phase shift φ appears between the switch voltage and output current as shown in Figure 6.11.

6.5.2   Class-E inverter

The class-E inverter was proposed by Sokals, which is known as a high-frequency high-efficiency inverter [6–9,15–20]. By satisfying the class-E ZVS/ZDS conditions at turn-ON instant, the switching losses are reduced to zero. Figure 6.12(a) shows a circuit topology of the class-E inverter, which consists of input voltage V_I, input inductance L_C, switching device S with shunt capacitance C_S, resonant filter $L{-}C$ and load resistance R. The most important component in the class-E inverter is the shunt capacitance. The shunt capacitance includes the drain-to-source parasitic capacitance of the switching device. By showing the shunt capacitance in the circuit topology explicitly, it is possible to discuss the circuit operation with parasitic capacitance effects.

Figure 6.12(b) shows example waveforms of the class-E inverter at the nominal operation. When an RF choke is used as the input inductance, the combination of the

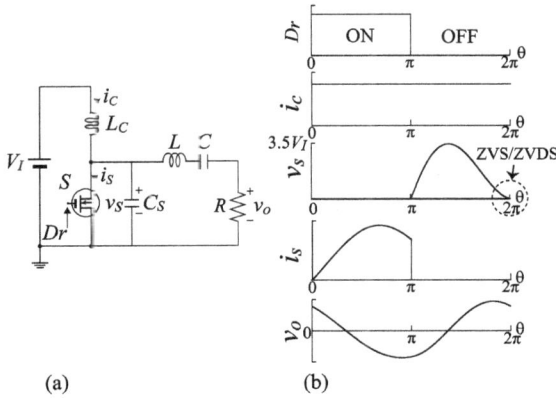

Figure 6.12   Class-E inverter. (a) circuit topology and (b) example waveforms

input voltage and the input inductance is regarded as a DC-current source. Therefore, the input current i_C is constant as shown in Figure 6.12(b). When the switch is in the ON state, the current flows through the switch, and the voltages across the switch and the shunt capacitance are zero. Conversely, the current flows through the shunt capacitance when the switch is in the OFF state. Therefore, the pulse-shape switch voltage appears across the switch as shown in Figure 6.12(b). The switch turns ON when the switch voltage is zero and the derivative of the switch voltage is also zero. Namely, the class-E ZVS/ZDS conditions are satisfied at turn-ON instant in the class-E inverter. We have to note that the class-E ZVS/ZDS conditions are not achieved automatically in the class-E inverter, which is different from the class-E rectifier. It is necessary to adjust and choose the component values for satisfying the class-E ZVS/ZDS conditions. The difficulty of the component-value decisions is one of the major problems of the class-E inverter. Actually, many design strategies have been proposed [9,16,17,19] since the class-E inverter was proposed.

6.5.3   Class-DE inverter

Generally, the peak value of the switch voltage is 3.5 times as high as the input voltage in the class-E inverter as shown in Figure 6.12(b) though it is the same as the input voltage in the class-D inverter. High peak switch voltage is also one of the problems of the class-E inverter.

The class-E ZVS/ZDS conditions can also be applied to the class-D inverter, which is called the class-DE inverter [9,21–24]. Figure 6.13 shows a circuit topology and example waveforms of the class-DE inverter. Compared with the class-D inverter, C_{S1} and C_{S2} appear in the circuit topology explicitly as shown in Figure 6.13(a), which are the shunt capacitances. By adjusting the shunt-capacitance and resonant-capacitance values properly, it is possible to satisfy the class-E ZVS/ZDS conditions at turn-ON instant on the class-D inverter as shown in

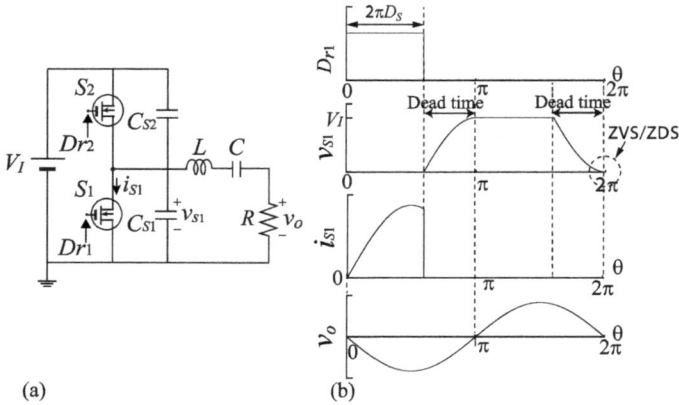

(a)   (b)

Figure 6.13   Class-DE inverter: (a) circuit topology and (b) example waveforms

Figure 6.13(b). It is regarded that the class-DE inverter is a special case of the class-D ZVS inverter.

The maximum switch voltage is the same as the input voltage in the class-DE inverter, which is an advantage of the class-DE inverter compared with the class-E inverter. It is, however, not easy to implement the high-side switch driver at high frequencies, in particular, which is a drawback of the class-DE inverter compared with the class-E inverter.

6.5.4   Class-E/F inverter

The purpose of the class-E/F inverter [9,25–27] is to reduce the peak value of the switch voltage compared with the class-E inverter, which can be achieved by subtracting harmonic component from the current flowing through the shunt capacitance.

Figure 6.14(a) shows a circuit topology of the class-E/F inverter. In the class-E/F inverter, the harmonic series-resonant filter L_h–C_h is added to the switching device of the class-E in parallel. Figure 6.14(b) shows example waveforms of the class-E/F$_3$ inverter. In the class-E/F$_3$ inverter, the third-harmonic current is subtracted. By adjusting the phase shift between the fundamental current and the harmonic one and the amplitude of the currents, the peak value of the switch voltage can be reduced. As shown in Figure 6.14(b), the square-shape type of switch voltage waveform can be obtained in the class-E/F inverter though the pulse-shape type of the switch voltage appears in the class-E inverter. Because the class-E ZVS/ZDS conditions are achieved, the class-E/F inverter also obtains high power-conversion efficiency at high frequencies. Of course, the circuit volume and implementation cost increase by adding the harmonic resonant filter, which are the drawbacks of the class-E/F inverter.

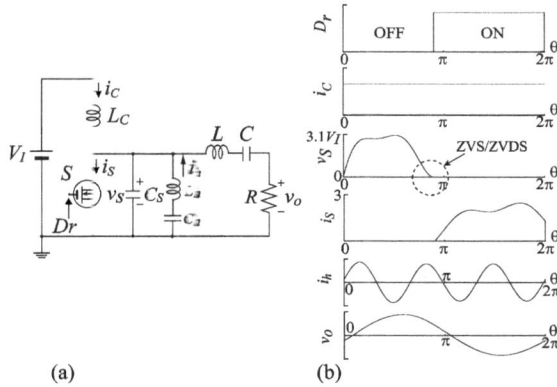

Figure 6.14 Class-E/F ZVS inverter: (a) circuit topology and (b) example waveforms of the class E/F₃ inverter

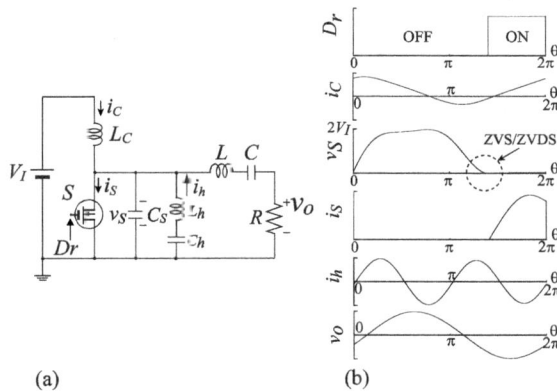

Figure 6.15 Class-Φ inverter: (a) circuit topology and (b) example waveforms of the class-Φ₂ inverter

6.5.5 Class-Φ inverter

The class-Φ inverter [28–30] is regarded as the extended version of the class-E/F inverter. Figure 6.15 shows a circuit topology and example waveforms of the class-Φ inverter. The circuit topology of the class-Φ inverter is the same as that of the class-E/F inverter as shown in Figure 6.15(a). In the class-Φ inverter, the resonant frequency of the LC filter is the same as the operating frequency. This concept is achieved by reducing the input inductance and by allowing partial input current to be regenerated to the input voltage source as shown in Figure 6.15(b). Because the resonant frequency of LC is the same as the operating frequency, the current through the resonant filter is independent of the load resistance. Namely, the class-E ZVS/ZDS conditions can be achieved in spite of load variations. This is the main advantage of the class-Φ inverter compared with the class-E/F inverter.

6.6 Design example of optimal WPT system

This section shows a design example of the optimal class-E^2 WPT system [10,46–49]. Figure 6.16(a) and (b) shows a circuit topology of the class-E^2WPT system, in

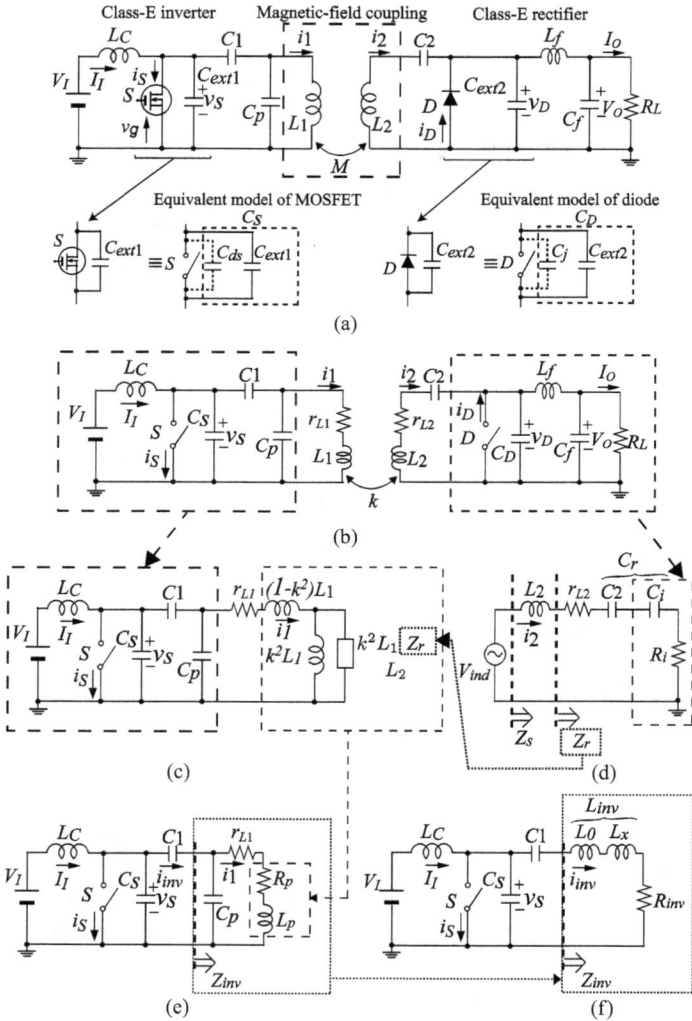

Figure 6.16 *Class-E^2 WPT system: (a) system topology, (b) equivalent circuit model, (c) equivalent circuit of the inverter part, (d) equivalent circuit of the rectifier part, and (e) and (f) equivalent circuits expressed by the class-E inverter*

which the class-E inverter and the class-E rectifier are applied to the transmitter part and the receiver one, respectively. The shunt capacitances C_S and C_D include the drain-to-source capacitance of the MOSFET C_{ds} and the junction capacitance of the diode C_j, respectively, as shown in Figure 6.16(a). Compared with the circuit topology in Figure 6.12(a), the impedance transformation capacitance C_p is newly added to the class-E inverter, which has a role in squeezing the transmission power to the receiver side. This is because all the component values of the rectifier are determined for satisfying the optimal value of the input resistance R_i. By adding C_p, it is possible to adjust the rated output power at the inverter part because of the increase in the degree of freedom in design. Additionally, C_p never affects to the power losses at the coupling part. From the previous discussions, it can be stated that C_p is a mandatory component for achieving the rated output power with optimal power-delivery efficiency.

In this design example, the design specifications are operating frequency $f = 1$ MHz, input voltage $V_I = 25$ V, output power $P_O = 13$ W, switch on-duty ratio of the class-E inverter $D_S = 0.5$, and load resistance $R_L = 50$ Ω. The optimal design is carried out under the physical-size limitation of the coils. In this design example, the primary and secondary coils are implemented in the enclosure of TAKACHI TWN-8-2-8W, whose inner size is 71 mm × 71 mm × 14 mm. The coil interval is set as 30 mm. The current density of the coil wire is less than 5 A/mm^2, which is also the constraint condition for optimization.

6.6.1 Optimal design for fixed coil parameters

Now, it is assumed that all the transformer parameters, namely, self-inductances and equivalent resistances of the primary and secondary coils and coupling coefficient are given. The class-E rectifier is transformed into the input resistance R_i and the input capacitance C_i as shown in Figure 6.16(d). The optimal value of R_i can be obtained from (6.29). It is necessary to design the class-E rectifier, whose input resistance is $R_i = R_{iopt}$.

6.6.1.1 Receiver-part design

From (6.36), the optimal shunt capacitance of the diode is

$$C_{Dopt} = \frac{1 - \cos(2\pi D_{dopt}) - 2\pi^2(1 - D_{dopt})^2 + \frac{[2\pi(1 - D_{dopt}) + \sin(2\pi D_{dopt})]^2}{1 - \cos(2\pi D_{dopt})}}{2\pi\omega R_L},$$

(6.39)

where D_{dopt} is the duty ratio of the diode for optimal operation, which can be obtained from (6.34) and (6.35).

The low-pass filter components L_f and C_f, which ensure the current ripple to be less than 10% of the output current, are selected in the range of

$$L_f > \frac{1 - D_{dopt}}{0.1 f R_L}$$

(6.40)

and

$$C_f > \frac{25}{\pi^2 f^2 L_f}. \tag{6.41}$$

The optimal value of the secondary resonant capacitance C_2 is expressed as

$$C_{2opt} = \frac{C_{iopt}}{\omega^2 L_2 C_{iopt} - 1}, \tag{6.42}$$

where C_{iopt} is the equivalent capacitance of the class-E rectifier for R_{iopt}, which can be obtained by substituting optimal values of C_D, D_d, and phi_d into (6.37).

6.6.1.2   Class-E inverter design

There are many analytical expressions of the class-E inverter. In this chapter, we use the analytical results in [6]. For optimal designs, the equivalent resistance and inductance seen from primary side, namely, R_p and L_p in Figure 6.16(e) can be expressed as (6.25) and (6.26).

By considering with the duty ratio of the class-E inverter is 0.5, the phase shift between the driving signal and the output current is expressed as

$$\phi_{inv} = \pi - \tan^{-1}\frac{2}{\pi}. \tag{6.43}$$

By adding impedance transformation capacitance C_p to primary coil in parallel as shown in Figure 6.16(a), the equivalent resistance and inductance of $C_p - L_{popt} - R_{popt}$ network, which appears in Figure 6.16(f), are

$$R_{inv} = \frac{8\cos^2\phi_{inv} V_I^2}{\pi^2 (R_{popt} + r_{L1}) I_1^2}, \tag{6.44}$$

and

$$L_{inv} = \frac{L_{popt}(1 - \omega^2 L_{popt} C_p) - C_p(R_{popt} + r_{L_1})^2}{\omega^2 C_p^2 \left[(R_{popt} + r_{L_1})^2 + \left(\omega L_{popt} - \frac{1}{\omega C_p}\right)^2 \right]}, \tag{6.45}$$

respectively, where

$$I_1 = \frac{V_o \sqrt{(R_i + r_{L_2})^2 + \left(\omega L_2 - \frac{1}{\omega C_r}\right)^2}}{\omega k R_L \sqrt{L_1 L_2} \sin \phi_d} \tag{6.46}$$

is the amplitude of AC current through the primary coil. By using R_{inv} and L_{inv}, the WPT system is modeled as the typical topology of the class-E inverter as shown in Figure 6.16(f).

By using R_{inv} in (6.44), the impedance transformation capacitance is

$$C_p = \frac{\omega L_{eq} R_{inv} - \sqrt{R_{inv}(R_{ec} + r_{L_1})\left[(R_{eq} + r_{L_1})(R_{eq} + r_{L_1} - R_{inv}) + \omega^2 L_{eq}^2\right]}}{\omega R_{inv}\left[(R_{eq} + r_{L_1})^2 + \omega^2 L_{eq}^2\right]}.$$

(6.47)

When C_p is obtained from (6.47), L_{inv} in (6.45) can be calculated.

In this design, L_{inv} is divided into L_0 and L_x virtually as shown in Figure 6.16(f), where L_0 satisfies

$$2\pi f \sqrt{L_0 C_1} = 1.$$

(6.48)

Namely, L_x means the inductive component of the resonant filter, which is expressed as

$$L_x = \frac{\pi R_{inv}(\pi^2 - 4)}{32\pi f}.$$

(6.49)

Therefore, we have

$$C_1 = \frac{1}{\omega^2 L_0} = \frac{1}{\omega^2 (L_{inv} - L_x)}.$$

(6.50)

The shunt capacitance is derived from

$$C_S = \frac{8}{\pi \omega R_{inv}(\pi^2 + 4)}.$$

(6.51)

Finally, the input inductance for satisfying 10% ripple of the input current should be in the range of

$$L_C > \frac{R_{inv}}{f}\left(\frac{\pi^2}{2} + 2\right).$$

(6.52)

It is possible to design the optimal class-E^2 WPT system by following the previous design procedure when the coil parameters are given.

6.6.2 Optimal WPT system design

The remained problem is how to design the primary and secondary coils. There are many parameters for coil designs, which are wire diameter, turn number, layer number, and coil diameter. In addition, the coils should be put in the specified enclosure.

The cost function of the optimization is a power delivery efficiency of the class-E^2 WPT system. It is assumed that the power losses at the class-E inverter and the class-E rectifier can be ignored because the class-E ZVS/ZDS conditions can be achieved at both the inverter and the rectifier. Of course, the conduction losses are generated in the inverter and the rectifier. It is, however, regarded that the conduction losses are independent of coil parameters. Therefore, only the power conversion efficiency of the coupling part in (6.23) is considered and is set as a cost function of the efficiency maximization. In this design example, the optimal combinations of the coil parameters for maximizing the cost function are searched by particle swarm optimization (PSO) algorithm [4,19,20,49], which is one of the well-known heuristic optimization algorithms.

As a result, it is possible to obtain all the component values as well as coil parameters of the optimal class-E^2 WPT system. Table 6.1 gives the obtained component values for optimal operation. The power-deliver efficiency is maximized when the primary-coil parameters are different from the secondary-coil ones. The maximum power-deliver efficiency can be obtained when $d_1 = 0.7$ mm, $d_2 = 0.5$ mm, $N_1 = 16$, $N_2 = 23$, and $N_{l_1} = N_{l_2} = 1$.

Figure 6.17 shows the photo of the implemented class-E^2 WPT system. Additionally, Figure 6.18 shows the experimental waveforms superimposed on the theoretical waveforms. The measurement results are also given in Table 6.1. It can

Table 6.1   Specifications and component values of the optimal class-E^2 WPT system

	Analytical	Measured	Difference (%)		Analytical	Measured	Difference (%)
L_C	169 μH	177 μH	4.6	k	0.109	0.111	1.8
L_f	219 μH	245 μH	12	r_{L_1}	1.24 Ω	1.26 Ω	1.6
C_S	1.20 nF	1.19 nF	−0.7	r_{L_2}	2.55 Ω	2.57 Ω	0.8
C_1	941 pF	935 pF	−0.6	R_L	50.0 Ω	50.1 Ω	0.2
C_p	97.5 pF	98.5 pF	1.0	f	1 MHz	1 MHz	0.0
C_L	482 pF	482 pF	−0.2	D_S	0.5	0.5	0.0
C_D	535 pF	538 nF	0.6	V_I	25.0 V	25.0 V	0.0
L_1	28.1 μH	28.5 μH	1.4	P_O	13.0 W	12.8 W	−1.5
L_2	58.4 μH	58.6 μH	0.3	η	81.5%	81.1%	−0.5

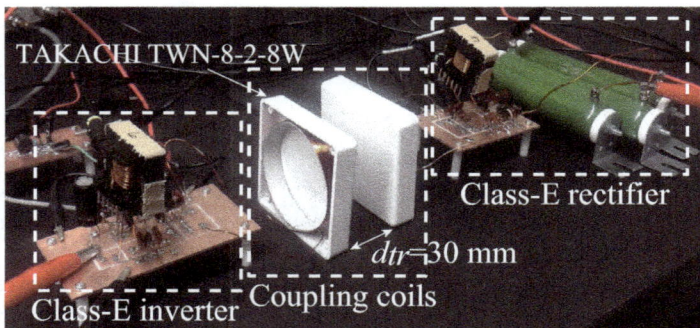

Figure 6.17   Overview of implemented WPT system

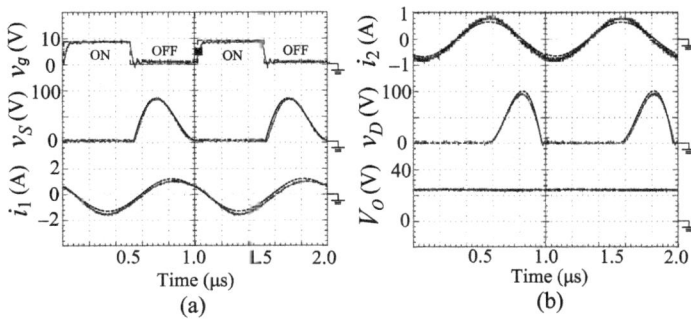

Figure 6.18 *Analytical (dashed line) and experimental (solid line) waveforms of the class-E^2 WPT system: (a) at inverter part and (b) at rectifier part*

be confirmed Figure 6.18 that the experimental waveforms achieved the class-E ZVS/ZDS conditions and the rated output power. Additionally, it is seen from Figure 6.18 and Table 6.1 that the experimental results agreed with the theoretical predictions quantitatively. These results showed the validity of the design procedure of the optimal WPT system as well as the analytical expressions. In the laboratory measurements, 81.1 % power delivery efficiency was obtained with 12.8 W output power and 1 MHz frequency.

6.7 Conclusion

This chapter has introduced the circuit topologies and fundamental operations of series-resonant inverters and current-driven rectifiers. By applying these inverters and rectifiers to a WPT system, it is possible to achieve high power-delivery efficiency even at high operating frequencies. The design theory of the WPT system is explained from a circuit-theory viewpoint. When the power losses of the inverter and rectifier can be sufficiently reduced, the coil parameters strongly influence the optimal design of the WPT system. For efficiency enhancement, it is essential to design the rectifier to satisfy the optimal equivalent resistance, and adjusting the output power in the inverter part is also beneficial. The chapter has presented a design example of the class-E^2 WPT system, demonstrating how the WPT system can be designed as an integrated system. Laboratory measurements showed high power-delivery efficiency at a 1-MHz operating frequency.

While the chapter did not focus on output controls, they remain another important design aspect of the WPT system. However, the control strategy can be established by using analytical expressions of the power-delivery efficiency, which are also applicable to the WPT system design theory introduced in this chapter.

References

[1]   A. Sagar, A. Kashyap, M. A. Nasab, *et al.*, "A comprehensive review of the recent development of wireless power transfer technologies for electric vehicle charging systems," *IEEE Access,* 2023, doi:10.1109/ACCESS.2023. 3300475

[2]   F. Musavi and W. Eberle, "Overview of wireless power transfer technologies for electric vehicle battery charging," *IET Power Electron.*, vol. 7, no. 1, pp. 60–66, 2014.

[3]   X. Yu, J. Feng, and Q. Li, "A planar omnidirectional wireless power transfer platform for portable devices," *2023 IEEE Applied Power Electronics Conference and Exposition (APEC)*, Orland, FL, USA, pp. 1654–1661, Mar 2023.

[4]   H. Sekiya, K. Tokano, W. Zhu, Y. Komiyama, and K. Nguyen, "Design procedure of load-independent class-E WPT systems and its application in robot arm," *IEEE Trans. Indust. Electron.*, vol. 70, no. 10, pp. 10014–10023, 2023.

[5]   M. Kiani, "Wireless power transfer and management for medical applications: Wireless power," *IEEE Solid-State Circuits Mag.*, vol. 14, no. 3, pp. 41–52, 2022.

[6]   M. K. Kazimierczuk and D. Czarkowski, *Resonant Power Converters,* 2nd Ed., Wiley, New York, 2011

[7]   M. K. Kazimierczuk, *RF Power Amplifiers* 2nd Ed., Wiley, West Sussex, 2014.

[8]   A. Grebennikov and M. J. Franco, *Switchmode RF and Microwave Power Amplifiers* 3rd Ed., Academic Press, Cambridge, MA, 2021.

[9]   A. Grebennikov, N. O. Sokal, and M. J. Franco, *Switchmode RF Power Amplifiers*, Newnes, Bulington, MA, 2011.

[10] T. Nagashima, X. Wei, E. Bou, E. Alarcon, M. K. Kazimierczuk, and H. Sekiya, "Analysis and design of loosely inductive coupled wireless power transfer system based on class-E^2 DC-DC converter for efficiency enhancement," *IEEE Trans. Circuits Syst.-I*, vol. 62, no. 11, pp. 2781–2791, 2015.

[11] O. Lucia, J. M. Burdio, I. Millan, J. Acero, and L. A. Barragan, "Efficiency-oriented design of ZVS half-bridge series resonant inverter with variable frequency duty cycle control," *IEEE Trans. Power Electron.*, vol. 25, no. 7, pp. 1671–1674, 2010.

[12] M. K. Kazimierczuk, "Class-D zero-voltage-switching inverter with only one shunt capacitor," *IEE Proc. B*, vol. 139, no. 5, pp. 449–456, 1992.

[13] X. Wei, H. Sekiya, T. Nagashima, M. K. Kazimierczuk, and T. Suetsugu, "Steady-state analysis and design of class-D ZVS inverter at any duty ratio," *IEEE Trans. Power Electron.*, vol. 31, no. 1, pp. 394–405, 2016.

[14] T. Osato, X. Wei, K. N. Asiya, and H. Sekiya, "Steady-state analysis and design of phase-controlled class-D ZVS inverter," *Nonlinear Theory and Its Applications, IEICE*, vol. 11, no. 2, pp. 189–205, 2020.

[15]   N. O. Sokal and A. D. Sokal, "Class E – A new class of high-efficiency tuned single-ended switching power amplifiers," *IEEE J Solid-State Circuits*, vol. 10, no. 3, pp. 168–176, 1975.

[16]   F. H. Raab, "Idealized operation of the class E tuned power amplifier," *IEEE Trans. Circuits Syst.*, vol. 24, no. 12, pp. 725–735, 1977.

[17]   H. Sekiya, I. Sasase, and S. Mori, "Computation of design values for class E amplifier without using waveform equations," *IEEE Trans. Circuits Syst.-I*, vol. 49, no. 7, pp. 966–978, 2002.

[18]   M. Hayat, A. Lotfi, M. K. Kazimierczuk, and H. Sekiya, "Generalized design considerations and analysis of class-E amplifier for sinusoidal and square input voltage waveform," *IEEE Trans. Ind. Electron.*, vol. 62, no. 1, pp. 211–220, 2015.

[19]   W. Zhu, Y. Komiyama, K. Nguyen, and H. Sekiya, "Comprehensive and simplified numerical design procedure for class-E switching circuits," *IEEE Access*, vol. 9, pp. 149971–149981, 2021.

[20]   A. Komanaka, W. Zhu, X. Wei, K. Nguyen, and H. Sekiya, "Generalized analysis of load-independent ZCS parallel-resonant inverter," *IEEE Trans. Indust. Electron.*, vol. 69, no. 1, pp. 347–356, 2022.

[21]   S. A. EI-Hamamsy, "Design of high-efficiency RF class-D power amplifier," *IEEE Trans. Power Electron.*, vol. 9, no. 3, pp. 297–308, 1994.

[22]   H. Koizumi, T. Suetsugu, M. Fujii, K. Shinoda, S. Mori, and K. Ikeda, "Class DE high-efficiency tuned power amplifier," *IEEE Trans. Circuits Syst. I*, vol. 43, no. 1, pp. 51–60, 1996.

[23]   H. Sekiya, X. Wei, T. Nagashima, and M. K. Kazimierczuk, "Steady-state analysis and design of class-DE inverter at any duty ratio," *IEEE Trans. Power Electron.*, vol. 30, no. 7, pp. 3685–3694, 2015.

[24]   A. Lotfi, A. Katsuki, F. Kurokawa, H. Sekiya, M. K. Kazimierczuk, and F. Blaabjerg, "Analysis of class-DE PA using MOSFET devices with non-equally grading coefficient," *IEEE Trans. Circts. Syst. I*, vol. 66, no. 7, pp. 2794–2802, 2019.

[25]   S. D. Kee, I. Aoki, A. Hajimiri, and D. Rutledge, "The class-E/F family of ZVS switching amplifiers." *IEEE Trans. Microwave Theor. Tech.*, vol. 51, no. 6, pp. 1677–1690, 2003.

[26]   Z. Kaczmarczyk, "High-efficiency class E, EF$_2$, and E/F$_3$ inverters." *IEEE Trans. Ind. Electron.*, vol. 53, no. 5, pp. 1584–1593, 2006.

[27]   J. Ma, Asiya, X. Wei, K. Nguyen, and H. Sekiya, "Analysis and design of generalized class-E/F2 and class-E/F3 inverters,' *IEEE Access*, vol. 8, pp. 61277–61288, 2020.

[28]   J. M. Rivas, Y. Han, O. Leitermann, A. D. Sagneri, and D. J. Perreault, "A high-frequency resonant inverter topology with low voltage stress," *in Proc. IEEE Power Electron. Spec. Conf.*, pp. 2705–2717, 2007.

[29]   J. M. Rivas, Y. Han, O. Leitermann, A. D. Sagneri, and D. J. Perreault, "A high-frequency resonant inverter topology with low-voltage stress," *IEEE Trans. Power Electron.*, vol. 23, no. 4, pp. 1759–1771, 2008.

[30] J. N. Rivas, O. Leitermann, Y. Han, and D. J. Perreault, "A very high fre-
      quency DC-DC converter based on a class Φ_2 resonant inverter," *IEEE
      Trans. Power Electron.*, vol. 26, no. 10, pp. 2980–2992, 2011.

[31] M. K. Kazimiercauk, "Class D current-driven rectifiers for resonant dc/dc
      converter applications," *IEEE Trans. Ind. Electron.*, vol. 38, no. 10, pp. 344–
      354, 1991.

[32] M. Mikotajewski, "Class D synchronous rectifiers," *IEEE Trans. Circuits
      Syst.-I*, vol. 38, no. 7, pp. 694–697, 1991.

[33] Y. Minami and H. Koizumi, "Analysis of class DE current driven low di/dt
      rectifier," *IEEE Trans. Power Electron.*, vol. 30, no. 12, pp. 6804–6816,
      2015.

[34] K. Fukui, and H. Koizumi, "Analysis of half-wave class DE low dv/dt rec-
      tifier at any duty ratio," *IEEE Trans. Power Electron.*, vol. 29, no. 1, pp. 234–
      245, 2014.

[35] M. K. Kazimierczuk, "Analysis of class E zero-voltage-switching rectifier,"
      IEEE Trans Circuits Syst.-I, vol. 37, no. 6, pp. 747–755, 1990.

[36] K. Jirasereeamornkul, M. K. Kazimierczuk, I. Boonyaroonate, and K.
      Chamnongthai, "Single-stage electronic ballast with class-E rectifier as
      power-factor corrector," *IEEE Trans. Circuits Syst.-I*, vol. 53, no. 1, pp. 139–
      148, 2006.

[37] Asiya, T. Osato, X. Wei, K. Nguyen, and H. Sekiya, "Analysis and design of
      generalized class-E rectifier," *Nonlinear Theor. Appl., IEICE*, vol. 11, no. 2,
      pp. 206–223, 2020.

[38] X. Wei, H. Sekiya, and T. Suetsugu, "New class-E rectifier with low voltage
      stress," *2016 IEEE Asia Pacific Conference on Circuits and Systems (APC-
      CAS2016)*, Oct. 2016.

[39] T. Murayama, X. Wei, and H. Sekiya, "Proposal and analysis of class-E/F$_3$
      rectifier," *2017 International Future Energy Electronics Conference
      (IFEEC2017)*, June 2017.

[40] X. Asiya, J. Wei, T. Ma, W. Osato, K. N. Zhu, and H. Sekiya, "Generalized
      analysis and performance investigation of the class-E/Fn rectifiers," *IEEE
      Access*, vol. 8, pp. 124145–124157, 2020.

[41] H. Nagaoka, "The inductance coefficients of solenoids," *Journal of the
      College of Science*, vol. 27, 1909.

[42] A. C. M de Queiroz "Mutual inductance and inductance calculations by
      Maxwell's method," *EE/COPPE*, Universidade Federal do Rio de Janerio,
      Brazil.

[43] T. Imura and Y. Hori, "Maximizing air gap and efficiency of magnetic
      resonant coupling for wireless power transfer using equivalent circuit and
      Neumann formula," *IEEE Trans. Ind. Electron.*, vol. 58, no. 2, pp. 544–554,
      2011.

[44] M. K. Kazimierczuk, *High-Frequency Magnetic Components*, John Wiley &
      Sons, New York, 2009.

[45] M. K. Kazimierczuk and H. Sekiya, "Design of AC resonant inductors using area product method," *2009 IEEE Energy Conversion Congress and Exposition (ECCE2009)*, pp. 994–1000, Sept. 2009.

[46] P. Luk, S. Aldhaher, W. Fei, and J. Whidborne, "State-space modelling of a class E^2 converter for inductive links," *IEEE Trans. Power Electron.*, vol. 30, no. 6, pp. 3242–3251, 2015.

[47] M. Liu, M. Fu, C. Ma "Parameter design for a 6.78-MHz wireless power transfer system based on analytical derivation of class E current-driven rectifier," *IEEE Trans. Power Electron.*, vol. 31, no. 6, pp. 4280–4291, 2016.

[48] M. Liu, C. Zhao, J. Song, and C. Ma, "Battery charging profile-based parameter design of a 6.78-MHz class E^2 wireless charging system." *IEEE Trans. Ind. Electron.*, vol. 64, no. 8, pp. 6169–6178, 2017.

[49] T. Noda, T. Nagashima, X. Wei, and H. Sekiya, "Design procedure for wireless power transfer system with inductive coupling-coil optimizations using PSO," *2016 IEEE International Symposium on Circuits and Systems (ISCAS2016)*, pp. 646–649, May 2016.

Chapter 7

Basic theory of wireless power transfer via radio waves

Naoki Shinohara[1]

7.1 Introduction

As described in Chapter 2, a magnetic field and an electric field are inductively generated in the space near a high-frequency power supply (e.g., in the kHz–MHz range). Power is transferred to a receiver wirelessly through the magnetic or electric fields. This form of wirelessly power transfer (WPT) is called inductive-coupling WPT; its effective distance is very short because wavelengths in the mi kHz–MHz range are too long to transfer power over long distances.

Combining the magnetic and electric fields, an electromagnetic field can WPT to a receiver at longer distances. The fields create an electromagnetic wave when the wavelength of the field or the wave becomes shorter. We often use electromagnetic waves with frequencies in the MHz–GHz range. When a high-frequency electromagnetic wave is created, power can be wirelessly transferred to a distant receiver. The electromagnetic waves and fields were originally based on Ampere's law and Faraday's law; however, they can be better described by Maxwell's equations as follows:

$$\nabla \times \mathbf{E} = -\frac{\partial \mathbf{B}}{\partial t} \quad \text{(Faraday's law of induction)}, \tag{7.1}$$

$$\nabla \times \mathbf{H} = \mathbf{J} + \frac{\partial \mathbf{D}}{\partial t} \quad \text{(Ampere's circuital law)}, \tag{7.2}$$

$$\nabla \cdot \mathbf{D} = \rho \quad \text{(Gauss' law)}, \tag{7.3}$$

$$\nabla \cdot \mathbf{B} = 0 \quad \text{(nonentity of magnetic charge)}, \tag{7.4}$$

where $\mathbf{H}$, $\mathbf{B} = \mu\mathbf{H}$, $\mathbf{E}$, $\mathbf{D} = \varepsilon\mathbf{E}$, $\mathbf{J} = \sigma\mathbf{E}$, and ρ indicate magnetic field (A/m), magnetic-flux density (T), electric field (V/m), electric-flux density (C/m²), current density (A/m²), and charge density (C/m³), respectively. μ, ε, and σ indicate magnetic permeability (H/m), permeability (F/m), and electrical conductivity (1/Ωm), respectively. Maxwell's equations indicate that electromagnetic waves propagate in a field based on relationships between the electric field, magnetic

[1]Research Institute for Sustainable Humanosphere, Kyoto University, Japan

field, time, and space. Maxwell's equations can describe all electromagnetic phenomena from electricity to light. "Radio waves" are electromagnetic waves with wavelengths longer than those of the infrared light.

Based on Maxwell's equations, power flow can be described as $\mathbf{E} \times \mathbf{H}$ (W/m²). The vector $\mathbf{E} \times \mathbf{H}$ is called the Poynting vector; it indicates that all electromagnetic waves are energy themselves along with the direction of propagation. Thus, power is wirelessly transferred even in common wireless-communication systems such as TVs and/or mobile phones. A modulated radio wave with information is transferred from a TV station or mobile phone station to users. We certainly receive radio waves as a power source in the wireless-communication system; however, we only use the radio waves as a carrier of information. In the WPT system, we can transfer and receive radio waves as a power source.

7.2   Propagation of radio waves

7.2.1   Radio waves in a far field

According to Maxwell's equations, a radio wave propagates non-preferentially in all directions from a point source (Figure 7.1). "Attenuation with distance" is often considered a key trait in wireless-communication systems; it means that the attenuation of a radio-wave power is inversely proportional to the square of the distance. However, this is not the correct word. When the radio wave propagates through lossless media, such as vacuum or dry air, the attenuation of power density of the radio wave is inversely proportional to the square of the distance, as described in the following equation:

$$P_{r\_density} = \frac{1}{4\pi d^2} P_t, \tag{7.5}$$

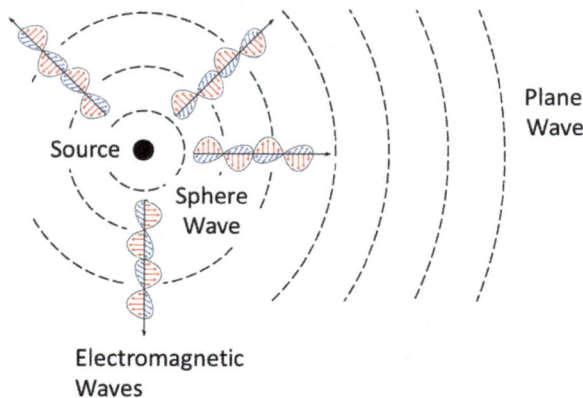

Figure 7.1   Propagation of electromagnetic waves

where $P_{r\_density}$, P_t, d are the power density at distance d, transmitted power, and distance from transmitting point, respectively. However, the radiated original power is not totally attenuated because there is no loss factor in lossless media.

A radio wave is radiated from a resonating antenna made mainly of metal. The antenna is also used as a receiver for radio waves. Power is transferred wirelessly among the antennas. In the resonance-coupling WPT system, LC or other resonators are used as transmitters and receivers of wireless power. The resonators for the resonance-coupling WPT are non-radiative, with power basically inside the resonators only. The wireless power in the resonance-coupling WPT is transferred via an evanescent-mode wave. In WPT systems using radio waves, the waves are radiated from the antenna, and power is wirelessly propagated among the antennas. The LC resonator and the antenna differ in that the former is non-radiative while the latter is radiative.

The antenna has the characteristics of an antenna gain G, a directivity, and an effective aperture A; these characteristics are mutually related. As mentioned earlier, the radio wave propagates equally in all directions from a source. However, when the source of the radio waves is an antenna, they do not propagate isotropically, and there is directivity in the radio wave and the antenna gain; in comparison with omnidirectional radiation, this directivity of the radio waves from antenna serves better in focusing the radio wave to our desired direction (Figure 7.2). The directivity and the antenna gain are decided by the initial current

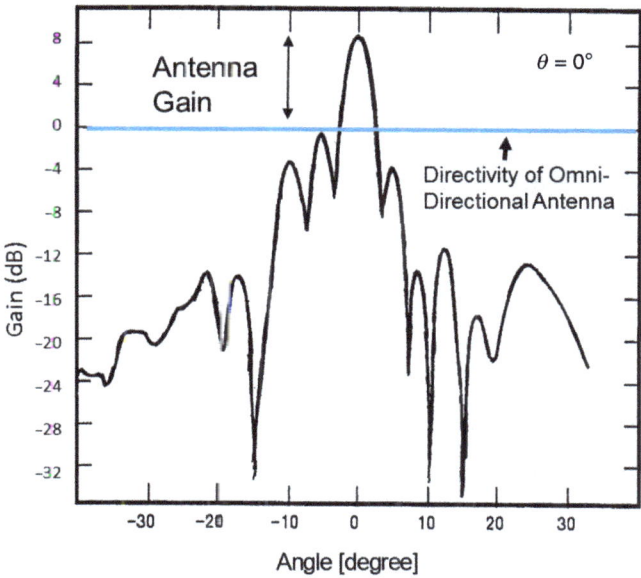

Figure 7.2   Typical directivity of an antenna

distribution on the antenna as a source of radio waves. The effective aperture of the antenna and the antenna gain are related as follows:

$$A = \frac{\lambda^2}{4\pi} G,$$
(7.6)

where λ is the wavelength of the radio wave. At the receiving antenna, the radio waves are received through the effective aperture area.

A received radio-wave power P_r at the effective aperture area A_r of the receiving antenna at a distance d from the transmitting antenna with transmitting radio-wave power P_t and gain G_t (Figure 7.3) is evaluated with (7.5) and (7.6) as follows:

$$P_r = \frac{G_t A_r}{4\pi d^2} P_t = \frac{A_t A_r}{(\lambda d)^2} P_t = \frac{G_t G_r}{\left(\frac{4\pi d}{\lambda}\right)^2} P_t.$$
(7.7)

Equation (7.7) is the Friis transmission equation. The $\left(\frac{4\pi d}{\lambda}\right)^2$ term is known as the propagation loss and is used in the design of wireless-communication systems. As mentioned before, this is not a real propagation loss but a diffusion factor of omnidirectional radio-wave propagation. Theoretically, we can obtain the beam efficiency, $\eta = P_r/P_t$, in the far field using (7.7). The beam efficiency in the WPT system based on radio waves corresponds to the transmission efficiency in an inductive- or resonance-coupling WPT system.

The most important assumption for the Friis transmission equation is that the radio waves are in far field where a plane wave is assumed. As mentioned before, the radio wave originally propagates spherically. For very long-distance transmission, a spherical wave becomes effectively planar. When the borderline between plane and spherical waves is considered, the phase difference between the center and edge of an aperture antenna is also taken into account. As a result, the distance d at which a plane wave can be assumed is $d > \frac{2D^2}{\lambda}$, which is called the far field or Fraunhofer region. D is the diameter of an aperture antenna. In the far field, a plane wave can be assumed, and the corresponding Friis transmission equation is viable (Figure 7.4).

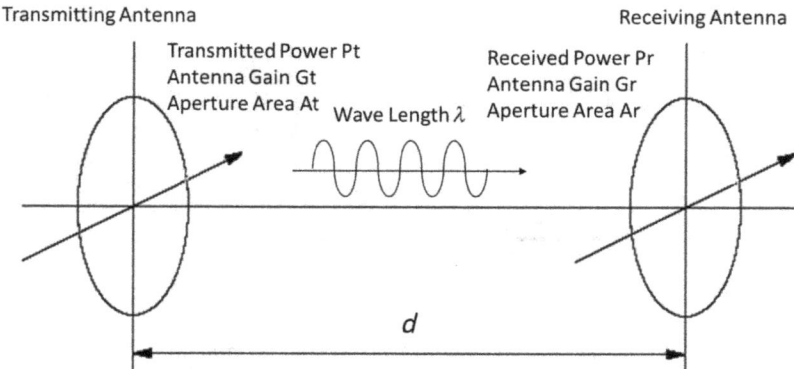

Figure 7.3   Antenna parameters for calculating beam efficiency

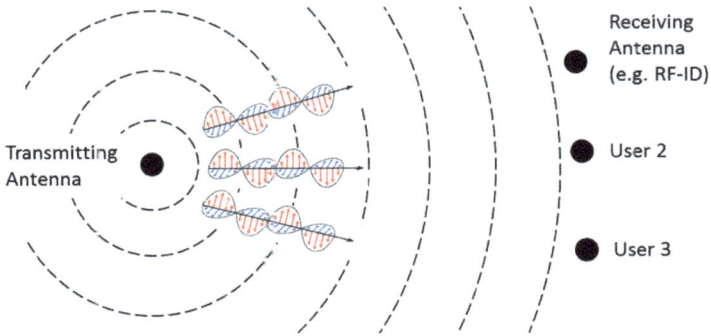

Figure 7.4   *WPT in the far-field region. Beam efficiency is very low, but all users can receive the same wireless power.*

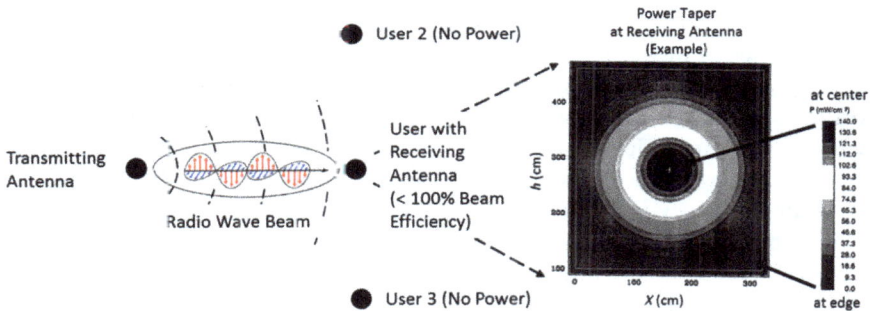

Figure 7.5   *WPT in the radiative near field. The beam efficiency is sufficiently high that the WPT system can be applied instead of the wired power system.*

7.2.2   Radio waves in the radiative near field

The far-field distance is not suitable for WPT because the theoretical beam efficiency η, which is obtained from the Friis transmission equation, is very low for a power system. Of course, the far-field WPT system can be used in RF–ID or weakly powered sensor-network applications. All users at any point in the far field can receive the same wireless power, making the system very convenient. However, beam efficiency is very low. In a wireless-communication system, efficiency is sufficient because an amplifier can be used to increase the received radio-wave power to obtain the information on the radio waves. On the contrary, if high efficiency is required instead of a wired power system, the beam efficiency in the far field is insufficient. To achieve high-beam efficiency for a WPT system based on radio waves, a short-distance WPT system with $d < \frac{2D^2}{\lambda}$ is often used as a "beam-type WPT" system. $d < \frac{2D^2}{\lambda}$ is named the near field or Fresnel region (Figure 7.5). For $d < \frac{2D^2}{\lambda}$, the Friis transmission equation is incorrect because radio waves cannot be assumed to be planar. This means that the power density in the vicinity of a receiving point is almost equal to that in the far field; however, the power density in the area of a receiving point is not equal to the tapered power density in the near field. The term "near field" is often used in inductive-coupling and resonance-coupling WPTs. However, the near field is not defined as $d < \frac{2D^2}{\lambda}$ for the coupling WPT system. There is no electromagnetic coupling between the transmitting and receiving antennas. Thus, the near field in $d < \frac{2D^2}{\lambda}$ is called the "radiative near field."

Instead of the Friis transmission equation, which requires a plane wave, the following equations must be used in the radiative near field:

$$P_r = \left(1 - e^{-\tau^2}\right)P_t, \tag{7.8}$$

$$\tau^2 = \frac{A_t A_r}{(\lambda d)^2}. \tag{7.9}$$

Equation (7.10) corresponds to the Friis transmission (7.7) in the radiative near field. When τ is sufficiently small (i.e., the antennas are in the far field), the power received in (7.8) is equal to that received in the Friis transmission (7.7). Based on (7.8), it can be inferred that, theoretically, the beam efficiency can reach 100% when τ is sufficiently large, meaning that the antennas are very close and are in the radiative near field. This beam-type WPT has high-beam efficiency. If the position of the antenna changes, the beam efficiency suddenly decreases. Thus, beam forming and target-position-detecting technologies are very important for beam-type WPT systems.

Beam efficiency is defined as the ratio of power radiated from a transmitting-antenna aperture to the power received at the receiving-antenna aperture. Figure 7.6

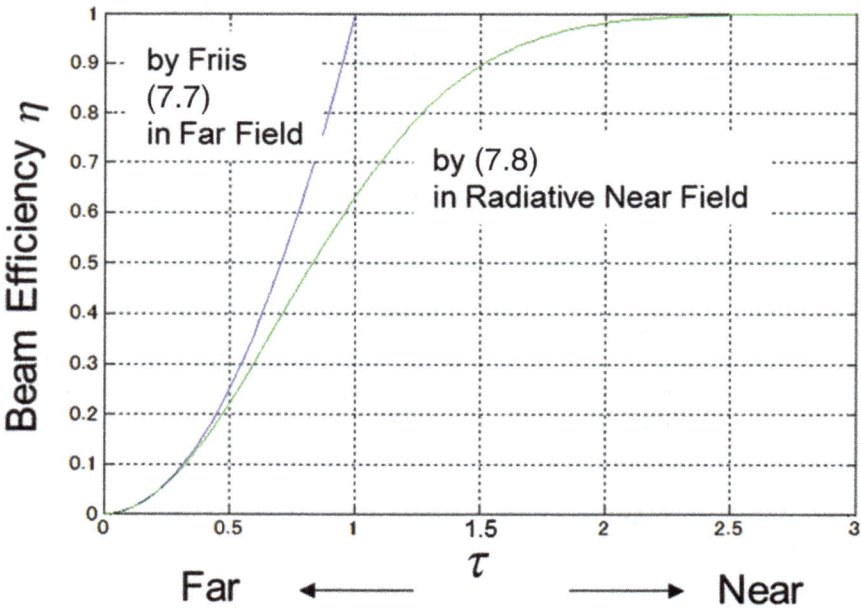

Figure 7.6   Theoretical beam efficiency

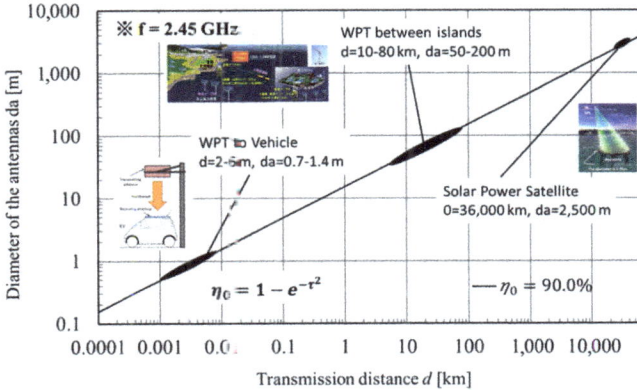

Figure 7.7   Theoretical relationship among beam efficiency, transmission distance, and diameter of an antenna

indicates the beam efficiency, which is obtained from (7.8) and (7.7). If the antennas are in the far field and τ is small, the beam efficiencies obtained from (7.8) and (7.7) are approximately the same because a plane wave can be assumed in the far field. If the antennas are in the radiative near field and τ is large, (7.8) for beam efficiency is correct; however, (7.7) for beam efficiency is wrong because plane waves cannot be assumed in the radiative near field.

Based on the definition of the borderline between the far field and the radiative near field, a distance of a few thousand kms can be considered to be within the radiative near field when large antennas are used. Figure 7.7 indicates the theoretical relationship of the beam efficiency, transmission distance, and diameter of transmitting and receiving antennas. When we use antennas whose diameters are 2,500 m and whose frequencies are 2.45 GHz, the beam efficiency reaches 90% even at a 36,000-km distance, meaning that the antennas are in the radiative near field. Even at 36,000 km, we cannot assume a plane wave.

7.2.3   Radio waves in the reactive near field

In general, a transmitting antenna and a receiving antenna are not electromagnetically coupled, even in the radiative near field. Thus, unlike in inductive-coupling WPT system, circuit parameters, such as impedance and resonance frequency, are not changed if the positions of the antennas or the distance between them change in the WPT system via radio waves. When the distance, d, between transmitting and receiving antennas is shorter than $\frac{\lambda}{\pi}$, even the antennas are electromagnetically coupled like resonators. This distance is in real near field that we call the "reactive near field." Within this distance, we have to consider the changes in the circuit impedance and the resonance frequency, which depends upon the position or the distance of the antennas [1]. Figure 7.8 indicates beam efficiency η as a function of frequency [1]. When frequency changes, the wavelength-normalized

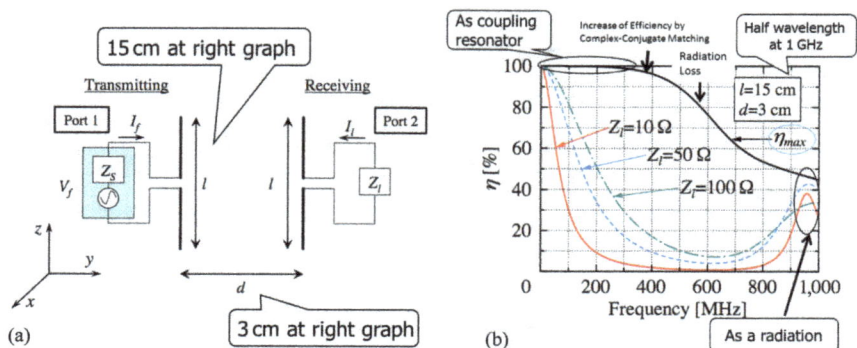

Figure 7.8   (a) Analytical parameters of antennas in the reactive near-field, and (b) beam efficiency among antennas in the reactive near field [1]

distance is changed in relation. Figure 7.8(a) indicates antenna parameters for analysis; dipole antennas are assumed. Figure 7.8(b) indicates the beam efficiency in the case where the load impedance, Z_L, is matched in all frequencies and where impedance, Z_L, is fixed at 10, 50, and 100 Ω. Figure 7.8(b) indicates that the antenna impedance changes when the distance between the transmitting and receiving antennas change. Thus, the beam efficiency decreases when the load impedance is fixed and kept higher, close to 100%, and when the load impedance is matched at each distance. In Figure 7.8(b), the beam efficiency rises again around 1,000 MHz, even when the load impedance is fixed. This indicates that the antenna impedance is matched at around the distance.

Figure 7.9 shows the electromagnetic-field-simulation result of beam efficiency in the reactive near field according to the Finite Difference Time Domain (FDTD) method [2]. Figure 7.9(a) indicates the simulation parameters in which a slot-antenna array is used for transmission and a microstrip-antenna array is used for receiving. Figure 7.9(b) indicates the simulation results for the resonance frequency of each antenna. Originally, the resonance frequency was designed to be 2.45 GHz in free space. But when the antennas are put in front, mutual coupling occurs and the resonance frequency moves to the other frequency. As a result, the fluctuation of the beam efficiency shown in Figure 7.9(c) appears. Peak efficiency is obtained in the case of impedance matching and is estimated by (7.8). In the other distance case, the antenna impedance is changed and is not matched, meaning that even radiating antennas are electromagnetically coupled in the reactive near field and that circuit parameters, such as impedance, are changed by the electromagnetic coupling effect.

Although we often divide inductive-coupling WPT and WPT via radio wave, with one being "coupling WPT" using coils and the other being "uncoupled WPT" using antennas, there is no strict borderline between them. Maxwell's equations cover all WPT systems. The borderline between the coupled and uncoupled WPT is decided based on the wavelength and distance between the transmitting and receiving elements. Resonance-coupled WPT fills the space between coupled and

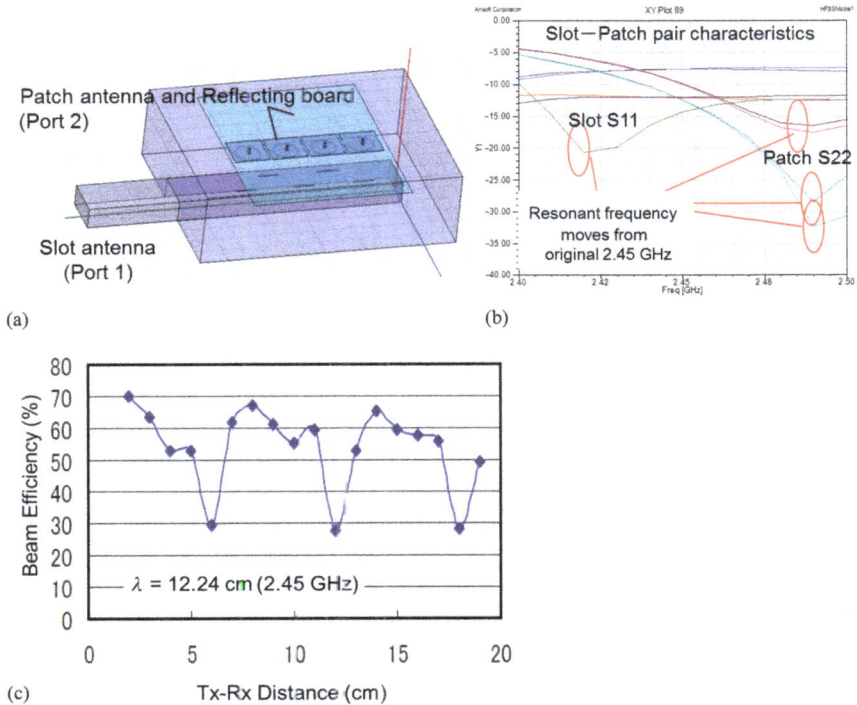

(a)

(b)

(c)   Tx-Rx Distance (cm)

Figure 7.9   (a) Simulation parameters of antennas in the reactive near-field; (b) S_{11} of the transmitting and receiving antennas that are put in front and whose resonance frequency is 2.45 GHz in free space, and (c) beam efficiency among antennas in the reactive near field according to the FDTD method

uncoupled WPT. For example, a bandpass filter (BPF) with coupled dielectric resonators is often used in mobile phones. The basic theory and technology of the BPF with coupled dielectric resonators are the same as those of the resonance-coupled WPT in the MHz and kHz bands. However, the BPF for mobile phones is in the GHz band. Some resonance-coupling WPT techniques in the GHz band have been developed in Japan [3,4].

7.2.4   Radio waves from a dipole antenna

This explanation is mainly based on aperture antennas. We would like to explain the reactive near field, radiative near field, and far field from another viewpoint. A dipole antenna, which is a line antenna of length $l = \lambda/2$, as shown in Figure 7.10, is considered. The current $I(Z)$ in the dipole antenna is assumed based on the following equation:

$$I(z) = I_0 \sin k_0(l - |z|)e^{j\omega t}, \tag{7.10}$$

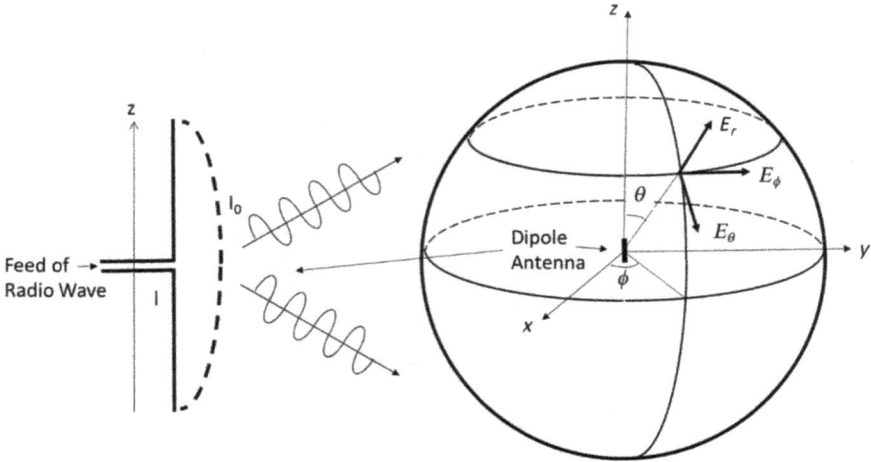

Figure 7.10   Dipole antenna and parameters

where I_0 is the current on the dipole, $k_0 = 2\pi/\lambda$ is the wave number, and ω is the angular frequency. A radio wave, whose components consist of electric and magnetic fields, is transferred from the dipole antenna. The electric and magnetic fields have components (r, θ, and ϕ). Each component in the electric field of the $\lambda/2$ dipole antenna is described as follows:

$$E_r = -j60I_0e^{-jk_0r}\left[\frac{j\frac{\lambda}{4}\sin\left(\frac{\pi}{2}\cos\theta\right)}{r^2} + \frac{\frac{\lambda^2}{16}\left\{\cos\theta\cos\left(\frac{\pi}{2}\cos\theta\right) + \frac{\pi}{4}\sin^2\theta\sin\left(\frac{\pi}{2}\cos\theta\right)\right\}}{r^3}\right],$$

$$(7.11)$$

$$E_\theta = j60I_0e^{jk_0r}\frac{\cos\left(\frac{\pi}{2}\cos\theta\right)}{\sin\theta}\left(\frac{1}{r} - j\frac{\pi}{16}\lambda\frac{\sin^2\theta}{r^2} - \frac{\lambda^2}{32}\frac{\sin^2\theta}{r^3}\right), \tag{7.12}$$

$$E_\phi = 0. \tag{7.13}$$

In (7.5), (7.7), and (7.9), r corresponds to d. When $r \gg \frac{\lambda}{\pi}$, which is considered the borderline of reactive near field and radiative near field, then the condition $\frac{1}{k_0r} \gg \frac{1}{(k_0r)^2}$, $\frac{1}{(k_0r)^3}$ occurs. Thus, the terms of $\frac{1}{(k_0r)^2}$ and $\frac{1}{(k_0r)^3}$ can be neglected at $r \gg \frac{\lambda}{\pi}$. As a result, at $r \gg \frac{\lambda}{\pi}$, the electric field of the $\lambda/2$ dipole antenna is described as follows:

$$E_r = 0, \tag{7.14}$$

$$E_\theta = j60I_0e^{jk_0r}\frac{\cos\left(\frac{\pi}{2}\cos\theta\right)}{r\sin\theta}, \tag{7.15}$$

$$E_\phi = 0. \tag{7.16}$$

The electric field described by (7.14)–(7.16) can be applied in the far field and in the radiative near field. Thus, we must divide the near field into "reactive near field" and "radiative near field."

7.3 Directivity control and beam formation using phased-array antenna

Antenna gain is usually defined as the maximum value, and the maximum value of the antenna gain is often the front gain of the antenna (see Figure 7.2). Beam efficiency can be calculated by (7.7) and (7.8) only when the receiving antenna is in front (i.e., in the maximum-gain direction) of a transmitting antenna (Figure 7.11 (a)). When the antenna position is not in front (Figure 7.11(b)), the oblique antenna gain and the beam efficiency decrease. In other words, the directivity of the antenna shown in Figure 7.2 affects the beam efficiency. It is similar to the transmission efficiency between coils in an inductively coupled WPT and a resonance-coupled WPT. In the case of the coupled WPT system, the transmission efficiency changes with the positions of the transmitting (coil or resonator) or receiving elements because of the circuit parameters (e.g., circuit impedance or resonance frequency).

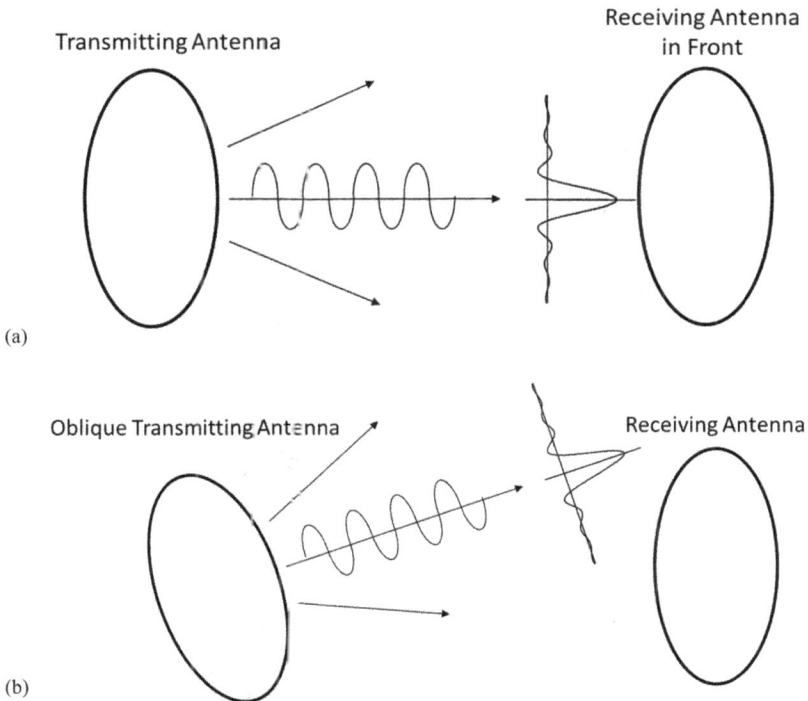

Figure 7.11   Direction and position of the transmitting and receiving antennas

In the WPT system via radio waves, instead of electromagnetic coupling, the directivity of the antenna affects the beam efficiency (transmission efficiency).

In all WPT systems, a high transmission efficiency is desirable. In the case of the coupled WPT system, we control circuit impedance and resonance frequency, sometimes actively. In the case of the WPT system via radio waves, we control directivity, again sometimes actively. The easiest way to maintain a high-beam efficiency of the WPT system via radio waves is to adjust the direction and position of the antenna optimally. When the receiving antenna moves, we must change its direction and the position mechanically like a parabolic antenna. However, there might be some problems with the accuracy of positioning and the control speed.

Instead of mechanically controlling the direction of the antenna, a phased-array antenna is often used to control directivity actively and electronically. The phased-array antenna is composed of a number of antenna elements, including an amplitude-control circuit, a phase-control circuit (phase shifter), and beam-forming network circuits. An exemplary phased-array antenna is shown in Figure 7.12. Electromagnetic waves (including radio waves) interfere with each other. In the phased-array antenna, the interference of the radio waves is used to control the directivity and to form the beam by controlling the amplitude and the phase of the radio wave from each antenna element. Figure 7.13(a) presents the direction control of the radio wave by the phased-array antenna. The phased-array antenna does not need to move mechanically to control the directivity. Figure 7.13(b) indicates a typical controlled directivity of the phased-array antenna.

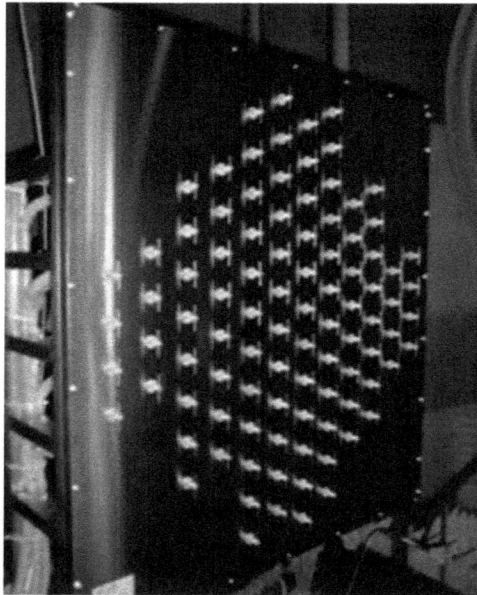

Figure 7.12   Example of a phased-array antenna

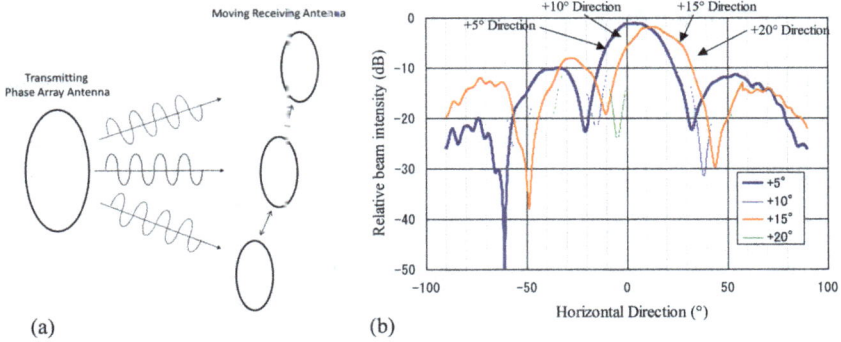

Figure 7.13 *(a) Image of direction control of radio waves by the phased-array antenna, and 'b) typical controlled directivity of the phased-array antenna*

Target-position detection is an important technology for maintaining high-beam efficiency with a phased-array antenna. There are various target-position-detection methods, including the Global Positioning System, supersonics, optics, and others. The direction of arrival (DOA) of radio waves is often used to detect the target position. A pilot or beacon signal is transmitted from targets to detect the target position first. At the transmitter, the pilot signal is received and the DOA of the pilot signal is estimated by a computational algorithm. The following algorithms are often used for the DOA:

- Capon method
- Linear-prediction method
- Minimum-norm method
- Multiple-signal-classification method
- Estimation of signal parameters via a rotational-invariance technique

With such DOA algorithms, we can estimate not only one target but multiple targets. However, it takes time to estimate the target position.

A retrodirective-target-detecting system, in which a pilot signal from a receiver is used, is often employed for WPT via radio waves [5]. The retrodirective target detecting system is based on the Van Atta array [6]. To detect the target position, a phased-array antenna is used as a duplexer antenna and phase-conjugate circuits are used instead of phase shifters for the retrodirective system. As a result of phase conjugation on the phased-array antenna, the radio-wave power from the trans-mitting antenna is transmitted in the direction of the pilot signal from the target. The retrodirective system can be developed only using analog circuits. Thus, it does not take time to estimate the target position. In the retrodirective system, infor-mation on the shape of the transmitting antenna is included. Thus, if the transmitting-antenna plane actively fluctuates, the radio-wave power can be trans-mitted to the target automatically and with high accuracy.

7.4   Receiving-antenna efficiency

To maintain high transmission efficiency in inductively coupled and resonance-coupled WPT systems, the impedance of circuits and resonance frequency must be matched. The beam efficiency in the WPT system via radio waves must also be considered. Of course, in this system, there is no electromagnetic coupling among antennas, and we do not need to consider coupling effects. However, we must consider the resonance frequency of each antenna to be the same to maintain high-beam efficiency. We also consider the optimal position, including directivity, of each antenna. Additionally, we must consider impedance matching between the antennas and space.

When the impedances of each antenna in an infinite array are matched, considering the antenna-element spacing and the DOA of a radio wave, the receiving efficiency, η_{recept}, can theoretically reach 100% [7–9]. η_{recept} is defined as the ratio of the input power at an aperture of the array antennas to the power absorbed into circuits through the antennas; η_{recept} is obtained as the following equation:

$$\eta_{recept} = \frac{\pi^2 a^2}{Z_0 G L_x L_y \cos \theta} \{J_0(\rho) - J_2(\rho)\}^2, \tag{7.17}$$

$$J_\alpha(x) = \sum_{m=0}^{\infty} \frac{(-1)^m}{m!(m + \alpha + 1)} \left(\frac{x}{2}\right)^{2m+\alpha} \text{ (Bessel function of the first kind)},$$

$$\tag{7.18}$$

where $\rho = ka \sin \theta$, k is wave number, a is the diameter of a microstrip antenna, Z_0 is the impedance of space, G is the active conductance, and L_x and L_y are element spacing. In this case, a single TM_{110} mode is considered.

Figure 7.14 indicates the receiving efficiency of the infinite array with circular microstrip antennas. The direction of arrival of the radio wave is θ. Considering the

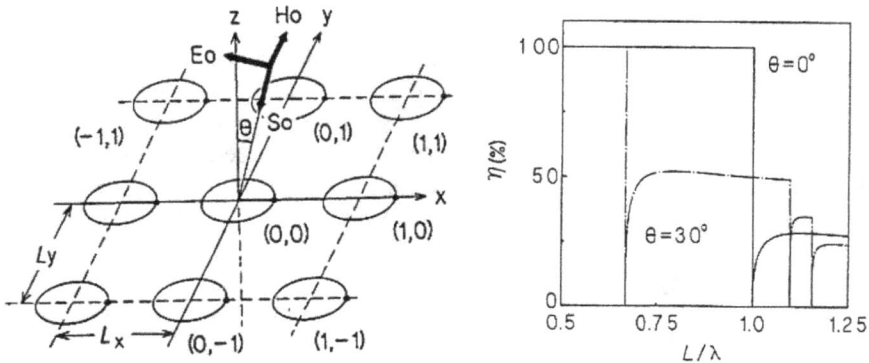

Figure 7.14   Receiving efficiency with element spacing L (=L_x=L_y) and incident angle θ [9]

element spacing and the DOA of the radio wave, when we matched the impedance of the antennas as given by (7.17), input radio wave to the infinite array antenna can be, theoretically, 100% absorbed. In the case of a finite array antenna, the receiving efficiency can reach up to 100% [10].

References

[1] Q. Chen, K. Ozawa, Q. Yuan, and K. Sawaya, "Antenna design for near-field wireless power-transfer (in Japanese)", *Technical Report of IEICE*, pp. 5–9, WPT2010-05, 2010.

[2] N. Shinohara, "Beam efficiency of wireless power transmission via radio waves from short range to long range", *Journal of the Korean Institute of Electromagnetic Engineering and Science*, Vol. 10, No. 4, pp. 224–230, 2011.

[3] Y. Fujiyama, "Field intensity measurement of the wireless electric power transmission equipment using a dielectric resonator (in Japanese)", *Technical Report of IEICE*, WPT2013-42, 2014.

[4] K. Nishikawa and T. Ishizaki, "Study on propagation-mode of wireless power transfer system using ceramic dielectric resonators *(in Japanese)*", *Technical Report of IEICE*, Vol. 113, No. 70, pp. 41–45, MW2013-17, 2013.

[5] N. Shinohara, "Beam control technologies with a high-efficiency phased array for microwave power transmission in Japan", *Proceeding of IEEE*, Vol. 101, No. 6, pp. 1448–1463, 2013.

[6] L. G. Van Atta, "Electromagnetic reflector", U.S. patent No. 2,908,002; Oct. 6, 1959.

[7] B. L. Diamond, "A generalized approach to the analysis of infinite planar array antennas", *Proceeding of IEEE*, Vol. 56, pp. 1837–1851, 1968.

[8] L. Stark, "Microwave theory of phased array antenna – A review", *Proceeding of IEEE*, Vol. 62, pp. 1661–1701, 1974.

[9] K. Itoh, T. Ohgane, and Y. Ogawa, "Rectenna composed of a circular microstrip antenna", *Space Power*, Vol. 6, pp. 193–198, 1986.

[10] N. Shinohara and Y. Tsukamoto, "Antenna absorption efficiency and beam efficiency of a microwave power transmission system", *Proceeding of APCAP2017*, 2017.

Chapter 8

Technologies of antenna and phased array for wireless power transfer via radio waves

A. Massa[1,2,3,4,5], *G. Oliveri*[1,2], *P. Rocca*[1,2,6], *N. Anselmi*[1,2], *A.A. Salas-Sanchez*[1,2] *and M. Salucci*[1,2]

The idea of transferring power over long distances via radio waves has emerged shortly after the development of high-power microwave amplifiers. The potential applications and impact of wireless power transmission (*WPT*) have motivated an always increasing interest from the academic and industrial communities in the research and development of effective *WPT* transmitting and receiving devices. Within such a framework, this chapter is aimed at introducing and formulating the problem of designing *WPT* antennas and arrays for power transmission via radio waves, and at illustrating the theoretical motivations, fundamentals, and technological guidelines behind state-of-the-art strategies for the synthesis of such systems. A discussion on the current trends and envisaged development within this research area will also be provided.

8.1 Introduction and rationale

The synthesis, configuration, and control of the transmitting/receiving (*TX/RX*) antennas have a fundamental importance in the development of effective systems for long-distance wireless power transmission (*WPT*) via radio waves [1–6]. As a matter of fact, *TX/RX* antennas are the unique components in *WPT* system that are responsible for the radiation and collection of the wireless power, and therefore they represent the fundamentally enabling technology in many envisaged

[1]DICAM – Department of Civil, Environmental, and Mechanical Engineering, ELEDIA Research Center (ELEDIA @ UniTN – University of Trento), Italy
[2]CNIT—'University of Trento' ELEDIA Research Unit, Italy
[3]School of Electronic Science and Engineering, ELEDIA Research Center (ELEDIA @ UESTC—UESTC), China
[4]Department of Electronic Engineering, ELEDIA Research Center (ELEDIA @ TSINGHUA—Tsinghua University), China
[5]School of Electrical Engineering, Tel Aviv University, Israel
[6]School of Mechano-Electronic Engineering, ELEDIA Research Center (ELEDIA @ XIDIAN—Xidian University), China

applicative scenarios of *WPT*, including unmanned aerial vehicles and platforms [7], electric vehicles [8], pervasive sensors and actuators [9], and mobile phones/laptops [10].

Unfortunately, standard strategies and technologies for the design and fabrication of antennas/phased arrays have not been conceived to meet the objectives of *WPT* applications [6,11]. As a matter of fact, state-of-the-art antenna/array synthesis methods are usually aimed at optimizing figures of merit that are relevant for radar/communications systems (such as the gain, the sidelobe level, and the half-power beamwidth) [11]. On the contrary, the key objective in the development of a *WPT* system is to maximize the power transfer efficiency (i.e., the ratio between the received and transmitter powers at the output/input of the *WPT* chain—Figure 8.1) [6,12,13].

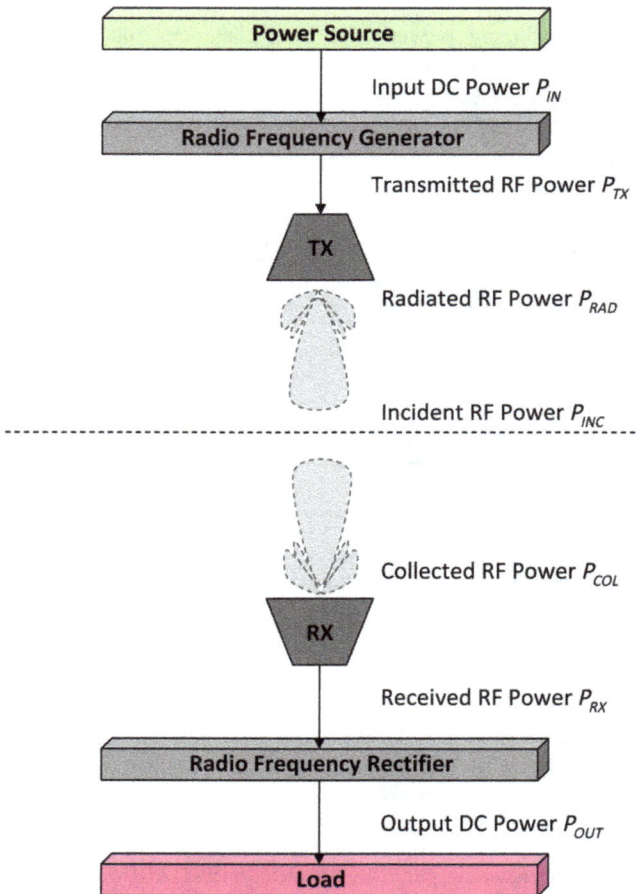

Figure 8.1   WPT Problem—functional scheme of the transmission-reception WPT chain.

Consequently, new classes of methodological approaches are actually required to handle the design of effective antennas/arrays for radio-wave *WPT* [6,12,13].

From the antenna architectural viewpoint, the unusual constraints/objectives arising in *WPT* antenna design have a major impact especially in the development of the phased arrays used in transmission [6]. Indeed, the *TX* antenna system is responsible for the power focusing capabilities of the *WPT* system, and therefore it must exhibit a carefully controlled beam shape (i.e., to guarantee high power transfer efficiencies) while minimizing the architectural complexity and resulting costs [6]. On the contrary, simple architectures are often adopted in *RX* [6,14]. In fact, most popular strategies for the design of the in *WPT* receiving array feature *non-coherent* reception schemes in which each antenna is responsible for the reception and rectification of the incident radio power, which is then combined at the direct current (*DC*) level [4,6,14]. This solution (unique to *WPT* receiving systems) is widely preferred over phased-array ones because of (*a*) its intrinsic redundancy [14], (*b*) its simplicity (i.e., it allows to avoid expensive, heavy, and complex radio-frequency feed networks in *RX* [15–17]), and (*c*) its modularity and stability against the frequency, the input power level, the noise, and the possible interferers [4,17].

Accordingly, and owing to the focus of the current chapter, the introduction and formulation of the problem of designing *WPT* arrays for power transmission via radio waves will be discussed in the following, and the theoretical motivations, fundamentals, and applicative guidelines of state-of-the-art strategies for antenna and phased array synthesis for *WPT* will be presented. Moreover, a discussion on the current trends and envisaged development within this research area will be also included.

The chapter is thus organized as follows. After an introduction on the concept of end-to-end *WPT* efficiency (Section 8.1), the problem of designing the *Transmitting WPT* antenna will be mathematically formulated (Section 8.2). Afterward, several classes of design strategies for *WPT* phased arrays will be presented in detail (Section 8.3), also discussing numerical and experimental state-of-the-art results. Some conclusions and final remarks will follow (Section 8.4).

8.2 Design of antenna and phased arrays for *WPT*: problem formulation

8.2.1 The end-to-end *WPT* efficiency

Let us consider the sketch model of a generic *WPT* system operating over long distances as illustrated in Figure 8.1 [1,3,5,6,18,19]. In such a scenario, a *DC* power source feeds the system input power P_{IN} to a radio-frequency (*RF*) generator, which then converts it to a transmitted *RF* power P_{TX} (Figure 8.1). The *RF* power is then transduced by the *TX* antenna system (usually, an array of relatively simple elements including dipoles [1], waveguide slots [20], horns [21,22], and microstrip patches [23,24]) into a total radiated power P_{RAD} (Figure 8.1), whose spatial distribution is controlled by the shape of the *TX* antenna beam [6,19,21,25–27]. Accordingly, only a portion P_{INC} of the total radiated power actually impinges on the *RX* antenna, which is then responsible for acquiring it according to the *RX* beam shape into the actually collected *RF* power P_{COL} (Figure 8.1). The *RX* antenna then transduces such a power into the received RF power (P_{RX}), a portion of which (i.e., P_{OUT}) is finally fed to the system load after the *RF-to-DC* conversion performed by the rectifier (Figure 8.1) [16,18,28].

The objective of any long-distance *WPT* system is naturally to maximize the end-to-end power transfer efficiency, which, according to the above description, is defined as [6,18,19]

$$\eta \triangleq \frac{P_{OUT}}{P_{IN}} \qquad (8.1)$$

and it is equal to [6,19]

$$\eta = \eta_{DC-RF} \times \eta_{TA} \times \eta_{BT} \times \eta_{BC} \times \eta_{RA} \times \eta_{RF-DC}, \qquad (8.2)$$

where the efficiencies of each subsystem of the *WPT* chain are defined as (Figure 8.1)

- the *DC-to-RF* conversion efficiency $\eta_{DC-RF} \triangleq \frac{P_{TX}}{P_{IN}}$,
- the transmission antenna efficiency $\eta_{TA} \triangleq \frac{P_{RAD}}{P_{TX}}$,
- the beam transmission efficiency $\eta_{BT} \triangleq \frac{P_{INC}}{P_{RAD}}$,
- the beam collection efficiency $\eta_{BC} \triangleq \frac{P_{COL}}{P_{INC}}$,
- the receiving antenna efficiency $\eta_{RA} \triangleq \frac{P_{RX}}{P_{COL}}$,
- the *RF-to-DC* conversion efficiency $\eta_{RF-DC} \triangleq \frac{P_{OUT}}{P_{RX}}$.

Accordingly, the end-to-end efficiency η of a long-range *WPT* system is mainly related to (*a*) the efficiency of the *DC*-to-*RF* conversion in transmission and radiation (i.e., $\eta_{DC-RF} \times \eta_{TA}$), (*b*) the capability to suitably shape the radiated beam toward the collecting area (i.e., η_{BT}), (*c*) the capability to gather and convert the power impinging on the receiver antenna (i.e., $\eta_{BC} \times \eta_{RA} \times \eta_{RF-DC}$). Solving a generic *WPT* problem thus requires to address several independent design subproblems [6].

Owing to the focus of this chapter, attention will be paid to the strategies and synthesis algorithms for the maximization of the "beam transmission efficiency" (i.e., key point (*b*)—Section 8.2). Indeed, most of the current studies on *WPT* via radio waves are focused on the synthesis and fabrication of receiving antennas and *RF-to-DC* and *DC-to-RF* conversion circuits and power management units [5,15,28–30]. On the contrary, the development of design methodologies and architectures aimed at addressing the η_{BT} maximization has been quite limited [6,19]. The interested reader should refer to Chapters 7 and 9 for a more detailed discussion on (*a*) and (*c*).

8.2.2 *The transmitting WPT antenna design problem*

The key objective of the transmitting array in a *WPT* system is to maximize η_{BT} by focusing the power associated with the radiated waves in the angular region corresponding to the *RX* collecting area, while also taking into account suitable technological constraints/guidelines (e.g., to minimize the fabrication and maintenance costs) [2,18,21,25–27]. Accordingly, a static* *WPT* transmitting array design strategy must be able to define the optimal tradeoff excitations so that (*i*) the power sent toward the receiver aperture Ψ (i.e., the angular area occupied by the *RX* antenna) is maximized, and (*ii*) the resulting tapering window is *feasible*.

To achieve this goal, the *WPT* beam shaping problem is usually formulated as a constrained *power-efficiency* maximization one [18,25–27,31,32]. More specifically, let us consider a phased array arrangement with N ideal radiating elements displaced either along a line or on a planar grid. Under these assumptions, the *WPT* *TX* antenna design problem can be mathematically formulated as that of finding the set of excitation coefficients $\mathbf{w} = \{w_n; n = 1, ..., N\}$ such that [27,31–33]

$$\mathbf{w}_{opt}^{D} \triangleq \arg\left[\max_{\mathbf{w}} \left(\eta_{BT}^{D}(\mathbf{w})\right) \; s.t. \, \mathbf{w} \in \mathcal{F}, \quad \mathcal{D} \in \{1D, 2D\}\right. \tag{8.3}$$

where $\mathcal{F}$ is the excitation "feasibility set", $\eta_{BT}^{D}(\mathbf{w})$ is the beam transmission efficiency, which is equal to

$$\eta_{BT}^{D}(\mathbf{w}) \triangleq \frac{P_{INC}^{D}(\mathbf{w})}{P_{RAD}^{D}(\mathbf{w})}, \quad \mathcal{D} \in \{1D, 2D\}, \tag{8.4}$$

where $P_{INC}^{D}(\mathbf{w})$ and $P_{RAD}^{D}(\mathbf{w})$ are the powers radiated in the region corresponding to the receiver and the total radiated power, which can be computed as

$$P_{INC}^{D}(\mathbf{w}) = \begin{cases} \int_{\Psi} \left|W^{1D}(u)\right|^2 du & \mathcal{D} = 1D \\ \iint_{\Psi} \left|W^{2D}(u,v)\right|^2 du\,dv & \mathcal{D} = 2D \end{cases} \tag{8.5}$$

and

$$P_{RAD}^{D}(\mathbf{w}) = \begin{cases} \int_{\Omega} \left|W^{1D}(u)\right|^2 du & \mathcal{D} = 1D \\ \iint_{\Omega} \left|W^{2D}(u,v)\right|^2 du\,dv & \mathcal{D} = 2D, \end{cases} \tag{8.6}$$

respectively, (u,v) are the direction cosines, $W^{1D}(u) = \sum_{n=1}^{N} w_n \exp[jkuz_n]$ and $W^{2D}(u,v) = \sum_{n=1}^{N} w_n \exp[jk(ux_n + vy_n)]$ are the 1D and 2D array factors,[†] $\Omega \triangleq \{(u,v) : u^2 + v^2 \leq 1\}$ is the overall visible range, $k = \frac{2\pi}{\lambda}$ is the wavenumber, and z_n and (x_n,y_n), $n = 1, ..., N$, are the element locations.[‡] It is worth remarking

---

[*]Beam steering in *TX*, which is required for mobile *WPT* applications, is discussed in detail in [6,10,18].
[†]The array factor expressions could be modified to account for element directivity and mutual coupling effects [34]. For the sake of notation simplicity, such effects will be neglected in the following. The reader is referred to [35] for an exhaustive discussion on such a topic.
[‡]Without loss of generality and following the most commonly adopted convention [33], the linear and planar arrangements are assumed hereinafter to be located in the *z* axis and in the *x–y* plane, respectively.

that $\mathcal{F}$ can mathematically encode a multiplicity of constraints/guidelines regarding the desired simplicity, weight, reliability, compactness, and robustness of the resulting feeding network.

Given the nature of the synthesis problem (8.3), which aims at finding a tradeoff between performance and complexity of the *WPT* layout, no general-purpose solution strategy exists, which is optimal regardless of the considered $\mathcal{F}$. Accordingly, many different design strategies have been proposed in the literature that are suitable to handle different classes of constraints on the size, target efficiency, and costs for the specific application.

8.3   *WPT*-phased array synthesis techniques

8.3.1   Uniform excitations in WPT

The design of architectures able to yield very low/possibly trivial realization costs is one of the key focus in the scientific literature dedicated to *WPT* transmitting arrangements because of their importance from the practical viewpoint [20,23,36,37]. In this framework, many design methods, realizations, and prototypes are conceived assuming very strong limitations on the allowed tapering windows, such as enforcing [20,23,36,37]

$$\mathcal{F} \triangleq \{|w_n| = 1, n = 1, ..., N\}, \tag{8.7}$$

where normalized excitation magnitudes have been assumed. The exploitation of *uniform arrays* in *TX*, as dictated by (8.7), is motivated by the fact that using *TX* modules with fixed and equal gain considerably simplifies the fabrication, maintenance, and replacement of the feed network, as well as the complexity of its calibration process [20,23]. Such features are of fundamental importance especially in those scenarios where the costs of the array architecture and of its maintenance must be mandatorily minimized even at the expenses of a reduction of η_{BT} [18,23].

Several experimental demonstrators of *WPT* systems have been actually obtained following the guideline (8.7). As an example, a *TX*-phased array featuring $N=256$ antennas operating at 5.8 GHz and fed with a 1.5-kW input power has been designed, fabricated, and experimentally validated in [23]. To this end, a regular triangular lattice has been adopted, and each element (i.e., a microstrip patch with circular polarization) has been fed with a 6-W high-power amplifier and a 5-bit phase shifter [23]. Despite its minimal complexity, the resulting prototype has been shown to guarantee an excellent beam steering accuracy [23]. An analogous strategy has been adopted for the design and realization of the active integrated *WPT TX* antenna shown in [20], which has been based on the combination of a waveguide power divider and a slot-coupling feeding network [20].

In order to further reduce the technological complexity of the array architecture, uniformly excited *WPT* layouts have been realized by using subarrayed feed structures [38]. In such a case, the same design rule (i.e., (8.7)) has been adopted, but an additional fabrication simplification has been achieved by feeding a set of radiating elements by means of a single *RF TX* module (comprising a phase shifter and a power amplifier) [38]. Thanks to such an approach, the same broadside performance of a fully populated uniformly fed layout has been achieved using $Q \ll N$ active transmitting modules, thus considerably reducing the prototype costs [38]. Within this framework, the design and fabrication of an $N = 128$ *WPT TX* antenna has been demonstrated in [38] by combining $Q = 2$ subarrays, each one comprising $\frac{N}{Q} = 64$ circularly polarized truncated microstrip patch antennas. The capability to radiate a 40-W beam at 5.8 GHz by means of only $Q = 2$ power amplifiers and two phase shifters has been success-fully demonstrated [38] following such a strategy. Unfortunately, according the theory of clustered arrays [11,39], the main drawback of this class of approaches is the reduction of their field-of-view with respect to their fully populated coun-terparts. As an example, a maximum steering angle of ±5 deg has been experi-mentally demonstrated for the prototype in [38].

The success of the uniform tapering rule (8.7), which is often adopted because of the minimum cost and complexity of the arising architecture despite its suboptimal η_{BT}, is demonstrated by its popularity in the fabrication of *WPT* demonstrators [22,40–42]. For instance, the *TX* array designed and fabricated for the long-distance *WPT* experiments demonstrated in [41,42] has been based on a uniform tapering window. In such an example, a large transmitting array able to convey 20 W at about 150 km at competitive costs has been shown [41,42]. Moreover, the same design rule has been adopted also for the implementation of arrangements with few antennas [22,43]. As an example, a *WPT TX* layout fea-turing $N = 5$ horn antennas operating at 5.8 GHz has been demonstrated to allow 4 W to be delivered to a micro-aerial vehicle via radio waves through an efficient and cost-effective feed network based on modular phase shifters and amplifiers [22]. Furthermore, an $N = 4 \times 8$ *WPT TX* planar layout with uniform tapering has been designed and fabricated (Figure 8.2) to feed a rover [43]. In such a demonstrator, a fixed feeding network (Figure 8.2(a)) or a single high-efficiency radiating module (Figure 8.2(b)) has been deployed to further reduce the system complexity.

8.3.2 Heuristic tapering methods

The solutions based on uniform tapering (Section 8.3.1) by definition guarantee the minimum architectural complexity for the *WPT TX* array. Unfortunately, the arising beams are not shaped to guarantee power focusing and thus η_{BT} max-imization. In order to enhance the beam transmission efficiency, pattern control capabilities have to be implemented in the *TX* feeding architecture, and suitable tapering strategies for *WPT* applications are required [6,11,25]. Standard excitation design methods based on elementary tapering functions (often derived from other applicative scenarios [11]) have been among the first strategies to be investigated [25,26,44,45].

In this framework, weighting techniques based on Gaussian functions have been employed to yield improved η_{BT} in *WPT* applications via radio waves [45,46]. More specifically, the excitation synthesis has been carried out assuming the

(a)

(b)

Figure 8.2   Prototypes of transmitting systems for WPT-fed rovers [43]. (a) The
4×8 element active integrated antenna (AIA) with no steering
capabilities. (b) The 4×8 active integrated phased-array antenna
(AIPAA). (courtesy of Prof. S. Kawasaki)

following technological constraint (in the 2D case) [6,45,46]:

$$\mathcal{F} \triangleq \left\{ |w_n| = \exp\left[\frac{\lambda^2 \ln(A_0)(x_n^2 + y_n^2)}{R^2}\right], n = 1, ..., N \right\},$$

(8.8)

where $A_0 \geq 0$ is a user-defined parameter controlling the tapering level, and R is the *TX* aperture size, in meters (analogous strategies have been adopted also in $1D$ scenarios). This choice is actually motivated by the capabilities of *continuous* Gaussian apertures to yield η_{BT} values close to the theoretical optimal solutions of (8.3) [47,48]. As a matter of fact, an aperture featuring a continuous Gaussian taper with Fresnel number equal to 5.0 has been shown to yield $\eta_{BT}^{Gauss} \approx 99.31\%$, which is very close to the ideal optimum $\eta_{BT}^{opt} \approx 99.53\%$ for such an aperture [48]. This feature has been demonstrated to guarantee significant beam transmission improvements with respect to equally tapered layouts (e.g., $\eta_{BT}^{Unif} \approx 83.8\%$ vs. $\eta_{BT}^{Gauss} \approx 99.7\%$ has been shown in [46] [6]. Moreover, design methods based on Gaussian tapers do not require any optimization and can be adopted even for very large apertures with trivial computational efforts (i.e., (8.8)) [6].

The fundamental technological drawback of Gaussian tapers is related to the *smooth* spatial variations of the resulting excitation magnitudes (8.8). In fact, these smooth variations can be implemented only by using fully populated architectures (complex and very expensive to fabricate and maintain) featuring a different magnitude control for each *WPT TX* element [25,44]. Alternative excitation distributions based on the concept of *edge tapering* have been then proposed to simplify the *WPT* array fabrication while still yielding high η_{BT} values [25,44]. This paradigm combines (a) a set of uniformly fed elements (located in the innermost part of the *WPT* array aperture) with (b) a small portion of antennas that are actually tapered and displaced only in the vicinity of the aperture border [25,44]. Within this line of reasoning, the following normalized tapering window has been proposed (for the $1D$ array case) [25]:

$$\mathcal{F} \triangleq \left\{ |w_n| = A_0 + \frac{n(1 - A_0)}{\alpha_0(N - 1)} \text{ if } n \in [1, N\alpha_0), \right.$$

$$|w_n| = 1 \text{ if } n \in [N\alpha_0, N(1 - \alpha_0)],$$

$$\left. |w_n| = 1 + (1 - A_0)\frac{(1 - \alpha_0)(N - 1) - n}{\alpha_0(N - 1)} \text{ if } n \in ((1 - \alpha_0)N, N] \right\} \quad (8.9)$$

where $A_0 \geq 0$ is the tapering level, α_0 is (normalized) user-defined size of the tapering region ($\alpha_0 \in [0, 0.5]$), and d is the *WPT* array lattice spacing [25,44]. By comparing the performance of the arising isosceles-trapezoidal distribution (*ITD*) (i.e., see (8.9)) with those obtained by Gaussian and uniform tapers, it has been shown that the *ITD* guarantees improved η_{BT} values in most design scenarios [25]. As an example, an $N = 25 \times 25$ *ITD* arrangement (operating frequency 5.8 GHz; $d = 0.75\lambda$, $\alpha_0 = 0.25$, $A_0 = -18$ dB; *RX* aperture with area 9.4×10 m displaced at 1 km) is shown in [25] to allow a 3% and 10% improvement in terms of η_{BT} over the corresponding Gaussian and uniform tapering, respectively. These power focusing capabilities, together with the easy computability of the tapering rule (8.9) that is suitable also for very large apertures, have made *ITD* a popular design strategy in *WPT* applications [25,26,44].

The combination of nonuniform array geometries with *ITD*-inspired tapering has been considered a possible approach to further improve η_{BT} with moderate architectural complexity [26]. More specifically, 1D *WPT TX* arrays based on the excitation design rule (8.9) and featuring an element displacement based on

$$
z_n = \begin{cases} nd \times \dfrac{\text{sinc}\left[A_1 + \dfrac{n(1-A_1)}{\alpha_1(N-1)}\right]}{0.8415} & n \in [1, \alpha_1(N)) \\[4mm] nd & n \in [\alpha_1 N, (1-\alpha_1)N] \\[4mm] nd \times \dfrac{\text{sinc}\left[1 + (1-A_1)\dfrac{(1-\alpha_1)(N-1)-n}{\alpha_1(N-1)}\right]}{0.8415} & n \in ((1-\alpha_1)N, N] \end{cases}
$$

(8.10)

(d being the basic element spacing, $\alpha_1 \in [0, 0.5]$ the size of "nonuniform" region, and $A_1 \geq 0$ is the associated distortion factor) have been proposed in [26]. The resulting *isosceles-trapezoidal distribution with unequal element spacing (ITDU)* methodology has been demonstrated to guarantee considerably enhanced η_{BT} values with respect to both standard *ITD* and Gaussian layouts (e.g., $\eta_{BT}^{ITD} \approx 98.76\%$, $\eta_{BT}^{Gauss} \approx 98.33\%$, $\eta_{BT}^{ITDU} \approx 99.69\%$ when $N \approx 100$ [26]). This improvement is actually expected from the theoretical viewpoint because of the increase in the complexity (and associated costs) of the geometrical layout, owing to its nonuniform nature (i.e., (8.9)) [26].

8.3.3 Designs based on optimization strategies

The *WPT TX* array synthesis strategies illustrated in Sections 8.3.1 and 8.3.2 are all based on explicit and closed-form design rules that can be computed very efficiently regardless of the aperture size [27]. Unfortunately, due to their heuristic nature, the arising radiated beams are not guaranteed to yield the maximum η_{BT}. Accordingly, the exploitation of optimization methods [49,50] has been proposed as a possible strategy to achieve higher beam efficiencies in *WPT TX* systems [27,51–53].

In this framework, a local optimization technique based on the *Coordinate Descent (CD)* method has been proposed in [53] to compute the tapering window that maximizes η_{BT} by matching a reference "optimal" field distribution $W_{REF}(u, v)$ in the Ψ region, and minimizing the power "lost" in the rest of the visible range [53]. To this end, the following cost function [53]

$$
\Phi^{CD} = \iint_\Psi |W^{2D}(u, v) - W_{REF}(u, v)|^2 \, du \, dv
$$

$$
+ \chi \left(\frac{\max_{(u,v) \notin \Psi+\Delta}\left(|W^{2D}(u,v)|^2\right)}{\max_{(u,v) \in \Psi}\left(|W^{2D}(u,v)|^2\right)} \right),
$$

(8.11)

(being Δ the *safety area* surrounding Ψ, and χ the associated weighting coefficient) has been minimized through an iterative $1D$ line-search strategy based on the following procedure [53]: (*a*) initialize $w_n = 1$, $n = 1, \dots, N$, iteration number $h = 0$, and element number $p = 0$; (*b*) minimize Φ_{CD} only with respect to $w_n|_{n=p}$ through a line search technique ($1D$ minimization); (*c*) if $p = N-1$ goto d, otherwise update $p \leftarrow p + 1$ and goto b; (*d*) if $\Phi_{CD} \leq \tilde{\Phi}$ terminate ($\tilde{\Phi}$ being the user-defined threshold value for termination), else, update $h \leftarrow h + 1$, $p = 0$, and goto b. The CD-based method has been shown to guarantee improved η_{BT} values over heuristic approaches, although at the cost of a non-monotone variation of the excitation magnitudes toward the TX aperture ends [53].

Within this line of reasoning, the exploitation of stochastic algorithms has been considered in the *WPT* scenario to synthesize wide arrangements with maximum η_{BT} and controlled sidelobes [51,52]. More specifically, a *Combined Stochastic Algorithm* (*CSA*) has been introduced for synthesizing linear *WPT* TX arrays with standard tapering windows (i.e., uniform tapering: $w_n = 1$, $n = 1, \dots, N$; squared cosine tapering: $w_n = \left[\cos\left(\frac{\pi x_n}{2}\right)\right]^2$, $n = 1, \dots, N$) but featuring antennas displaced according to an irregular lattice [51,52]. To achieve this goal, a small (with respect to d) random perturbation δ_n is applied to each antenna position as follows:

$$z_n = nd + \delta_n \quad n = 1, \dots, N \tag{8.12}$$

which is stochastically optimized to minimize the sidelobe level (*SLL*) and to maximize η_{BT} [52]. Thanks to such a nonuniform spacing, the *CSA* method has been shown to guarantee considerable improvements in terms of η_{BT} and *SLL* control over standard uniform arrangements (e.g., $SLL^{UNI} = -13.5$ dB, $SLL^{CSA} = -35$ dB when $N = 16\,000$—[52]), while also allowing the optimization of very large layouts (i.e., $N > 10^4$) with trivial computational efforts [51,52].

An improvement of the *CSA* method has been achieved by introducing in (8.12) two user-defined constants (ν_1 and ν_2) aimed at enhancing the degrees-of-freedom (*DoF*s) available in the design [52]. The resulting "Improved CSA" (*ICSA*) method, which features the following antenna positioning strategy [52]

$$z_n = \nu_1 \times nd + \nu_2 \times \delta_n \quad n = 1, \dots, N, \tag{8.13}$$

has demonstrated to enable a reduction of the average sidelobe level (*ASLL*) from $ASLL^{CSA} = -42$ dB to $ASLL^{ICSA} = -58$ dB when $N = 16\,000$ [52]. Moreover, it is worthwhile to notice that simple and robust tradeoff guidelines for the configuration of the control parameters ν_1 and ν_2 in (8.13) have been derived in the literature (i.e., $\nu_1 = 0.93$; $\nu_2 = 0.1$ [52]).

Still within the framework of optimization strategies, the maximization of η_{BT} subject to a requirement on the *SLL* sidelobe level has been addressed by an iterative method in [27]. To this end, the application of an evolutionary programming (*EP*) approach [54] to design a *WPT* TX layout matching a radiation mask $W_{REF}(u, v)$ and with a maximum η_{BT} has been demonstrated in [27]. The adopted *EP* algorithm, which features a "Levy" and a "Gaussian" mutation operator [54], has been customized and exploited to minimize the following cost function [27]:

$$\Phi^{EP} = -\eta_{BT}$$

$$+ \frac{1}{Area(\Omega)} \iint_{\Omega} \{\mathcal{H}[W(u,v) - W_{REF}(u,v)] \times [W(u,v) - W_{REF}(u,v)]\}dudv$$

$$(8.14)$$

($\mathcal{H}(\cdot)$ being the Heaviside function). The *EP* method has been show to yield excellent η_{BT} values and *SLL* control capabilities (e.g., $\eta_{BT}^{EP} = 0.963$, $SLL^{EP} = -40$ dB [27]) for circular arrangements.

In [55], the maximization of η_{BT} has been instead addressed by means of an *EP* technique aimed at minimizing the sidelobe level of the *TX* antenna pattern in the angular region outside the *RX* antenna, referred to as complementary sidelobe level (*CSL*). More specifically, the use of the *Chaotic Particle Swarm Optimization* (*CPSO*) [56] has been adopted for the case of uniformly excited and unequally spaced planar arrays with constraints on the number of elements, the aperture, the minimum element spacing and the *CSL*. To this end, the following cost function has been minimized [55]:

$$\Phi^{EP} = -\eta_{BT} + w_{CSL} \times \{\mathcal{H}[CSL(X) - CSL_{REF}] \times [CSL(X) - CSL_{REF}]\},$$

$$(8.15)$$

(X being the coordinates of the antenna elements, $w_{CSL} = 10^6$ a real-valued weighting coefficient, and CSL_{REF} the desired *CSL* level). The *EP* method has been show to yield a good compromise between η_{BT} values and *CSL* control capabilities (e.g., $\eta_{BT}^{EP} = 0.900$, $CSL = -12.30$ dB [55]) for square arrangements.

Unfortunately, because of the global optimization nature of the *EP* techniques [49,50], the resulting design strategies turn out computationally expensive when large apertures are at hand (as an example, several thousand computations of (8.15) were necessary to achieve convergence in the examples in [27]).

8.3.4   *Optimal WPT-phased array synthesis*

Despite their interesting features and performance, the *WPT TX* design methods discussed in the previous sections (which are based either on heuristic tapering or on local/global optimization techniques) do not mathematically guarantee to achieve the theoretical optimal η_{BT} value for a given array geometry. Indeed, it has been demonstrated that, at least under suitable assumption, a closed-form optimization of η_{BT} is possible [31–33,57–61]. Accordingly, the possibility to analytically design the tapering windows that maximize the beam transmission efficiency of *WPT TX* antennas has been investigated from the theoretical and practical viewpoint [31–33,57–61]. Because of the computationally expensive nature of the arising synthesis strategies and the smoothness of the optimal tapering windows (which can result in high fabrication costs) [6,33], theoretically optimal designs are mostly adopted when compact layouts are at hand. Nevertheless, the existence of the theoretical optimal solutions of the *WPT* design problem (8.3) has a fundamental importance for the design of any actual *WPT* systems since (*i*) they provide the theoretical upper bound for η_{BT}, which can be used to assess the effectiveness of other (suboptimal) methodologies [6]; (*ii*) the resulting tapering windows can be employed as a starting point/reference for other design methodologies [19].

In this framework, early approaches aimed at η_{BT} maximization have been formulated for continuous antenna apertures rather than for array layouts [60,61]. Under this assumption, an eigenvalue problem formulated with an integral equation similar to those adopted for *Generalized Confocal Resonators* has been solved to compute the theoretically optimal current distribution for *2D* apertures with arbitrary shape [60]. Furthermore, a solution in the form of a product of prolate spheroidal wave functions has been explicitly obtained as the optimal illumination for the rectangular aperture case [60]. Conjugate-gradient approaches have been introduced more recently to take into account *SLL* constraints in the same mathematical framework [32]. Moreover, the *Matrix Method* has been effectively adopted to synthesize continuous antenna current distributions comprising sub-apertures [62].

With reference to the antenna array design problem, the tapering window that theoretically maximizes η_{BT} has been obtained for the *1D* case in [57,58] by exploiting the so-called discrete prolate spheroidal sequences (DPSSs) [63]. Such a method has been then generalized to the *2D* case by demonstrating that the design problem can be formulated in the *1D*/*2D* case as a (generalized) eigenvalue one, regardless of the aperture geometry, the antenna element displacement, and the shape/size of the receiver region Ψ [33]. In fact, (8.4) can be actually rewritten according to [33] as

$$\eta_{BT}^{\mathcal{D}}(\mathbf{w}) = \frac{\mathbf{w}^H \mathcal{A}^{\mathcal{D}} \mathbf{w}}{\mathbf{w}^H \mathcal{B}^{\mathcal{D}} \mathbf{w}} \quad \mathcal{D} \in \{1D, 2D\} \tag{8.16}$$

where $\mathcal{A}^{1D} = \int_{\Psi} \mathbf{s}^{1D}(u)[\mathbf{s}^{1D}(u)]^H du$, $\mathcal{B}^{1D} = I_{N \times N}$ is the $N \times N$ identity matrix, $\mathcal{A}^{2D} = \iint_{\Psi} \mathbf{s}^{2D}(u,v)[\mathbf{s}^{2D}(u,v)]^H dudv$, $B^{2D} = \iint_{\Omega} \mathbf{s}^{2D}(u,v)[\mathbf{s}^{2D}(u,v)]^H dudv$, ^{H} stands for conjugate transpose, $\mathbf{s}^{1D}(u) \triangleq \{\exp[-j2\pi u z_n]; n = 1, ..., N\}$ and $\mathbf{s}^{2D}(u,v) \triangleq \{\exp[-j2\pi(ux_n + v_{yn})]; n = 1, ..., N\}$ are the steering vectors. Therefore, closed-form solution to (8.3) can be computed as

$$\mathbf{w}_{opt}^{\mathcal{D}} = \arg\left[\max_{\mathbf{w}} \left(\frac{\mathbf{w}^H \mathcal{A}^{\mathcal{D}} \mathbf{w}}{\mathbf{w}^H \mathcal{B}^{\mathcal{D}} \mathbf{w}}\right)\right] \quad \mathcal{D} \in \{1D, 2D\} \tag{8.17}$$

which is known to correspond to the eigenvector associated with the maximum eigenvalue ξ^* of the following generalized eigenvalue problem [33,64]:

$$\mathcal{A}^{\mathcal{D}} \mathbf{w} = \xi \mathcal{B}^{\mathcal{D}} \mathbf{w} \quad \mathcal{D} \in \{1D, 2D\}. \tag{8.18}$$

It is worthwhile to remark that ξ^* turns out to be also the theoretical limit of the beam transmission efficiency, since

$$\eta_{BT}^{\mathcal{D}}\Big|_{opt} \triangleq \eta_{BT}^{\mathcal{D}}(\mathbf{w}_{opt}) = \frac{\mathbf{w}_{opt}^H \mathcal{A}^{\mathcal{D}} \mathbf{w}_{opt}}{\mathbf{w}_{opt}^H \mathcal{B}^{\mathcal{D}} \mathbf{w}_{opt}} = \frac{\xi^* \mathbf{w}_{opt}^H \mathcal{B}^{\mathcal{D}} \mathbf{w}_{opt}}{\mathbf{w}_{opt}^H \mathcal{B}^{\mathcal{D}} \mathbf{w}_{opt}} = \xi^* \quad \mathcal{D} \in \{1D, 2D\}. \tag{8.19}$$

As can be noticed, the entries of $\mathcal{A}^{\mathcal{D}}$ and $\mathcal{B}^{\mathcal{D}}$ in (8.17) can be computed in explicit form (for canonical Ψ shapes [33]) or through numerical integration (in all the other cases) both in $1D$ and $2D$ cases [33]. Accordingly, the tapering window that maximizes η_{BT} as well as the corresponding beam transmission efficiency (i.e., theoretical upper bound) can be retrieved as the solution of (8.18), which can be implemented by several algorithms [65] and can be carried out by many popular software tools [66] even for large N values (planar arrangements comprising up to $N = 1600$ antennas—affording $\eta_{BT} \approx 99.98\%$—have been synthesized in about 10^2 s on a laptop PC [33]).

To visually show the results of such an approach, the optimal tapering window obtained when addressing an $N = 100$ half-wavelength spaced array design is reported in Figure 8.3. More specifically, the optimal beams obtained through the previous technique have been computed by assuming different Ψ regions (i.e., $\Psi = \{u \in [-\psi, \psi]\}$, $\psi \in \{1.6 \times 10^{-2}, 3.2 \times 10^{-2}, 4.8 \times 10^{-2}\}$—Figure 8.3) to point out the dependency of the mainlobe size, sidelobe level, and beam shape on the size of the receiver region (Figure 8.3(b)). For completeness, the associated array weights are reported as well (Figure 8.3(a)). As can be noticed and anticipated from the theoretical viewpoint, the optimal $\mathbf{w}$ exhibit smooth spatial variations (Figure 8.3(a)), which can potentially impact the realization complexity of the corresponding *WPT TX* arrangement.

In the same line of reasoning, constrained synthesis problems formulated as the maximization of η_{BT} subject to sidelobe level, directivity, or signal-to-noise ratio constraints have been solved through a modified *Eigenvalue Method* which has been deduced starting from the above techniques [31]. Moreover, approximate solution strategies (e.g., based on *Kaiser* tapering) have been presented to address the computation of the optimal $\mathbf{w}$ for very wide apertures (for which the solution of (8.18) is computationally unfeasible) [58].

8.3.5   Unconventional architectures for WPT-phased arrays

As previously pointed out, the fundamental limitation of the theoretically optimal solutions for *WPT TX* problems is the smooth spatial variations of the resulting tapering windows [19,33]. This feature can significantly increase the fabrication costs of the associated array architectures because of the need to realize fully populated arrangements with several antennas and a different control point for each radiating element [6,19,33]. In order to address such technological/fabrication issues, the use of unconventional array architectures [39] has been recently proposed in the *WPT* scenario [19].

In such a case, the objective is to employ unconventional arrangements to mitigate the *WPT TX* array complexity of while still affording η_{BT} performance as close to the state-of-the-art theoretical optimum as possible [19,33]. To achieve such a goal, two different architectures have been specifically considered in the state-of-the-art [19]: (*i*) clustered *WPT* layouts [67–69], in which the antenna elements are gathered into sub-arrays and the magnitude of the excitation coefficients are controlled at the sub-array level (Section 8.3.5.1); (*ii*) maximally sparse *WPT*

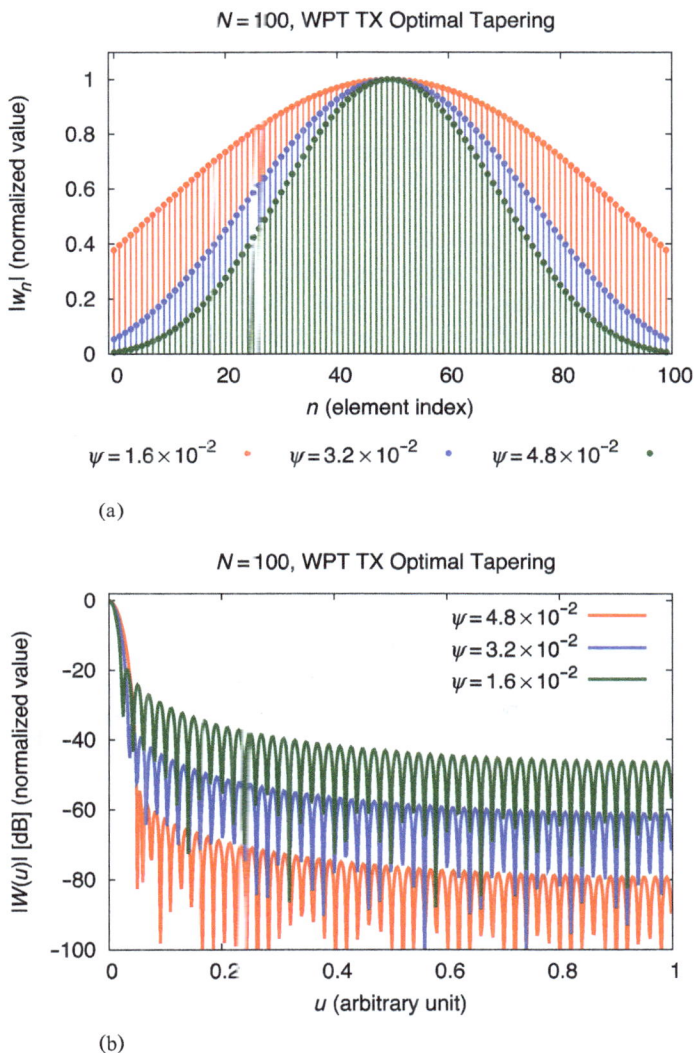

Figure 8.3 Optimal WPT TX Design [1D array, N = 100,
$\psi \in \{1.6 \times 10^{-2}, 3.2 \times 10^{-2}, 4.8 \times 10^{-2}\}$] Optimal tapering window
(a) and associated radiation patterns (b).

arrangements [70–72], in which the number of antennas is actually reduced with respect to the corresponding fully populated layout and they are displaced in a nonuniform fashion (Section 3.3.5.2). These architectures are derived from two fundamentally different strategies from the technological viewpoint. In fact, clustered arrays comprise the same number of radiating elements of the associated fully populated layouts, but they include only few control points [39]. On the other hand, sparse arrangements require a different control point for each antenna, but they are intrinsically lighter since fewer elements are comprised with respect to standard designs [39]. Owing to their unconventional nature, the design of such arrangements has been formulated as an "optimal-solution matching problem" either in the excitation (Section 8.3.5.1) or in the pattern (Section 8.3.5.2) domains, rather than as the direct maximization of $\eta_{BT}^{\mathcal{D}}(\mathbf{w})$, $\mathcal{D} \in \{1D, 2D\}$ [19].

8.3.5.1 Clustered *WPT* Layouts

The design of *clustered WPT* arrays has been addressed in [19] with reference to both 1D and 2D apertures. In such a case, a *WPT* architecture in which the N antennas are clustered in Q (logically contiguous [67]) sub-arrays according to the grouping vector $\mathbf{g} \overset{\Delta}{=} \{g_n \in [1,Q]; n = 1, ..., N\}$, $q = 1, ..., Q$, each one with associated excitation magnitude ω_q, $q = 1, ..., Q$ [67], has been assumed as a reference. Under such assumptions, the excitation coefficient of the nth element can be expressed as [19]

$$w_n = \sum_{q=1}^{Q} \delta_{qg_n} \omega_q, \quad n = 1, ..., N \tag{8.20}$$

where δ_{qg_n} is the Kronecker delta (i.e., $\delta_{qg_n} = 1$ if $q = g_n$, $\delta_{qg_n} = 0$ otherwise). The objective of the synthesis procedure presented in [19] is to define the architecture descriptors (i.e., Q, $\mathbf{g}$, $\mathbf{w} \overset{\Delta}{=} \{\omega_q; q = 1, ..., Q\}$) such that its radiation features and $\eta_{BT}^{\mathcal{D}}(\mathbf{w})$ are as close as possible to those of the theoretically optimal solution computed by solving the eigenvalue problem (8.17). To achieve such a goal, the contiguous-partition method (*CPM*) [67,73] has been adopted and customized to the *WPT* scenario in [19]. Accordingly, the design problem is formulated as *excitation matching* one as follows:

$$\mathbf{g}_{opt} \overset{\Delta}{=} \arg\left\{ \min_{\mathbf{g}} \left[\sum_{n=1}^{N} \left(w_{n,opt}^{\mathcal{D}} - \sum_{q=1}^{Q} \delta_{qg_n} \omega_q \right)^2 \right] \right\} \quad \mathcal{D} \in \{1D, 2D\}, \tag{8.21}$$

where Q is a user-defined synthesis parameter, and the entries of $\mathbf{w}$ are computed as [67,73]

$$\omega_q = \frac{\displaystyle\sum_{n=1}^{N} \delta_{qg_n} w_n}{\displaystyle\sum_{n=1}^{N} \delta_{qg_n}}, \quad q = 1, ..., Q. \tag{8.22}$$

The size of the search space arising if trying to directly solve (8.21) through standard search/optimization methods turns out huge (i.e., Q^N) [19]. As a consequence, the problem (8.21) is apparently computationally unfeasible to solve [19]. Thanks to the *CPM* approach, the clustering algorithm actually requires to scan only the so-called essential solution space (the solutions corresponding to

logically contiguous clusters). Since the size of this latter set is $\begin{pmatrix} N-1 \\ Q-1 \end{pmatrix} \ll Q^N$
[67], the synthesis procedure introduced in [67,73] has been adopted to successfully solve (8.21) [19].

The arising *CPM* technique has been applied seamlessly to 1D and 2D apertures with arbitrary size [19,74] because of its high numerical efficiency even when handling very wide apertures [19]. In fact, very high beam transmission efficiencies (almost identical to the theoretical optimum $\eta_{BT}^{2D}(\mathbf{w}_{opt}) - \eta_{BT}^{2D}(\mathbf{w})\big|_{Q/N=2.0\times 10^{-2}} \leq 0.9\%$ [19]) have been obtained by *CPM*-clustered architectures with low $\frac{Q}{N}$ (i.e., $\frac{Q}{N} \in [2.0 \times 10^{-2}, 2.5 \times 10^{-2}]$ [19]) if applied to sufficiently wide apertures (e.g., $N > 256$), and *WPT*-clustered arrangements have been demonstrated to exhibit a higher degree of regularity (and thus, lower fabrication costs) when the beam transmission efficiency increases [19]. Such results are of particular interest from the methodological viewpoint since they have been obtained despite the fact that the *CPM* minimizes the mismatching between the ideal optimum and the clustered arrangements in the excitation domain, thus only indirectly matching the corresponding η_{BT} [19,67,73,74].

The design of a set of *WPT* clustered arrangements synthesized by the strategy discussed in [19] is shown in Figure 8.4 for illustrative purposes. More in detail, the same $N = 100$ linear layout considered in Figure 8.3 has been assumed, and the synthesis of clustered arrangements matching a reference optimal layout with $\psi = 1.6 \times 10^{-2}$ has been performed [19]. The results obtained by employing $\frac{Q}{N} \in \{0.10, 0.15\}$ clusters (Figure 8.4) show that a significant reduction of the architecture complexity can be achieved (Figure 8.4(a)) with radiation performance almost identical to the ideal case (Figure 8.4(b)), as it is also confirmed by the corresponding beam efficiencies (i.e., $\eta_{BT}^{opt} \approx 95\%$, $\eta_{BT}^{\frac{Q}{N}=0.1} \approx 94.94\%$, $\eta_{BT}^{\frac{Q}{N}=0.1} \approx 94.97\%$—Figure 8.4(b)).

Alternative approaches for *clustered WPT arrays* have been considered in [75,78,79], where the design of planar subarrays has been developed by keeping the same synthesis objective, namely, defining a structure whose radiation features and $\eta_{BT}^{D}(\mathbf{w})$ are as close as possible to those of the theoretically optimal solution computed by solving the eigenvalue problem (8.17).

In [75], the *K*-means algorithm [76,77] has been proposed for addressing the excitation matching problem (8.21). In this work, different apertures and receiver aperture shapes have been analyzed, showing a fast convergence rate of the proposed procedure with respect to the *CPM*. As this method can lead to a suboptimal solution being a local optimization technique sensitive to the initial clustering, a combined strategy integrating the *particle swarm optimization* (*PSO*) algorithm has been proposed in [78]. In this work, the design of *WPT* clustered arrays with high η_{BT}^{2D} performance has been addressed by analyzing the impact of varying the number of array elements, the number of clusters, as well as the shape of the receiving antenna aperture. In all cases, the results have been compared with the performance of the *K*-means algorithm and the theoretical maximum value.

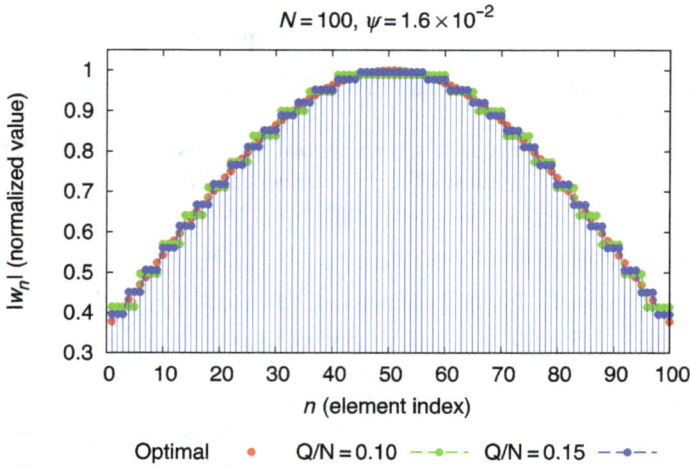

(a)

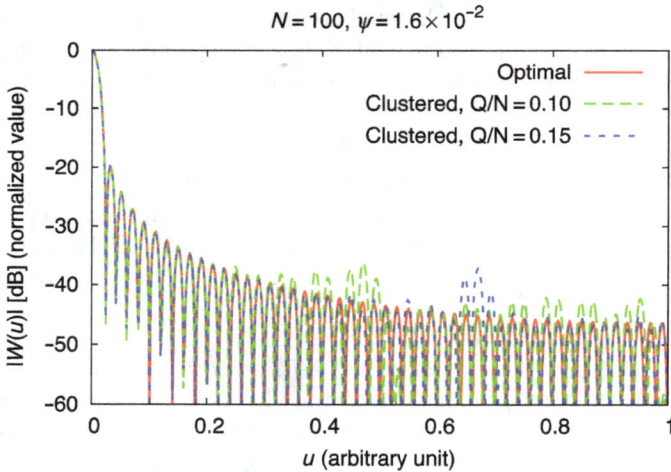

(b)

Figure 8.4   Clustered WPT TX Design [1D array, N = 100, $\psi = 1.6 \times 10^{-2}$, $\frac{Q}{N} \in \{0.1, 0.15\}$] Synthesized tapering window (a) and associated radiation patterns (b).

Finally, the synthesis of an innovative modular architecture for 2D layouts of *WPT TX* array antennas has been proposed in [79,80]. This methodology determines both the array layout and the corresponding subarray beamforming weights by optimizing the η_{BT}^{2D} performance through the use of simple domino-shaped tiles. To be more precise, starting from an optimal tapering of a reference fully populated array, the array aperture is divided into sections using domino tiles through an exhaustive or, depending on the array size, an optimization-based approach. This optimization ensures complete coverage of the antenna aperture and maximizes power transfer efficiency. The numerical results shown in [79] demonstrate that the *BCE* degradation in the synthesized tiled arrays, compared to the reference fully populated arrays, is consistently below 1%, even when using only half the number of control points in the domino-tiled array architectures as compared to the fully populated counterparts. At the same time, it can be concluded from the reported examples that these tiled arrays maintain high *BCE* and low *SLL* within limited scan ranges, even when steering the beam along diagonal directions, thanks to the optimized aperiodic arrangements of domino-shaped tiles. Additionally, real antenna models have been included in the analysis, showing that the performance remains high. This confirms the robustness, the practical applicability, and relevance of the proposed solutions.

8.3.5.2   Sparse *WPT* arrangements

Sparse WPT TX layouts have been designed in [19] as well by formulating the associated problem as the design of a linear/planar irregular *S*-element layout which (*i*) radiates an array factor matching that associated with the ideal optimal solution $\mathbf{w}_{opt}^{\mathcal{D}}$, $\mathcal{D} \in \{1D, 2D\}$, and (*ii*) features a minimum number of antennas *S*. According to such assumptions, the *WPT TX* design is stated as a pattern matching problem [19], and the maximization of the beam efficiency is indirectly obtained by approximating $\eta_{BT}^{\mathcal{D}}\left(\mathbf{w}_{opt}^{\mathcal{D}}\right)$ [19].

More in detail, and following the line of reasoning of [70,72], the *S* antenna positions are assumed to be selected from a set of *C* candidate locations (with $C \gg S$) [19]. The *WPT TX* antenna design can be then mathematically formulated as a sparse recovery problem [19]:

$$\gamma_{opt}^{\mathcal{D}} = \arg\left[\min_{\gamma^{\mathcal{D}}}\left(\|\gamma^{\mathcal{D}}\|_{\ell_0}\right)\right] \text{ s.t. } \mathbf{W}_{opt}^{\mathcal{D}} - \Xi^{\mathcal{D}}\gamma^{\mathcal{D}} = \mathbf{e} \quad \mathcal{D} \in \{1D, 2D\} \quad (8.23)$$

where $\mathbf{W}_{opt}^{\mathcal{D}}$ is the vector containing the optimum array factor samples:

$$\mathbf{W}_{opt}^{\mathcal{D}} \triangleq \begin{cases} \left\{W_{opt}^{1D}\left(u_m^{1D}\right); m = 1, ..., M\right\} & \mathcal{D} = 1D \\ \left\{W_{opt}^{2D}\left(u_m^{2D}, v_m^{2D}\right); m = 1, ..., M\right\} & \mathcal{D} = 2D, \end{cases} \quad (8.24)$$

$u_m^{1D} \in \Omega$ and $\left(u_m^{2D}, v_m^{2D}\right) \in \Omega$, $m = 1, ..., M$, are the matching directions in the 1D and 2D cases, respectively, $\gamma^{\mathcal{D}} \triangleq \{\gamma_c^{\mathcal{D}}; c = 1, ..., C\}$, $\mathcal{D} \in \{1D, 2D\}$, is the *fictitious* vector of the unknown excitations (i.e., $\gamma_c^{\mathcal{D}} = 0$ means that the *c*th candidate location is *empty*; $w_n \equiv \gamma_c^{\mathcal{D}}$ otherwise), $\mathbf{e} \triangleq \{e_m; m = 1, ..., M\}$ is the tolerance vector (with zero-mean Gaussian entries), $\Xi^{\mathcal{D}}$ is the *problem kernel* matrix

$$\Xi^{\mathcal{D}} = \begin{cases} \left\{ \exp\left[j2\pi\left(u_m^{1D}z_c\right)\right]; c = 1, ..., C, m = 1, ..., M \right\} & \mathcal{D} = 1D \\ \left\{ \exp\left[j2\pi\left(u_m^{2D}x_c + v_m^{2D}y_c\right)\right]; c = 1, ..., C, m = 1, ..., M \right\} & \mathcal{D} = 2D, \end{cases}$$

$$(8.25)$$

and z_c and (x_c, y_c), $c = 1, ..., C$, are the candidate element positions in the 1D/2D cases, respectively [19].

To solve the arising sparse linear problem (which turns out to be a *Compressive Sensing* one [19]), a Bayesian formulation has been adopted in [19]. The solution of the synthesis problem has been then obtained as

$$\gamma_{opt}^{\mathcal{D}} = \frac{\Phi_R^{\mathcal{D}}\left[\mathcal{R}\{\Xi^{\mathcal{D}}\}, \mathcal{I}\{\Xi^{\mathcal{D}}\}\right]^H \mathbf{T}_R^{\mathcal{D}}}{\hat{\nu}_R} + j\frac{\Phi_I^{\mathcal{D}}\left[\mathcal{R}\{\Xi^{\mathcal{D}}\}, \mathcal{I}\{\Xi^{\mathcal{D}}\}\right]^H \mathbf{T}_I^{\mathcal{D}}}{\hat{\nu}_I}, \qquad (8.26)$$

where $\mathcal{R}\{\cdot\}/\mathcal{I}\{\cdot\}$ stand for real/imaginary-part, $\mathbf{T}_R^{\mathcal{D}} \triangleq [\mathcal{R}\{\Xi^{\mathcal{D}}\}\mathcal{R}\{\mathbf{w}_{opt}^{\mathcal{D}}\}, \mathcal{I}\{\Xi^{\mathcal{D}}\}\mathcal{R}\{\mathbf{w}_{opt}^{\mathcal{D}}\}]$, $\mathbf{T}_I^{\mathcal{D}} \triangleq [\mathcal{R}\{\Xi^{\mathcal{D}}\}\mathcal{I}\{\mathbf{w}_{opt}^{\mathcal{D}}\}, \mathcal{I}\{\Xi^{\mathcal{D}}\}\mathcal{I}\{\mathbf{w}_{opt}^{\mathcal{D}}\}]$, $\Phi_O^{\mathcal{D}} = \left(\frac{[\mathcal{R}\{\Xi^{\mathcal{D}}\}, \mathcal{I}\{\Xi^{\mathcal{D}}\}][\mathcal{R}\{\Xi^{\mathcal{D}}\}, \mathcal{I}\{\Xi^{\mathcal{D}}\}]}{\hat{\nu}_O^{\mathcal{D}}} + diag(\hat{\mathbf{a}}_O^{\mathcal{D}})\right)^{-1}$, $O \in \{R, I\}$, and $(\hat{\mathbf{a}}_O^{\mathcal{D}}, \hat{\nu}_O^{\mathcal{D}})$ are the hyper-parameters of the Bayesian priors, which can evaluated as $(\hat{\mathbf{a}}_O^{\mathcal{D}}, \hat{\nu}_O^{\mathcal{D}}) \triangleq \arg\{\max_{(\mathbf{a}_O^{\mathcal{D}}, \nu_O^{\mathcal{D}})} \mathcal{L}^{\mathcal{D}}(\mathbf{a}_O^{\mathcal{D}}, \nu_O^{\mathcal{D}}; \beta_1, \beta_2)\}$, $\mathcal{L}^{\mathcal{D}}(\mathbf{a}_O^{\mathcal{D}}, \nu_O^{\mathcal{D}}; \beta_1, \beta_2)$, $\mathcal{D} \in \{1D, 2D\}$, $O \in \{R, I\}$, being the marginal likelihood of the associated Bayesian problem, and β_1, β_2 the design control parameters [70,72].

Thanks to the effectiveness and numerical efficiency of the resulting *BCS*-based *WPT* sparse array design method (which is motivated by the exploitation of fast relevance vector machines to compute the hyperparameter vectors [70,72,81]), large linear, and planar arrangements with effective power transmission capabilities have been synthesized in [19]. Indeed, it has been demonstrated that such class of *WPT* architectures can guarantee a considerable complexity/cost saving with respect to theoretically optimal fully populated arrangements computed by (8.17) [33] despite achieving very close power focusing values [19]. As an example, sparse 2D *WPT* layouts with element saving factor $\frac{S}{N} \in [50\%, 60\%]$ have been shown to yield an η_{BT} almost identical to the theoretical optimum (i.e., $\eta_{BT}^{2D}(\mathbf{w}_{opt}) - \eta_{BT}^{2D}(\mathbf{w})|_{S/N=0.55} \leq 0.09\%$ [19]).

A set of *WPT* sparse layouts designed by the method introduced in [19] is reported in Figure 8.5 for the sake of completeness. The results obtained employing the optimal $N = 100$ arrangement considered above (i.e., $\psi = 1.6 \times 10^{-2}$, $\eta_{BT}^{opt} \approx 95\%$—Figure 8.4) as a benchmark show that a significant element reduction (i.e., $\frac{S}{N} \in \{0.6, 0.7\}$—Figure 8.5(a)) can be achieved by means of the illustrated unconventional *WPT* design strategy [19] while affording excellent power focusing capabilities (i.e., $\eta_{BT}^{\frac{S}{N}=0.6} \approx 92.46\%$, $\eta_{BT}^{\frac{S}{N}=0.7} \approx 94.99\%$—Figure 8.5(b)).

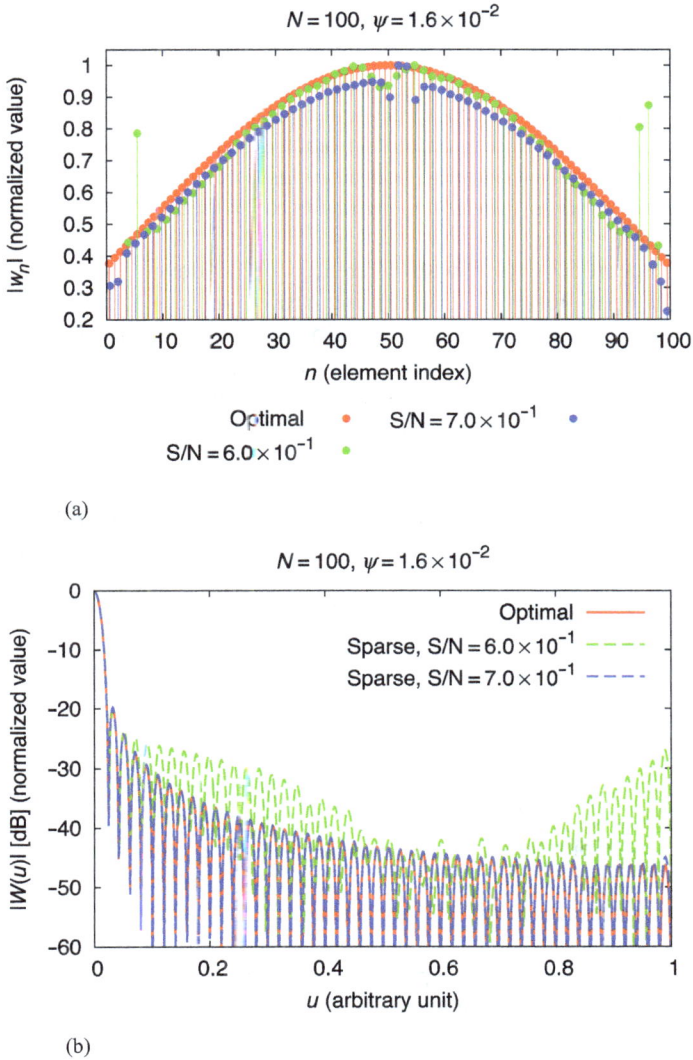

Figure 8.5 *Sparse WPT TX Design [1D array, N = 100, ψ= 1.6 × 10⁻²,* $\frac{S}{N} \in \{0.6, 0.7\}$*] Synthesized tapering window (a) and associated radiation patterns (b).*

8.4 Final remarks, current trends, and future perspectives

WPT via radio waves is an interesting technology for those scenarios in which the power/energy must be delivered over non-negligible distances in the absence of a standard electrical network. Several research activities have been recently focused on the development and experimental demonstration of *TX/RX* antennas for *WPT* systems, because of their potential impact in current and future wireless/mobile systems. Indeed, the unique challenges and requirements of *WPT* systems prevent the use of standard strategies and technologies developed for the design and fabrication of antennas/phased arrays, and new classes of methodological approaches are actually required to handle the synthesis of effective radiating systems for radio-wave *WPT*.

This is specifically true with reference to the development of innovative synthesis strategies for *WPT* transmitting arrays, as it is demonstrated by the number of contributions and involved groups worldwide. Despite such recent advancements, many methodological and technological challenges still need to be addressed in the design and fabrication processes in order to reduce the complexity and weight, increase the beam transmission efficiency, and mitigate cost and maintenance issues of *WPT TX* systems [6].

Accordingly, a review of the state-of-the-art techniques for the design of radio-wave *WPT* transmitting systems has been illustrated in this chapter. More in detail, the fundamental objectives in the design of a *WPT* system have been mathematically formulated, and the most recent advances and paradigms adopted in the synthesis of *TX* array architectures have been discussed. Moreover, some of the current trends and frontiers of this research area have been illustrated.

The following key aspects and open challenges regarding the design of *WPT TX* systems have been illustrated in the chapter:

- The design and fabrication of actual *WPT* prototypes must take into account several conflicting requirements regarding power focusing and beam steering capabilities, architecture modularity, power consumption, and maintenance complexity. Consequently, design techniques able to yield different tradeoffs between performance and cost of the overall architectures will have an increasing importance in the synthesis of *WPT TX* arrays.
- Despite their apparent simplicity, strategies based on uniform excitations are often selected for current experimental validations because of their reduced cost and complexity and high modularity, although they provide sub-optimal η_{BT} performance.
- Significant enhancements of the transmission efficiency can be achieved by means of aperture tapering based on heuristic/optimization techniques, which often yield also low complexity architectures.
- The design of the tapering window which theoretically maximizes η_{BT} is often possible by solving a generalized eigenvalue problem. However, the resulting excitations may correspond to expensive fully populated layouts requiring accurate magnitude control in each antenna element.
- The use of unconventional (i.e., clustered or sparse) architectures has been recently proposed to considerably mitigate the *WPT TX* array costs/complexity while achieving η_{BT} almost identical to the theoretical optimal methods.

Despite such successful results and efforts, many open challenges still exist concerning the design and fabrication of *WPT* transmitting arrays. Indeed, future efforts are expected to deal with (*i*) the experimental validation of theoretically optimal *WPT* arrangements, especially when addressing mobility and reconfigurability constraints; (*ii*) the extension of the unconventional design methodologies and architectures to three-dimensional/conformal scenarios; (*iii*) the possibility to combine sparsening and clustering techniques to further simplify *WPT* transmitting arrays; (*iv*) the integration of low-complexity *WPT* transmitting systems in cooperative *TX-RX* schemes able to address both power transfer efficiency and safety constraints/interference minimization objectives in realistic scenarios.

Acknowledgments

This work benefited from the networking activities carried out within the Project "AURORA—Smart Materials for Ubiquitous Energy Harvesting, Storage, and Delivery in Next Generation Sustainable Environments" funded by the Italian Ministry for Universities and Research within the PRIN-PNRR 2022 Program. Moreover, it benefited from the networking activities carried out within the Project "SEME @ TN—Smart ElectroMagnetic Environment in TrentiNo" funded by the Autonomous Province of Trento (CUP: C63C22000720003), the Project SPEED (Grant No. 61721001) funded by National Science Foundation of China under the Chang-Jiang Visiting Professorship Program, the Project "Electromechanical Coupling Theory and Design Method for Uncertain Factors of Electronic Equipment" (Grant No. 2022-JC-33) funded by Department of Science and Technology of Shaanxi Province under the Natural Science Basic Research Program, the Project "Research on Design Method of Efficient Microwave Wireless Energy Transfer Antenna Array for Space Power Stations' (Grant No. 2023-GHZD-35) funded by the Department of Science and Technology of Shaanxi Province under the Key Research and Development Program, and the Project "National Centre for HPC, Big Data and Quantum Computing (CN HPC)" funded by the European Union—NextGenerationEU within the PNRR Program (CUP: E63C22000970007). Views and opinions expressed are however those of the author(s) only and do not necessarily reflect those of the European Union or the European Research Council. Neither the European Union nor the granting authority can be held responsible for them. A. Massa wishes to thank E. Vico for her never-ending inspiration, support, guidance, and help.

References

[1]   Brown, W. C. "The technology and application of free-space power transmission by microwave beam". *Proc. IEEE.* 1974; **62**(1): 11–25.
[2]   Dickinson, R. M. "The beamed power microwave transmitting antenna". *IEEE Trans. Microw. Theory Tech.* 1978; **26**(5): 335–340.

[3]   Brown, W. C. "The history of power transmission by radio waves". *IEEE Trans. Microw. Theory Tech.* 1984; **MTT-32**(9): 1230–1242.

[4]   Brown, W. C. and Eves, E. "Beamed microwave power transmission and its application to space". *IEEE Trans. Microw. Theory Tech.* 1992; **40**(6): 1239–1250.

[5]   McSpadden, J. O., Yoo, T., and Chang, K. "Theoretical and experimental investigation of a rectenna element for microwave power transmission". *IEEE Trans. Microw. Theory Tech.* 1992; **40**(12): 2359–2366.

[6]   Massa, A., Oliveri, G., Viani, F., and Rocca, P. "Array designs for long-distance wireless power transmission—State-of-the-art and innovative solutions". *Proc. IEEE.* 2013; **101**(6): 1464–1481.

[7]   Gavan, J. and Tapuch, S. "Microwave wireless-power transmission to high-altitude-platform systems". *Radio Sci. Bull.* 2010; **334**: 25–42.

[8]   Oida, A., Nakashima, H., Miyasaka, J., Ohdoi, K., Matsumoto, H., and Shinohara, N. "Development of a new type of electric off-road vehicle powered by microwaves transmitted through air". *J. Terramech.* 2007; **44**(5): 329–338.

[9]   Shams, K. M. Z. and Ali, M. "Wireless power transmission to a buried sensor in concrete". *IEEE Sensors J.* 2007; **7**(12): 1573–1577.

[10]   Li, Y. and Jandhyala, V. "Design of retrodirective antenna arrays for short-range wireless power transmission". *IEEE Trans. Antennas Propag.* 2012; **60**(1): 206–211.

[11]   Mailloux, R. J. *Phased Array Antenna Handbook*, 2nd ed. (Norwood, MA: Artech House, 2005).

[12]   Borges-Carvalho, N., Georgiadis, A., Costanzo, A., *et al.* "Wireless power transmission: R&D activities within Europe". *IEEE Trans. Microw. Theory Tech.* 2014; **62**(4): 1031–1045.

[13]   COST Action IC1301 Team, "European contributions for wireless power transfer technology". *IEEE Microw. Mag.* 2017; **18**(4): 56–87.

[14]   Nishida, K., Taniguchi, Y., Kawakami, K., *et al.* "5.8 GHz high sensitivity rectenna array". *Proc. IEEE International Microwave Workshop Series on Innovative Wireless Power Transmission: Technologies, Systems, and Applications.* Kyoto, Japan, May 2011; 19–22.

[15]   Shinohara N. and Matsumoto, H. "Experimental study of large rectenna array for microwave energy transmission". *IEEE Trans. Microw. Theory Tech.*, 1998; **46**(3): 261–268.

[16]   Ren Y.-J. and Chang, K. "5.8 GHz circularly polarized dual-diode rectenna and rectenna array for microwave power transmission". *IEEE Trans. Microw. Theory Tech.* 2006; **54**(4): 1495–1502.

[17]   Olgun, U., Chen, C.-C., and Volakis, J. L. "Investigation of rectenna array configurations for enhanced RF power harvesting". *IEEE Antennas Wireless Propag. Lett.* 2011; **10**: 262–265.

[18]   McSpadden J. O. and Mankins, J. C. "Space solar power programs and microwave wireless power transmission technology". *IEEE Microw. Mag.* 2002; **3**(4): 46–57.

[19]   Poli, L., Oliveri, G., Rocca, P., Salucci, M., and Massa A. "Long-distance WPT unconventional arrays synthesis". *J. Electromagn. Waves Appl.* 2017; **31**(14): 1399–1320.

[20]   Nanokaichi, K., Shinohara, N., Kawasaki, S., Mitani, T., and Matsumoto, H. "Development of waveguide-slot-fed active integrated antenna for microwave power transmission". *URSI General Assembly*. New Delhi, India. Oct. 2005.

[21]   Takano, T., Sugawara, A., and Sasaki, S. "System considerations of onboard antennas for SSPS". *Radio Sci Bull.* 2004; **311**: 16–20.

[22]   Ishiba, M., Ishida, J., Komurasaki, K., and Arakawa, Y. "Wireless power transmission using modulated microwave". *Proc. IEEE International Microwave Workshop Series on Innovative Wireless Power Transmission: Technologies, Systems, and Applications*. Kyoto, Japan, May 2011; 51–54.

[23]   Takahashi, T., Mizuno, T., Sawa, M., Sasaki, T., Takahashi, T., and Shinohara, N. "Development of phased array for high accurate microwave power transmission". *Proc. IEEE International Microwave Workshop Series on Innovative Wireless Power Transmission: Technologies, Systems, and Applications*. Kyoto, Japan. May 2011; 157–160.

[24]   Kawasaki, S., Seita, H., Suda, T., Kaori, T., and Nakajima, K. "32-Element high power active integrated phased array antennas operating at 5.8 GHz". *Proc. IEEE Antennas and Propagation Society International Symposium*. San Diego, USA, July 2008; 1–4.

[25]   Baki, A. K. M., Hashimoto, K., Shinohara, N., Matsumoto, H., and Mitani, T. "Comparison of different kinds of edge tapering system in microwave power transmission". *IEICE Tech. Rep.* 2006; **SPS 2006-12**: 13–18.

[26]   Baki, A. K. M., Shinohara, N., Matsumoto, H., Hashimoto, K., and Mitani, T. "Isosceles-trapezoidal-distribution edge tapered array antenna with unequal element spacing for solar power station/satellite". *IEICE Trans. Commun.* 2008; **E91-B**(2): 527–535.

[27]   Jamnejad V. and Hoorfar, A., "Optimization of antenna beam transmission efficiency". *Proc. IEEE Antennas and Propagation Society International Symposium*. San Diego, USA; Jul. 2008: 1–4.

[28]   Strassner, B. and Chang, K. "5.8-GHz circularly polarized dual-rhombic-loop traveling-wave rectifying antenna for low power-density wireless power transmission applications". *IEEE Trans. Microw. Theory Tech.* 2003; **51**(5): 1548–1553.

[29]   Soares Boaventura, A. J., Collado, A., Georgiadis, A., and Borges Carvalho, N. "Spatial power combining of multi-sine signals for wireless power transmission applications". *IEEE Trans. Microw. Theory Tech.* 2014; **62**(4): 1022–1030.

[30]   Masotti, D., Costanzo, A., Del Prete M., and Rizzoli, V. "Time-modulation of linear arrays for real-time reconfigurable wireless power transmission". *IEEE Trans. Microw. Theory Tech.* 2016; **64**(2): 331–342.

[31] Sanzgiri, S. T. and Butler, J. K. "Constrained optimization of the perfor-
      mance indices of arbitrary array antennas". *IEEE Trans. Antennas Propag.*,
      1971; **AP-19**(4): 493–498.
[32] Uno T. and Adachi, S. "Optimization of aperture illumination for radio wave
      power transmission". *IEEE Trans. Antennas Propag.* 1984; **AP-32**(6): 628–
      632.
[33] Oliveri, G., Poli, L., and Massa, A. "Maximum efficiency beam synthesis of
      radiating planar arrays for wireless power transmission". *IEEE Trans.
      Antennas Propag.* 2013; **61**(5): 2490–2499.
[34] Balanis, C. A. *Antenna Theory: Analysis and Design*, 2nd ed. (New York:
      Wiley, 1997).
[35] Craeye, C. and Gonzalez-Ovejero, D. "A review on array mutual coupling
      analysis". *Radio Sci.* 2011; **46**(RS2012): 1–25.
[36] Shinohara, N. and Ishikawa, T. "High efficient beam forming with high
      efficient phased array for microwave power transmission". International
      Conference on Electromagnetics in Advanced Applications. Turin, Italy.
      Sept. 2011, 729–732.
[37] Ishikawa T. and Shinohara, N. "Study on optimization of microwave power
      beam of phased array antennas for SPS". *Proc. IEEE International Micro-
      wave Workshop Series on Innovative Power Transmission: Technologies,
      Systems, and Applications.* Kyoto, Japan. May 2011: 153–156.
[38] Hsieh, L. H., Strassner, B. H., Kokel, S. J., *et al.* "Development of a retro-
      directive wireless microwave power transmission system". *Proc. IEEE
      Antennas and Propagation Society International Symposium.* Columbus,
      USA. Jun. 2003; 2: 393–396.
[39] Rocca, P., Oliveri, G., Mailloux, R. J., and Massa, A. "Unconventional
      phased array architectures and design methodologies—A review". *Proc.
      IEEE.* 2016; **104**(3): 544–560.
[40] Kaya, N., Iwashita, M., and Mankins, J. C. "Hawaii project for microwave
      power transmission". *Proc. 54th International Astronautical Congress of the
      International Astronautical Federation.* 2003; 3: 163–168.
[41] Kaya, N., Iwashita, M., Little, F., Marzwell, N., and Mankins, J. C.
      "Microwave power beaming test in Hawaii". *Proc. 60th International
      Astronautical Congress.* 2009; **8**: 6128–6132.
[42] Kaya, N. and Mankins, J. "The second microwave power beaming experi-
      ment in Hawaii". *Proc. 61st International Astronautical Congress.* 2010; 12:
      9702–9707.
[43] Kawasaki, S. "Microwave WPT to a rover using active integrated phased
      array antennas". *Proc. 5th European Conference on Antennas and Propa-
      gation.* April 2011, 3909–3912.
[44] Baki, A. K. M., Shinohara, N., Matsumoto, H., Hashimoto, K., and Mitani,
      T. "Study of isosceles trapezoidal edge tapered phased array antenna for solar
      power station/satellite". *IEICE Trans. Commun.* 2007; **E90-B**(4): 968–977.
[45] Garmash, V., Katsenelenbaum, B. Z., Shaposhnikov, S. S., Tioulpakov, V.
      N., and Vaganov, R. B. "Some peculiarities of the wave beams in wireless

power transmission". *IEEE Aerospace Electron. Syst. Mag.* 1998; **13**(10): 39–41.

[46]   Shaposhnikov S. S. and Garmash, V. N. "Phase synthesis of antennas". *Proc. IEEE Aerospace Conference*. Big Sky, USA. 2002, **2**: 891–900.

[47]   Takeshita, S. "Power transfer efficiency between focused circular antennas with Gaussian illumination in Fresnel region". *IEEE Trans. Antennas Propag.* 1968; **16**(3): 305–309.

[48]   Goubau, G. "Microwave power transmission from an orbiting solar power station". *J. Microw. Power.* 1970; **5**(4): 223–231.

[49]   Rocca, P., Benedetti, M., Donelli, M., Franceschini, D., and Massa, A. "Evolutionary optimization as applied to inverse scattering problems". *Inverse Probl.* 2009; **25**(12): 1–41.

[50]   Rocca, P., Oliveri, G., and Massa, A. "Differential evolution as applied to electromagnetics". *IEEE Antennas Propag. Mag.* 2011; **53**(1): 38–49.

[51]   Shishkov, B., Shinohara, N., Hashimoto, K., and Matsumoto, H. "On the optimization of side lobes in large antenna arrays for microwave power transmission". *IEICE Tech. Rep.* 2006; **SPS 2006-11**: 5–11.

[52]   Shinohara, N., Shishkov, B., Matsumoto, H., Hashimoto, K., and Baki, A. K. M. "New stochastic algorithm for optimization of both side lobes and grating lobes in large antenna arrays for MPT". *IEICE Trans. Commun.* 2008; **E91-B** (1): 286–296.

[53]   Shaposhinikov, S. S., Vaganov, R. B., and Voitovich, N. N. "Antenna amplitude distributions for improved wireless power transmission efficiency". *Proc. IEEE AFRICON Conference*. George, South Africa., Oct. 2002, 559–562.

[54]   Hoorfar, A. "Evolutionary programming in electromagnetic optimization: A review". *IEEE Trans. Antennas Propag.* 2007; **55**(3): 523–537.

[55]   Li, X., Duan, B., Zhou, J., Song, L. and Zhang, Y. "Planar array synthesis for optimal microwave power transmission with multiple constraints". *IEEE Antennas Wirel. Propag. Lett.* 2017; **16**: 70–73.

[56]   Park, J. B., Jong, Y W., Shin, J. R. and Lee, K. Y. "An improved particle swarm optimization for nonconvex economic dispatch problems". *IEEE Trans. Power Syst.* 2010; **25**(1): 156–166.

[57]   Prasad, S. "On an index for array optimization and the discrete prolate spheroidal functions". *IEEE Trans. Antennas Propag.* 1982; **AP-30**(5): 1021–1023.

[58]   Van Trees, H. L. *Optimum Array Processing, Part IV of Detection, Estimation, and Modulation Theory*. (New York: Wiley, 2002).

[59]   Oliveri, G., Rocca, P., and Massa, A. "Antenna array architectures for solar power satellites and wireless power transmission". *Proc. XXXth URSI General Assembly and Scientific Symposium*. Istanbul, Turkey. Aug. 2011; 1–4.

[60]   Borgiotti, G. "Maximum power transfer between two planar apertures in the Fresnel zone". *IEEE Trans. Antennas Propag.* 1966; **14**(2): 158–163.

[61]   Heurtley, J. "Maximum power transfer between finite antennas". *IEEE Trans. Antennas Propag.* 1967; **15**(2): 298–300.

[62]   Garmash, V. and Shaposhnikov, S. S. "Matrix method synthesis of transmitting antenna for wireless power transmission". *IEEE Trans. Aerospace Electron. Syst.* 2000; **36**(4): 1142–1148.

[63]   Slepian D. and Pollack, H. O. "Prolate spheroidal wave functions, Fourier analysis, and uncertainty". *Bell Syst., Tech. J.* 1961; **40**: 43–64.

[64]   Gantamacher, F. R. *The Theory of Matrices.* (New York: Chelsea Publishing Company, 1959).

[65]   Hogben, L., (eds.). *Handbook of Linear Algebra.* (New York: Chapman and Hall, 2007).

[66]   Anderson, E., Bai, Z., Bischof, C., *et al. LAPACK Users' Guide.* (Philadelphia, PA: Society for Industrial and Applied Mathematics, 1999).

[67]   Manica, L., Rocca, P., Martini, A., and Massa, A. "An innovative approach based on a tree-searching algorithm for the optimal matching of independently optimum sum and difference excitations". *IEEE Trans. Antennas Propag.* 2008; **56**(1): 58–66.

[68]   Rocca, P., Manica, L., Azaro, R., and Massa, A. "A hybrid approach to the synthesis of subarrayed monopulse linear arrays". *IEEE Trans. Antennas Propag.* 2009; **57**(1): 280–283.

[69]   Manica, L., Rocca, P., Benedetti, M., and Massa, A. "A fast graph-searching algorithm enabling the efficient synthesis of sub-arrayed planar monopulse antennas". *IEEE Trans. Antennas Propag.* 2009; **57**(3): 652–663.

[70]   Oliveri, G. and Massa, A. "Bayesian compressive sampling for pattern synthesis with maximally sparse non-uniform linear arrays". *IEEE Trans. Antennas Propag.* 2011; **59**(2): 467–481.

[71]   Oliveri, G., Carlin, M., and Massa, A. "Complex-weight sparse linear array synthesis by Bayesian compressive sampling". *IEEE Trans. Antennas Propag.* 2012; **60**(5): 2309–2326.

[72]   Viani, F., Oliveri, G., and Massa, A. "Compressive sensing pattern matching techniques for synthesizing sparse planar arrays". *IEEE Trans. Antennas Propag.* 2013; **61**(9): 4577–4587.

[73]   Rocca, P., D'Urso, M., and Poli, L. "Advanced strategy for large antenna array design with sub-array-only amplitude and phase control". *IEEE Antennas Wireless Propag. Lett.* 2014; **13**: 91–94.

[74]   Moriyama, T., Poli, L., and Rocca, P. "On the design of clustered planar phased arrays for wireless power transmission". *IEICE Electronics Express.* 2015; **12**(4): 1–6.

[75]   Li, X., Duan, B., and Song, L. "Design of clustered planar arrays for microwave wireless power transmission". *IEEE Trans. Antennas Propag.* 2019; **67**(1): 606–611.

[76]   Celebi, M. E., Kingravi, H. A., and Vela, P. A. "A comparative study of efficient initialization methods for the K-means clustering algorithm". *Expert Syst. Appl.* 2013; **40**(1): 200–210.

[77]   Xiong, Z.-Y., Xu, Z.-H., Zhang, L., and Xiao, S.-P. "Cluster analysis for the synthesis of subarrayed monopulse antennas". *IEEE Trans. Antennas Propag.* 2014, **62**(4), 1738–1749.

[78] Zhang, Q. H., Zhang, Q. H., and Shen, Z. Y. "Planar array subarray division method in microwave wireless power transmission based on PSO&*k*-means algorithm". *IEEE Open J Antennas Propag.* 2023; **4**: 520–527.

[79] Anselmi, N., Polo, A., Hannan, M. A., Salucci, M., and Rocca, P. "Maximum BCE synthesis of domino-tiled planar arrays for far-field wireless power transmission". *J. Electromagn. Waves App.* 2020; **34**(17): 2349–2370.

[80] Anselmi, N., Rocca, P., Salucci, M., and Massa, A. "Irregular phased array tiling by means of analytic schemata-driven optimization". *IEEE Trans Antennas Propag.* 2017; **65**(9): 4495–4510.

[81] Massa, A., Rocca, P., and Oliveri, G. "Compressive sensing in electromagnetics—A review". *IEEE Antennas Propag. Mag.* 2015; **57**(1): 224–238.

Chapter 9

Transmitter/rectifier technologies in WPT via radio waves

Naoki Shinohara[1]

9.1 Introduction

As described in Chapter 6, an inverter circuit with lumped elements is always applied for coupling wireless power transfer (WPT) because its frequency is usually in the kHz–MHz range, and its wavelength is very low compared to the circuit size. As described in Chapters 7 and 8, radio waves with frequencies in the GHz range that are below 10s of cm^{-1} are used for WPT to extend the WPT distance. At frequencies above 1 GHz, the size of the circuit and the wavelength are essentially the same; therefore, distributed constant lines are often applied to the circuit at frequencies greater than GHz instead of the lumped elements (Figure 9.1). We must consider a phase and an amplitude at each point in the circuit and the characteristic impedance of the circuit given by the amplitude and the phase of the voltage and current in the circuit. We should additionally consider parasitic capacitances and inductances at GHz frequencies. While the inverter circuits and design method described in Chapter 6 can be applied for WPT via radio waves, other transmitter/receiver circuits and design methods with the distributed constant lines

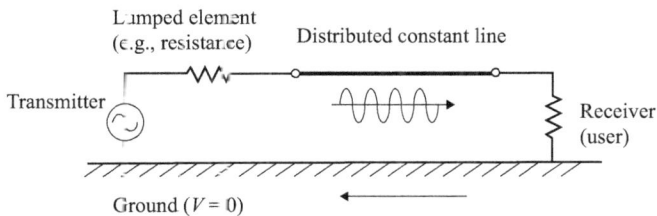

Figure 9.1   Typical circuit with distributed constant line for the GHz frequency range

[1]Research Institute for Sustainable Humanosphere, Kyoto University, Japan

are more suited for WPT via radio waves and should be used instead of the circuits described in Chapter 6.

"Mode" and "impedance (Z)" are the most important concepts for GHz circuit design. For boundary conditions with metal and dielectric materials in space such as those found in an antenna or in the case of distributed constant lines in a circuit, Maxwell's equations provide solutions for the electric and magnetic fields, which are the constituent elements of the electromagnetic wave. All electromagnetic waves are described by a linear combination of sinusoidal waves that change in space and in time. Since sinusoidal waves are periodic, Maxwell's equations have more than one solution, and each solution for the electromagnetic wave obtained using Maxwell's equations called a "mode." For example, TE_{10} mode, TM_{11} mode, and the other modes exist in a waveguide, which is a guiding tube made of metals that is often used in GHz systems.

For each mode, Maxwell's equations provide a solution for the electric field and for the magnetic field, and characteristic impedance in space is obtained from the ratio of the electric field and the magnetic field. Thus, each mode has its characteristic impedance, with different characteristic impedance values for different frequencies. In circuits, the characteristic impedance is obtained from the ratio of the voltage and the current and is considered to be constant. While the modes can coexist, in the circuits, only one mode is often considered. Therefore, for each circuit, one characteristic impedance of the circuit is considered, which is determined by the shape of the real distributed constant line. The concept of the "characteristic impedance" is most important at the GHz frequencies due to the requirement for impedance matching between the circuits. If the impedances of the connected circuits are different, radio wave reflection occurs giving rise to loss of the power transmitted to the user. This effect is the same as for the theory of radio waves in space.

As mentioned earlier, impedance matching is extremely important in resonance-coupling WPT. An inductive-coupling WPT can be described mainly by low-frequency theory with lumped circuits. A WPT via radio waves should be described by a high-frequency theory with distributed constant lines. It is very interesting that the resonance coupling WPT can be described both by low-frequency theory with lumped circuits and by high-frequency theory with distributed constant lines.

In a GHz circuit, we usually use S-parameters instead of impedance parameters Z with voltage and current because it is easy to consider impedance matching. We first define the normalized complex waves a_i and b_i that contain the amplitude and phase information, with a_i indicating the waves propagating toward the two-port and b_i indicating the waves traveling away from it (Figure 9.2). Index i indicates the port number. S-parameters are defined by the following matrix:

$$\begin{pmatrix} b_1 \\ b_2 \end{pmatrix} = \begin{pmatrix} S_{11} & S_{12} \\ S_{21} & S_{22} \end{pmatrix} \begin{pmatrix} a_1 \\ a_2 \end{pmatrix} \qquad (9.1)$$

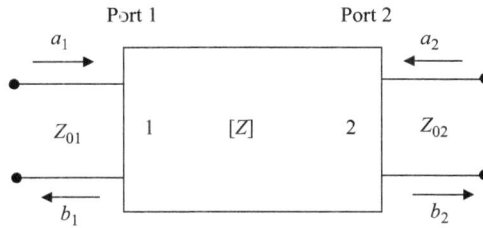

Figure 9.2 Two representations of two-port

where S_{11} is the reflection ratio at port 1, S_{12} is the power gain from port 2 to port 1, S_{21} is the power gain from port 1 to port 2, and S_{22} is the reflection ratio at port 2. Port 1 is usually defined as the input port and port 2 as the output port.

Figure 9.3 shows the typical circuits of a radio frequency (RF) amplifier and a rectifier for WPT via radio waves. In the RF amplifier at a GHz frequency, field-effect transistor (FET) or high-electron-mobility transistor (HEMT) or the other RF semiconductor is applied to amplify the RF source power for the RF transmitter. Impedance matching of the circuits before/after the semiconductor is very important for the realization of a high DC–RF conversion efficiency. Theoretically, 100% DC–RF conversion efficiency can be realized with wave forms of voltage and current shown in Figure 9.1 as an example.

Characteristics of the semiconductor, the impedance-matching circuits, and additionally management circuit for higher harmonics are very important for the realization of the theoretical 100% conversion efficiency in practice. Amplified RF power is transmitted from a transmitting antenna to space, while the receiving antenna receives the propagated radio waves from space. There is no electro-magnetic coupling between the transmitting antenna and the receiving antenna if the distance is sufficiently large and in radiative near field and in the far field. Therefore, the possible changes of circuit parameters (e.g., impedance) by the mutual coupling effect due to the changes of the distance or positions of the transmitter and the receiver do not need to be considered. After the receiving antenna, a rectifier with a diode is often installed and the received radio wave is converted from RF to DC.

9.2 RF transmitter

9.2.1 RF amplifier with semiconductor

A semiconductor is the most important element for determining the characteristics of RF amplifier circuits. Semiconductors are either p-type, in which positive holes are used as carriers of electric charges, or n-type, in which electrons are used as the carriers of electric charges. The choice of the semiconductor materials determines the efficiency and power of the circuit. Present materials for semiconductor devices are silicon (Si) for lower frequency below a few GHz and gallium arsenide (GaAs)

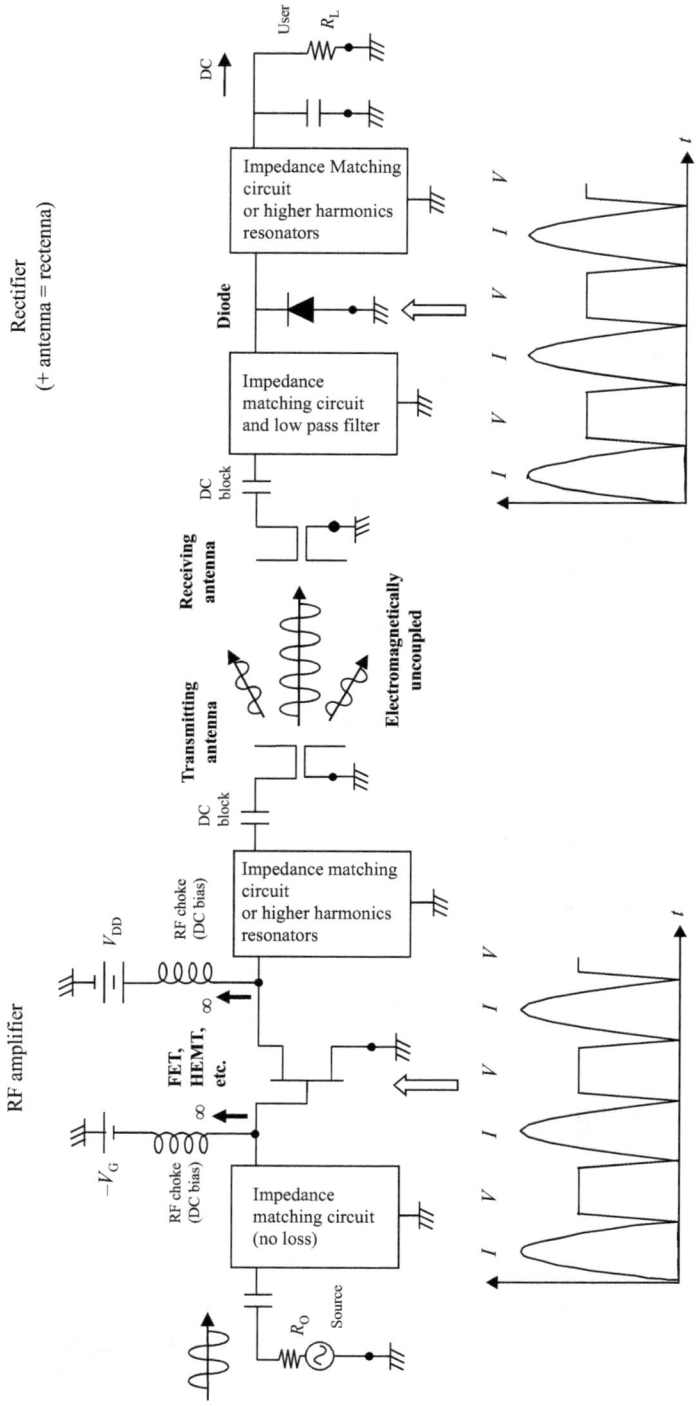

Figure 9.3 Typical RF amplifier and rectifier circuits for WPT via radio waves

for higher frequencies. Recently, the wide-band gallium nitride (GaN) and silicon carbide (SiC) semiconductors have been expected to be applied to high-power, high-efficiency RF circuits because these materials can accommodate high power and high frequency similar to the GaAs semiconductors.

A semiconductor device is composed of the p-type semiconductor, n-type semiconductor, and other materials (e.g., metals). For example, a semiconductor diode is a device that typically combines p- and n-type semiconductors. The p–n junction at the interface of the p- and n-type semiconductors creates nonlinearity in the current flow, with positive current only flowing in the diode. Thus, the diode can be used as a rectifier. A semiconductor transistor is a device typically combining p–n–p or n–p–n semiconductors as a three-terminal device and can be used for amplifying the RF signal. However, standard transistors are not suitable for amplification of waves with frequencies in the GHz range (microwave). Therefore, typical semiconductor devices for microwave amplifier circuits are the FET, heterojunction bipolar transistor (HBT), and HEMT, which are also three-terminal devices but with a form and shape that are different from those of the standard transistor. A combination of the semiconductor materials and the form of the three-terminal devices form the semiconductor device, e.g., GaAs FET, GaN HEMT, among others (Figure 9.4).

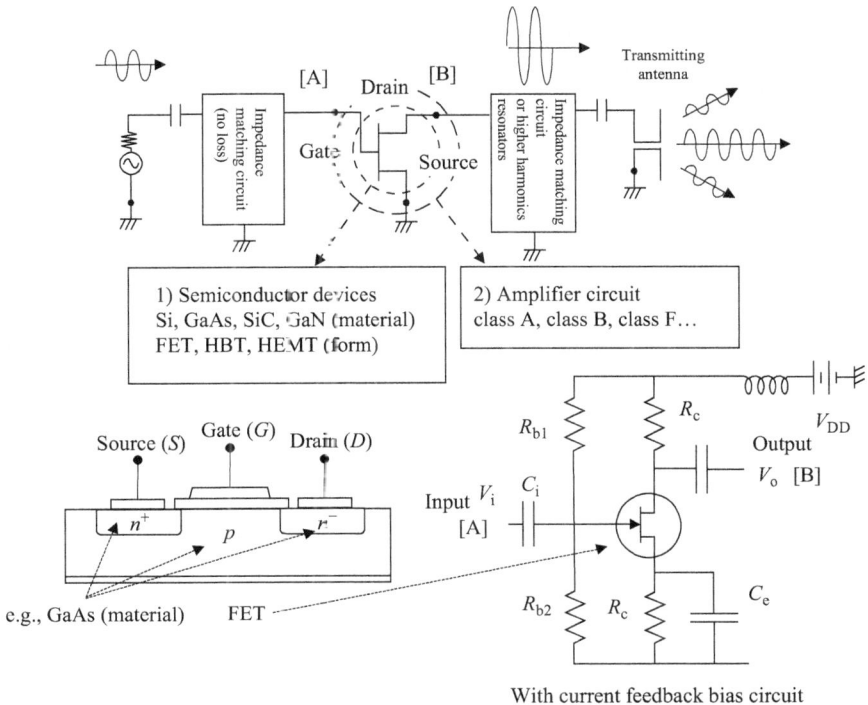

Figure 9.4 Elements of the RF amplifier: (1) semiconductor device and (2) amplifier circuit

The characteristics of the three-terminal semiconductor device determine the frequency, power, conversion efficiency, and amplifier gain. To realize a high-efficiency RF amplifier, next, we should consider circuit design with semi-conductor devices. There are three terminals in the three-terminal semiconductor device. For example, in an FET device, these terminals are referred to as "gate," "source," and "drain." It is important to decide which terminal is connected to ground and which terminals are used for signal input and output. In the micro-wave amplifier circuit, the source is usually connected to the ground, the gate is the input, and the drain is the output. In the RF amplifier circuit, a bias voltage is applied to amplify the input RF. Based on the bias voltage, the circuit design determines the theoretical efficiency and gain. Class A–C amplifiers are classi-fied by the bias voltage used in the device and in the same amplifier circuit. These classes can also be applied not only to GHz systems but also to kHz–MHz sys-tems. Theoretical RF amplifier efficiency is 50% for class A, 78.5% for class B, and <100% for class C. While efficiency is higher for classes B and C than for class A, the waveform is distorted and wave linearity cannot be maintained. For class C, the output power tends to be near zero when the efficiency is close to 100%. As a result, class B and C devices are not typically used for wireless communication systems that require high-wave linearity. Before and after the amplifier circuit, impedance-matching circuits must be installed to increase the efficiency, gain, and power.

Higher harmonics resonators are applied as the output impedance-matching circuit to increase efficiency with a combination of higher harmonics. Class D and E amplifiers can be applied not only to kHz–MHz systems but also to GHz sys-tems. Additionally, class F and class F^{-1} amplifiers are often applied to GHz systems that can theoretically realize 100% efficiency. As shown in Figure 9.5, special impedance-matching circuits controlling higher harmonics are added at the output, and a transistor is driven at class B bias [1]. The drain current becomes a half-wave-rectified waveform and higher harmonics occur in the class B bias cir-cuit. When the drain voltage comprises DC, and the fundamental frequency f_0, whose phase is different by 180° from that of the fundamental frequency of the drain current and odd harmonics, power consumption at the harmonics approaches zero and power with the power factor of −1 exists at the fundamental frequency. This means that DC power is converted to f_0 with theoretically 100% conversion efficiency if it is optimized by an impedance-matching circuit for f_0, and the bias voltage is adjusted. The theoretical maximum efficiency and the number of har-monics of the current and voltage are shown in Table 9.1. For higher efficiency, it is important to increase the number of harmonics. Some class F amplifiers have been developed for WPT via radio waves. The drain efficiency reached 80.1% and power-added efficiency (PAE) = $(P_{out}-P_{in})/P_{DC}$ is a maximum of 72.6% at 1.9 GHz in developed class F amplifier using GaN HEMT [2]. The same research group also developed a class F amplifier operating at 5.65 GHz using AlGaN/GaN HEMT with a drain efficiency of 90.7%, a maximum PAE of 79.5%, and a satu-rated power of 33.3 dB m [3].

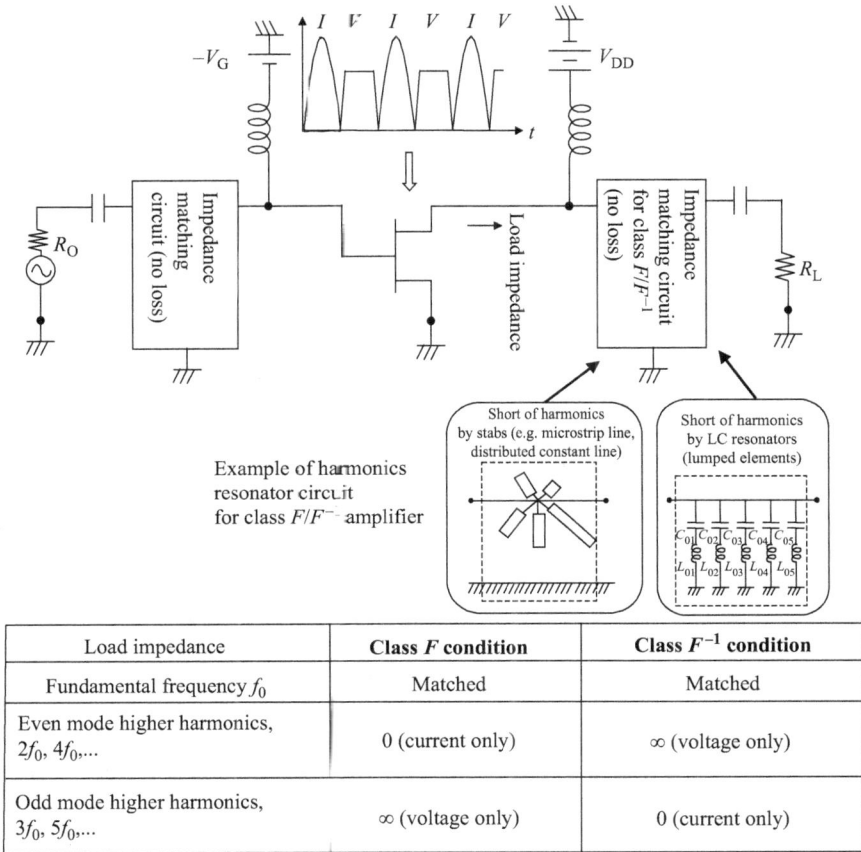

Load impedance	**Class *F* condition**	**Class *F*⁻¹ condition**
Fundamental frequency f_0	Matched	Matched
Even mode higher harmonics, $2f_0, 4f_0,...$	0 (current only)	∞ (voltage only)
Odd mode higher harmonics, $3f_0, 5f_0,...$	∞ (voltage only)	0 (current only)

Figure 9.5 Class F/F⁻¹ amplifier

Table 9.1 Relation between harmonics and efficiency [1]

Current even harmonics	Voltage odd harmonics (%)			
	f_0	$3f_0$	$5f_0$	**Infinity**
f_0	50 (class A)	57.7	60.3	63.7
$2f_0$	70.7 (real class B)	81.7	85.3	90.3
$4f_0$	75.0	86.6	90.5	95.5
Infinity	78.5 (ideal class B)	90.7	94.8	100

The PAEs were achieved in a high power amplifier (HPA). Depending on the gain of the PAE, the RF amplifier is generally composed of the HPA and driver amplifier. The DC-RF conversion efficiency of the RF amplifier is generally smaller than that of the HPA only. For example, new HPA with newly developed GaN HEMT with F class was developed in 2015, which is supported by Ministry of Economy, Trade, and Industry (METI) from 2009. The PAE 70%, with 7-W output power was successfully obtained in the HPA at 5.8 GHz. They developed a phased array module with 76 RF amplifiers, including the HPAs and phase shifters on one module. Module weight: 16.1 kg (specification was less than 19 kg), thickness of sub array: 25 mm (specification was less than 40 mm), average maximum power 449.8 W (specification was more than 400 W), the HPA efficiency 60.3% (specification was more than 60%, whose efficiency decreased by making a lot.), total efficiency of an RF amplifier with the HPA and the driver amplifier was 35.1% (specification was more than 30%) [4]. The METI's project continues in 2023 and the RF amplifier with the HPA and the driver amplifier is developed. In 2022, the RF amplifier with the integrated and the best adjusted GaAs HEMT driver stage to the HPA stage achieved the PAE of 71% with 35 dB gain and 33.9 dB m output power at 5.75 GHz [5]. For the WPT, total DC-RF efficiency is important.

The RF amplifier with a semiconductor is suitable for use in a phased array antenna. Figure 9.6 shows a phased array antenna with 256 class F RF high-power amplifiers (PAs) of 7 W with GaN HEMT at the Kyoto University. The module unit of the phased array antenna contains an RF amplifier, a driver amplifier, a phase shifter, and an isolator. At each module unit, a phase of 5.8 GHz microwave radiation can be controlled by the phase shifter to form a microwave beam in the phase array antenna.

9.2.2   Vacuum tube type microwave generator/amplifier

In kHz–MHz frequency range, vacuum tubes were initially used as three-terminal devices for signal amplification. For the signal frequency increased to GHz, vacuum tubes could no longer be used because of the long travel time of the electrons in the vacuum tube and the presence of unexpected parasitic inductance and capacitance. This has led to the recent replacement of vacuum tubes by semiconductor devices. But in the GHz frequency (microwave) range, the basic design that relied on the vacuum tubes for the amplification and generation of the microwave was changed and the conceptually different vacuum tubes that operate in the GHz range are still applied. These are known as microwave tubes and are high-power, high-efficiency devices. Klystron, magnetron, traveling wave tube (TWT), and TWT amplifier, gyrotrons, and other such devices are examples of such microwave tubes that widely applied for satellite broadcasting systems, remote sensing and radar systems, and microwave heating.

For a microwave WPT system, the magnetron is often used, especially high-power WPT systems. The magnetron is a vacuum tube used as a microwave generator, which is composed of a doughnut-shaped vacuum space with a high electric

(a)   (b)

(c)

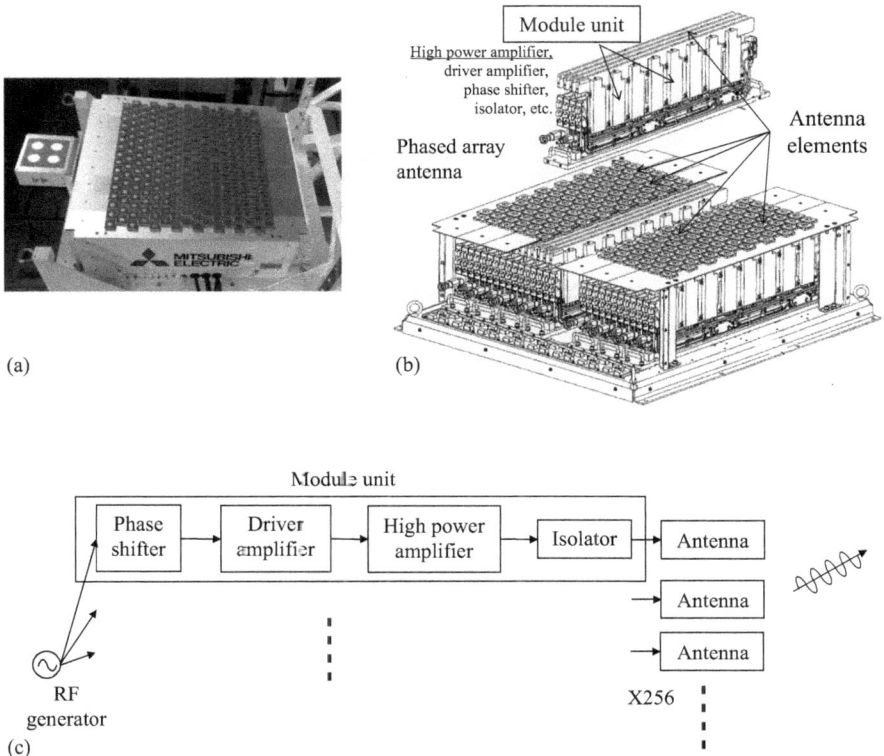

Figure 9.6   (a) Phase array antenna at Kyoto University; (b) schematic of phased array antenna; and (c) typical and simple module unit in phased array antenna

field **E** between the anode and cathode, and a perpendicular magnetic field **B** (Figure 9.7). Electrons from the cathode with heated filament are rotated in the doughnut-shaped vacuum space by **E**×**B** drift. Figure 9.8 shows the results of a computer simulation for a sample magnetron. Rotation and bunching of electrons are seen in the doughnut-shaped vacuum space. At the anode voltage, over 3,500 V is usually used. It is easy to generate high microwave power with high efficiency only by optimizing the structure, including resonance cavities and the electric and magnetic fields. Thus, the magnetron is a very cheap microwave generator and is always applied for microwave ovens at 2.45 GHz. However, the phase and amplitude of the generated microwave cannot be controlled and stabilized in the magnetron because the magnetron is only a microwave generator. Additionally, the magnetron is a noisy device and is therefore unsuitable for use in the WPT system as is (Figure 9.9). A low-noise device is required for the WPT system because the WPT system must coexist with the conventional wireless systems such as those used in mobile phones. Therefore, a phase/amplitude controllable device is

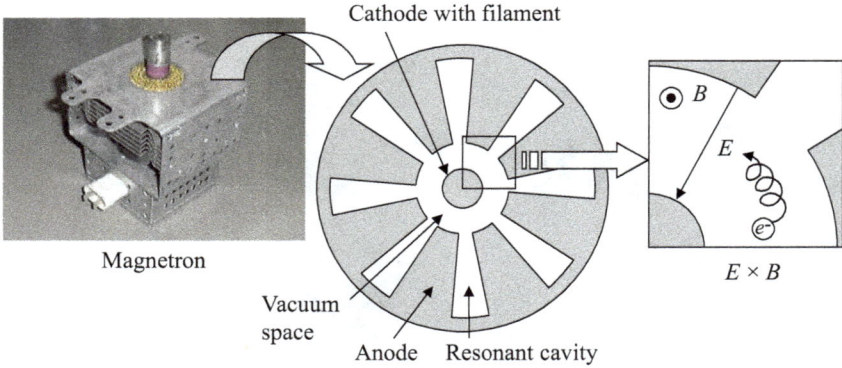

Figure 9.7   Picture and schematic of a magnetron

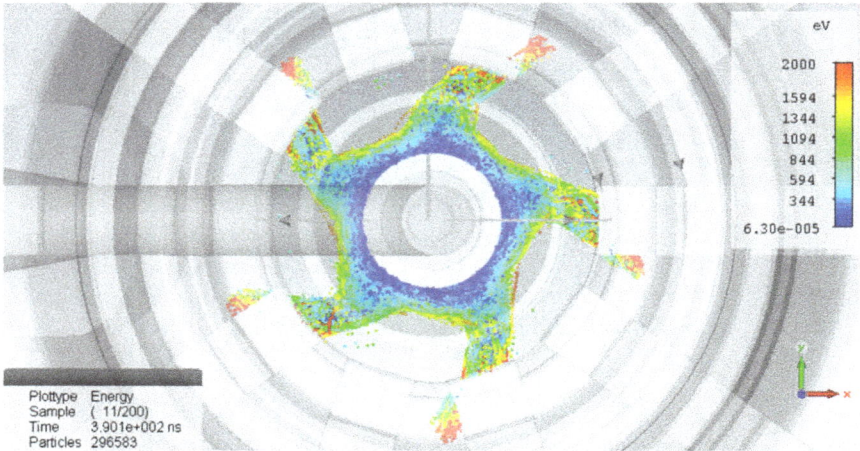

Figure 9.8   Computer simulation results for a magnetron

necessary for the development of a phased array antenna to maintain a high beam efficiency for the WPT system.

However, recently, the cause of the noise from the magnetron was identified as (1) thermal noise caused by the high temperature of the filament at the cathode used for electron generation, and (2) complex operation point of fluctuated anode voltage by using the half-wave rectified voltage source as a voltage boosting power circuit from AC 50/60-Hz power [6]. When the magnetron is driven by a different voltage, main frequency changes and noise occur. Due to the use of the half-wave rectified voltage source, which is suitable for a cheap microwave oven, the microwaves generated in the magnetron are very noisy. Therefore, to suppress the noise of the microwave from the magnetron, stabilized high DC

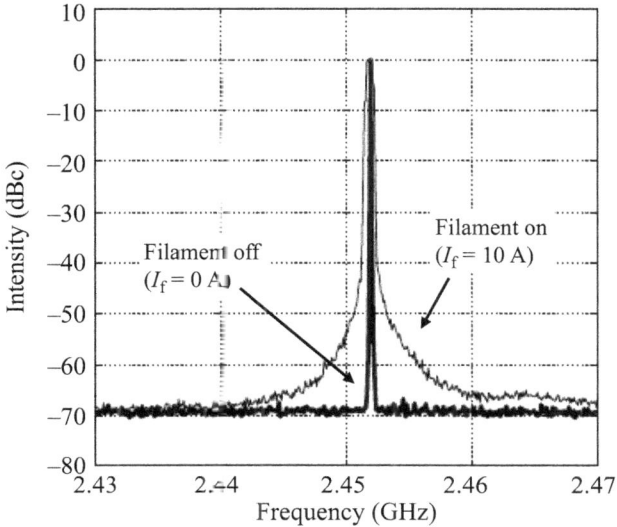

Figure 9.9 Frequency spectrum of a magnetron [6]

voltage should be used. As an additional effect of the stabilized DC high voltage, the excitation of the electrons can be maintained without the filament current after the first excitation of the electron occurs with the filament current in the usual manner because a high DC voltage maintains a sufficient filament temperature to keep the electrons sufficiently excited to generate the microwaves. If a half-wave rectified voltage source is used, a zero voltage time is obtained in the half-cycle of the half-wave rectification and a sufficiently high temperature of the filament cannot be maintained for the excitation of the electrons. The waveform of the microwave from the magnetron driven by the stabilized high DC voltage and with filament off technology is very pure and shows no noise (Figure 9.9).

After suppressing the noise and stabilizing the frequency of the magnetron, phase/amplitude control techniques must be applied to the magnetron [7–10].

To stabilize and control the phase/amplitude of the magnetron, an injection-locking technique and the phase-locked loop (PLL) feedback technique are applied. In the injection-locking technique, a weak signal is injected to the magnetron through an output antenna of the magnetron and then the frequency of the microwave from the magnetron is locked by the injected weak signal as defined by the Adler equation [11]. When the frequency of the injected signal is changed, the high-power microwave from the magnetron is matched to the injected signal. Additionally, the PLL feedback technique is applied to stabilize and control the phase of the microwave from the magnetron. The magnetron can be considered a voltage-controlled oscillator (VCO), which can change the frequency and phase of the VCO by changing the added voltage. Frequency and phase control are also obtained for the change of the magnetic field in the magnetron. Thus, using the

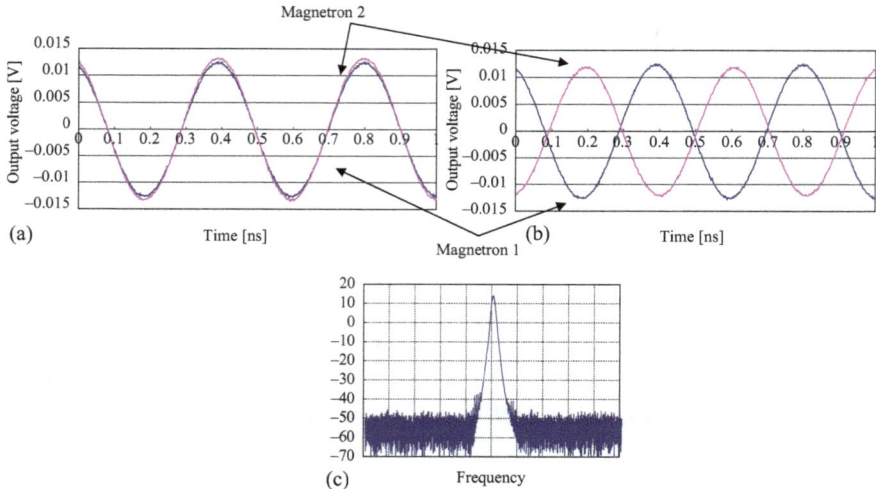

Figure 9.10   (a) Waveform of two PCMs for phase difference of 0°; (b) waveform of two PCMs with phase difference of 180°; and (c) frequency spectrum of PCM

injection locking and the PLL feedback techniques, the microwave from the magnetron can be stabilized and controlled by controlling the phase and the frequency of the weak injected signal. This is known as a phase-controlled magnetron (PCM) [9], and since it is very similar to the amplifier, it is also known as a magnetron amplifier [7]. Typical wave forms and the frequency spectrum of the PCM are shown in Figure 9.10.

A phased array antenna can be developed using the PCMs. Figure 9.11 shows the phased array antenna developed using the PCMs in Kyoto University [12]. Figure 9.10(a) shows a picture of the phased array antenna with 12 PCMs at the continuous wave frequency of 2.45 GHz and power of more than 340 W for each PCM and total power greater than 4 kW. After the PCM, a power divider is installed and a 1-bit low loss phase shifter is installed between an antenna and an output port of the power divider. This is necessary because the PCM power is very high, which is better to connect one high-power PCM with many antennas in order to disperse and decrease the microwave power to the antenna. If the antennas are directly connected with the output port of the power divider, a good phased array antenna cannot be obtained because the microwave beam theoretically cannot be controlled widely without grating lobes. Therefore, the 1-bit low loss phase shifter is installed between an antenna and an output port of the power divider to control the microwave beam widely without the grating lobes [13]. Figure 9.11(b) shows a picture of the phased array antenna developed in 2015 for use in a future Space Solar Power Satellite application by Mitsubishi Heavy Industries, Inc. An antenna with 2.45-GHz PCM is located behind each white panel. The total power of this phased array antenna was over 10 kW. In this system, the antenna is connected directly with the PCM.

We can apply well-made cooker-type magnetron at 2.45 GHz. However, the magnetron at 5.8 GHz, which is another often used frequency for the WPT, is not a mass-produced device and has varied performance. So the PCM is not the best system with 5.8-GHz magnetron. So the other phase and amplitude control system with the 5.8-GHz magnetron is proposed and developed [14]. It is named "Power-Variable Phase-Controlled Magnetron" (PVPCM) by Kyoto University. The magnetron in the PCM considered a VCO, and its frequency/phase is controlled by the feedback loop to voltage. On contrary, in the PVPCM, the feedback loop to control the phase is in the loop of an injection locking signal and the other feedback loop to control the power is installed in parallel (Figure 9.12). By the dual feedback loop,

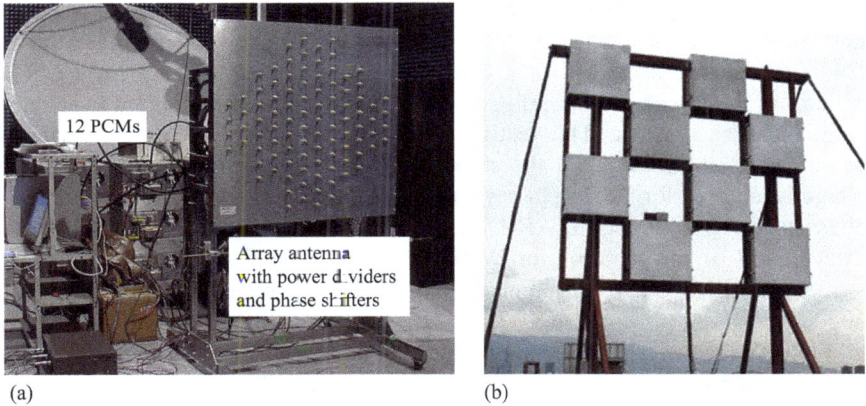

(a)   (b)

Figure 9.11   (a) Phased array antenna with PCMs in Kyoto University (2001); and (b) phased array antenna with PCMs. Developed by Mitsubishi Heavy Industries, Inc. (2015).

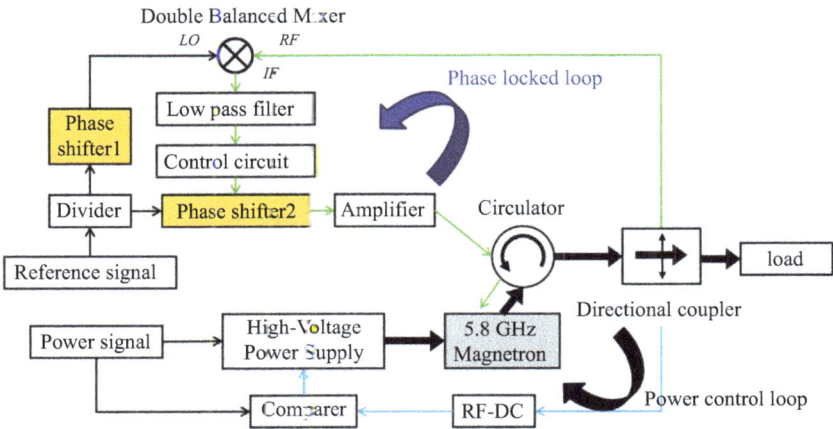

Figure 9.12   Power-variable phase-controlled magnetron with 5.8-GHz Magnetron

we don't have to consider the magnetron as the VCO and can apply any magnetron to control the phase.

9.3   RF rectifier

9.3.1   RF rectifier with semiconductor

While RF amplifiers are used not only for WPT but also in all wireless applications, the rectifier at a receiver is applied only for WPT. Therefore, the rectifier is one of the most important circuits for WPT. Instead of a three-terminal semiconductor device such as an FET, HEMT, or HBT, a diode as a two-terminal device semiconductor is usually applied as a rectifier at the receiver of a system for WPT via radio waves. Especially when a microwave at a GHz frequency is used as a carrier of wireless power, a theory for a circuit with distributed constant lines can be applied. As the circuit with the diode, a wave detector is often applied in wireless communication systems. The rectifier for WPT and the detector for wireless communications are basically the same. However, only a linearity of the output voltage is required for the detector whereas a high RF–DC conversion efficiency is not essential. By contrast, the RF–DC conversion efficiency is the most important characteristic of the rectifier for WPT. Therefore, the design approach for the rectifier for the WPT is different from that of the conventional detector with a diode.

First, the rectifier is designed without an additional management circuit for higher harmonics as an impedance-matching output. There are many diode circuits that are similar to an RF amplifier, e.g., a single series, single shunt, bridge, and double voltage (Figure 9.13). Considering the flow angle of the current and

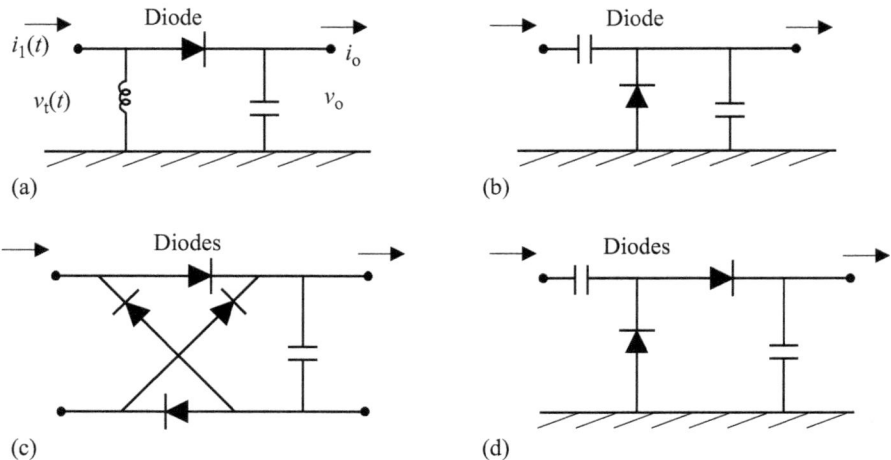

Figure 9.13   Rectifiers: (a) single series; (b) single shunt; (c) bridge; and (d) double voltage

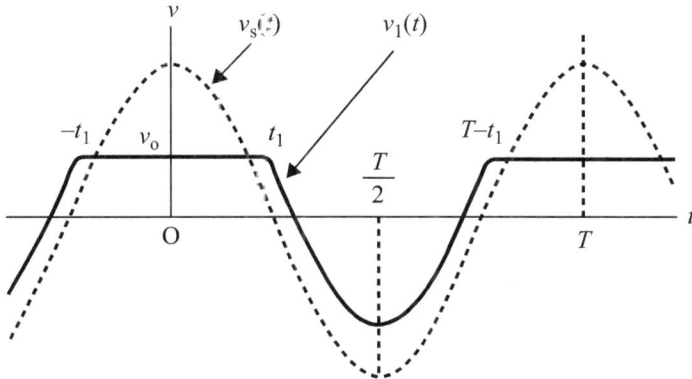

Figure 9.14   Waveform of voltage at single-series rectifier [15]

the voltage at the optimal load resistance, the theoretical RF–DC conversion efficiencies of the single series, the single shunt, the bridge, and the double voltage rectifier are 81.1%, 81.1%, 92.3%, and 92.3%, respectively [15]. The example of the waveform at the single series is shown in Figure 9.14. The waveforms are biased by the output voltage v_o. As a result, the theoretical efficiency of the single series is greater than 50% due to the effect of the capacitance and inductance. The rest power, e.g., 18.9% at the single series, of the input power dissipates in distortion. It is easy to understand why the single-series and the single-shunt rectifiers are operated in a class B like an amplifier. Additionally, when an inductance and capacitance (LC) resonator, which acts as open at the fundamental frequency and short in harmonics, is installed before the diode circuit at the single series or in the single shunt as an input matching, the rectifying circuit is operated as class C. As described before, the amplifier classes A–C are determined by the drain voltage added from a power source. On the contrary, the class of the rectifier is determined by the input power to the rectifier. As described later, the maximum efficiency is found for the operation in class B, so that we must add a near breakdown voltage of a diode at the diode, which has the same effect as the operation of a class B amplifier.

Next, similar to the design for increased efficiency of an amplifier, the rectifier is designed with an additional management circuit for higher harmonics as an impedance-matching output to use the dissipated power in distortion. It is easy to understand that here the class F output load should be used to use the harmonics to increase the RF–DC conversion efficiency in the single series or in the single shunt. Figure 9.15(a) shows a class F rectifier based on the single-shunt rectifier [16,17]. Due to the use of the harmonics, the rectifier can rectify an RF to DC at a theoretical efficiency of 100%. However, originally, the single-shunt rectifier, with the theoretical RF–DC conversion efficiency of 100%, is composed of a single-shunt diode, 1/4 distributed constant line, and a sufficiently high shunt capacitance [Figure 9.15(b)] [18]. Instead of the resonators at an output, the $\lambda/4$

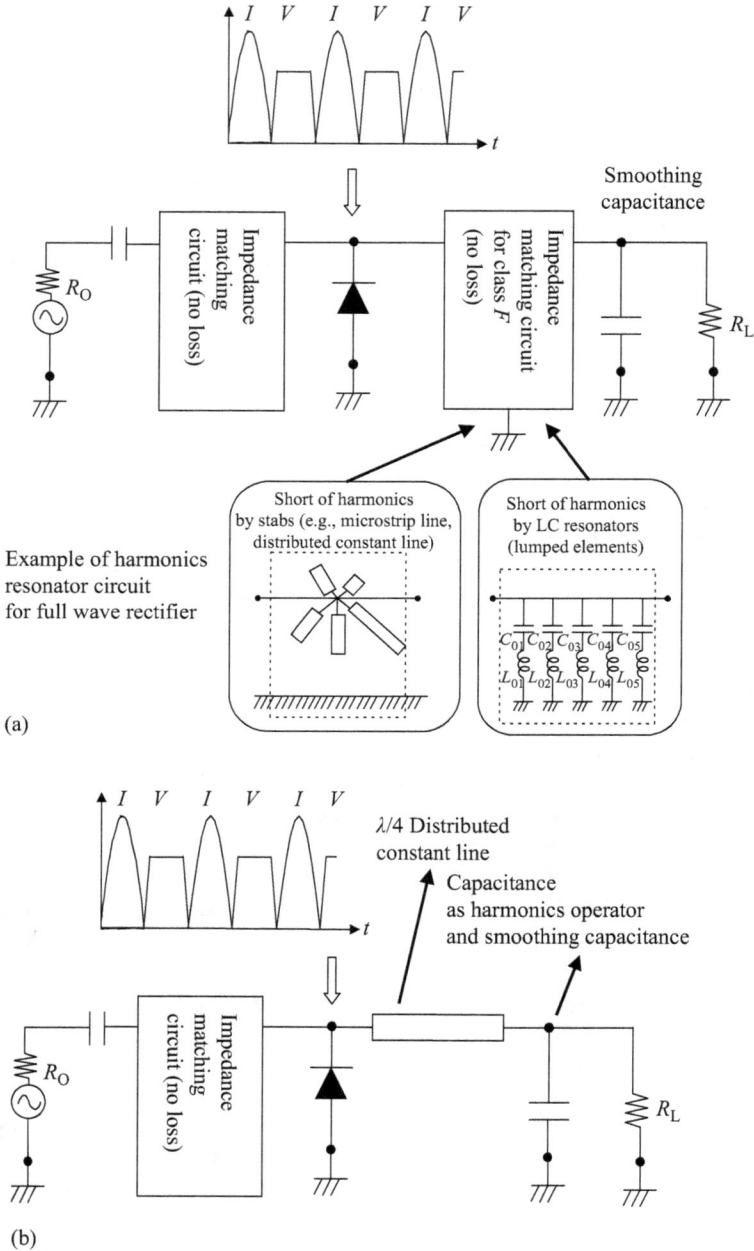

Figure 9.15   Rectifier with RF–DC conversion efficiency of 100%: (a) class F
rectifier; (b) single shunt rectifier with λ/4 distributed constant line
and shunt capacitance

distributed constant line and shunt capacitance create the relation of the harmonics such that impedance is open for odd harmonics and short for even harmonics, same as for a class F amplifier. As a result, the theoretical RF–DC conversion efficiency reaches 100% also in the single-shunt rectifier with the $\lambda/4$ distributed constant line and shunt capacitance. Instead of the class F rectifier, class D rectifier [19] and class E rectifier [20] can be developed. The theoretical RF–DC conversion efficiencies of the class D and the class E rectifier also reach 100%. We can thus consider the same design method for the rectifier and the amplifier. A low pass filter is often installed at an input circuit to suppress reradiation of higher harmonics and to reuse the higher harmonics to increase the RF–DC conversion efficiency.

While the theoretical RF–DC conversion efficiency of the single-shunt rectifier with class F load reaches 100%, loss factors are introduced in a real diode and a real circuit. The *I–V* characteristics of the diode, in particular, determine the actual RF–DC conversion efficiency [21,22]. Figure 9.16 shows the typical RF–DC conversion efficiencies of single-shunt rectifier with the $\lambda/4$ distributed constant line and shunt capacitance. Here, V_{J} is the junction voltage of the diode and V_{br} is the breakdown voltage of the diode, and their characteristic curves are shown in Figure 9.17 and imply that optimum input microwave power is required to realize the maximum actual RF–DC conversion efficiency. If the input power is too low or too high, the efficiency decreases even for a rectifier with the theoretical efficiency of 100%. The optimum input microwave power for the rectifier is found to be close to the breakdown voltage of the diode at the diode as described in the explanation of the operation of the class B rectifier. These characteristics also hold for a connected load instead of the input power, and

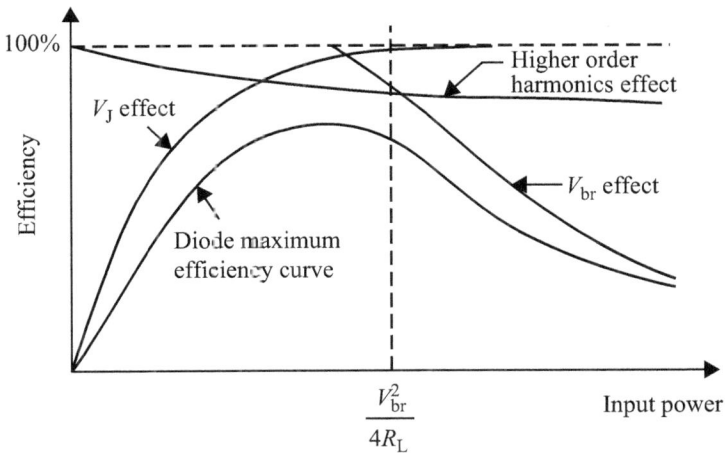

Figure 9.16   Typical relations between RF–DC conversion efficiency and input power [21]

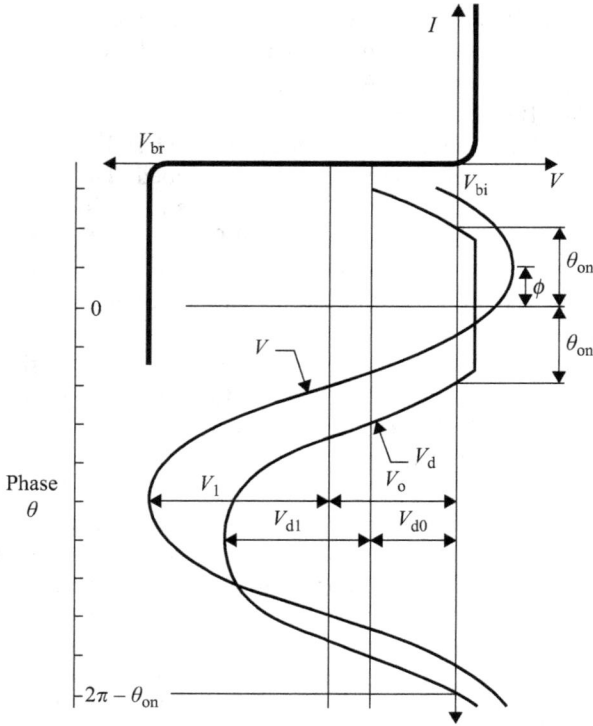

Figure 9.17   Rectification cycle represented by an input fundamental waveform and diode junction voltage waveforms impressed on the diode I–V curve [22]

optimum load is required to realize the highest possible RF–DC conversion efficiency because the added voltage at the diode changes with the changes in the load.

The waveform at a diode can expressed as

$$V = -V_0 + V_1 \cos \omega \tag{9.2}$$

$$V_d = \begin{cases} -V_{d0} + V_{d1} \cos(\omega t - \varphi) & \text{if diode is off} \\ V_{bi} & \text{if diode is on} \end{cases}, \tag{9.3}$$

where V_0 is the output self-bias DC voltage across the load and V_1 is the peak voltage amplitude of the input microwave. V_{d0} and V_{d1} are the DC and fundamental frequency components of the diode junction voltage, and V_{bi} is the diode's built-in voltage in the forward bias region. The diode model comprises a series resistance R_s, a nonlinear junction resistance R_j described by its DC I–V characteristics, and a nonlinear junction capacitance C_j as shown in Figure 9.18 [19,20]. A DC load resistance R_L is connected in parallel to the diode along the DC path represented by the dotted line to complete

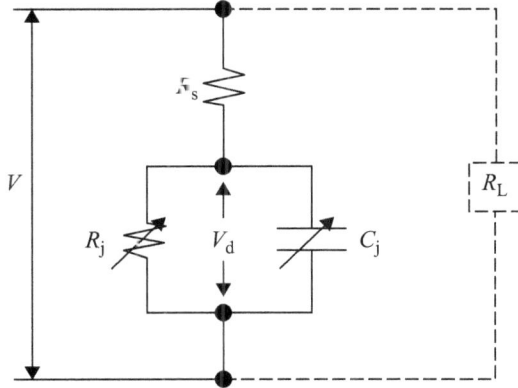

Figure 9.18   Equivalent circuit of a diode [22]

the DC circuit. The junction resistance is assumed to be zero for forward bias and infinite for reverse bias. By applying Kirchhoff's voltage law, closed-form equations for the diode's efficiency and input impedance are determined. The RF–DC conversion efficiency η is theoretically calculated from (9.4) [21,22]:

$$\eta = \frac{1}{1 + A + B + C} \tag{9.4}$$

$$A = \frac{R_L}{\pi R_s} \left(1 + \frac{V_{bi}}{V_o}\right)^2 \left[\theta_{on}\left(1 + \frac{1}{2\cos^2\theta_{on}}\right) - \frac{3}{2}\tan\theta_{on}\right]$$
[Power loss on R_s at diode on]

$$B = \frac{R_s R_L C_j^2 \omega^2}{2\pi}\left(1 + \frac{V_{bi}}{V_o}\right)\left(\frac{\pi - \theta_{on}}{\cos^2\theta_{on}} + \tan\theta_{on}\right) \text{ [Power loss on } R_s \text{ at diode off]}$$

$$C = \frac{R_L}{\pi R_s}\left(1 + \frac{V_{bi}}{V_o}\right)\frac{V_{bi}}{V_o}(\tan\theta_{on} - \theta_{on}) \text{ [Power loss on } R_j \text{ at diode on]}$$

where θ_{on} is defined as follows

$$\tan\theta_{on} - \theta_{on} = \frac{\pi R_s}{R_L\left(1 + \frac{V_{bi}}{V_o}\right)}, \tag{9.5}$$

and C_j is nonlinear junction capacitance described using zero bias junction capacitance C_{j0} as

$$C_j = C_{j0}\sqrt{\frac{V_{bi}}{V_{bi} - |V_o|}}. \tag{9.6}$$

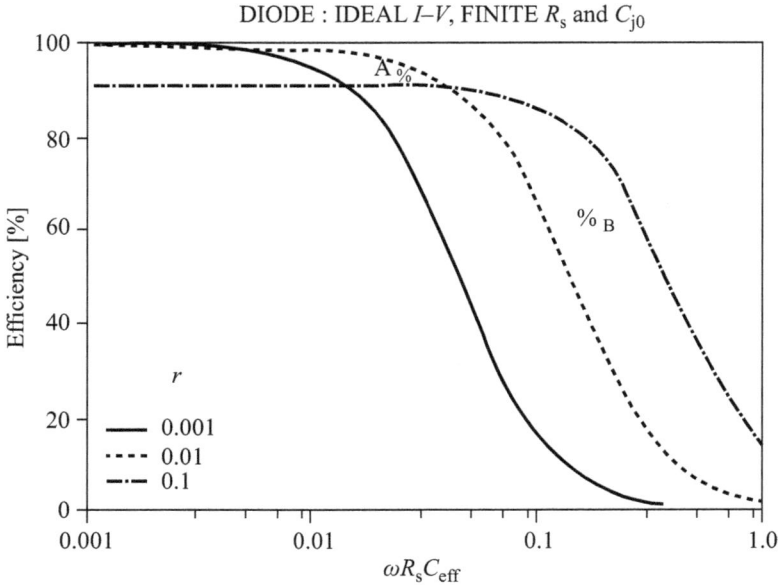

Figure 9.19   *RF–DC conversion efficiency of a diode estimated by frequency, internal resistance, and capacitance ($r = R_s/R_L$) [22] (point A: $R_s = 0.5\ \Omega$, $C_{j0} = 3$ pf, 2.45 GHz, $R_L = 100\ \Omega$; point B: $R_s = 4.85\ \Omega$, $C_{j0} = 0.13$ pf, 35 GHz, $R_L = 100\ \Omega$)*

The estimated RF–DC conversion efficiency of a diode calculated using (9.4)–(9.6) for various circuit parameters is shown in Figure 9.19.

To increase the RF–DC conversion efficiency, a diode with low R_s and low C_{j0} should be chosen. Therefore, a Schottky barrier diode is often used in the rectifier for WPT. The junction voltage of the Schottky barrier diode is very low, making it suitable for use as a high-efficiency rectifier. GaAs was found to be a better diode material than Si [23], and GaN diodes are expected to be used due to their high-efficiency, high-power rectification. Since the diode gives rise to loss factors of R_s and C_{j0}, the number of diodes in the rectifier should be as small as possible. Therefore, the single-shunt and the single-series rectifier are often chosen as the rectifier for WPT. Figure 9.19 also shows that the RF–DC conversion efficiency decreases with increasing WPT frequency. The RF–DC conversion efficiency of the rectifier reached the values of 91% at 920 MHz, 32 dB m with a diode [24], over 90% at 2.45 GHz with a GaAs diode at 5–10 W [23,25,26], 92.8% with high impedance dipole (500 Ω) at 5.8 GHz, 1 W [27], over 80% at 5.8 GHz with a diode [22], 77.9% with GaAs E-pHEMT double voltage rectifier MMIC at 5.8 GHz, 37.1 dB m [28], 61.5% with new developed GaN diode at 94 GHz, 3.6 kW/m^2 [29], and 37.7% at over 90 GHz with complementary metal–oxide–semiconductor [30].

9.3.2 Vacuum tube-type microwave rectifier

A vacuum tube can be applied as a microwave rectifier or as a microwave amplifier/generator. A cyclotron wave converter (CWC) is used as the vacuum tube rectifier. The most studied CWC comprises an electron gun, a microwave cavity with the uniform transverse electric field in the interaction gap, a region with a symmetrically reversed (or decreasing to zero) static magnetic field, and a collector with depressed potential, as shown in Figure 9.20. Microwave power from an external source is converted by this coupler into the energy for the electron beam rotation, and the latter is transformed into additional energy for the longitudinal motion of the electron beam by a reversed static magnetic field. Then, the energy is extracted by decelerating the electric field of the collector and appears as a DC signal at the load resistance of this collector. At Moscow State University, a variant of the CWC was tested and its efficiency was found to be 70%–74% at 2.5–25 W. The TORIY Corporation and Moscow State University collaborated to create several high-power CWCs with efficiencies of 60%–83% at 10–20 kW [31–33].

A vacuum microwave rectifier based on multipactor discharge was also proposed in the 1960s in the United States [34]. The term "multipactor" is derived from the words "multiple electron impact." A multipactor discharge consists of a thin electron cloud that is driven back and forth across a gap in response to an RF field applied across the gap. Multipactor discharges can be used to provide both half- and full-wave rectification. A full-wave rectifier utilizes a reentrant microwave cavity with the electric field concentrated at the center of the cavity where the secondary emitting electrodes are located. If the DC load voltage is set to be equal to the electron impact voltage, the theoretical RF–DC conversion efficiency of the single cavity full-wave rectifier is 59%, and that of the dual cavity two-phase rectifier approaches 92% or approximately $\pi/2$ times larger than that of the single cavity full-wave rectifier.

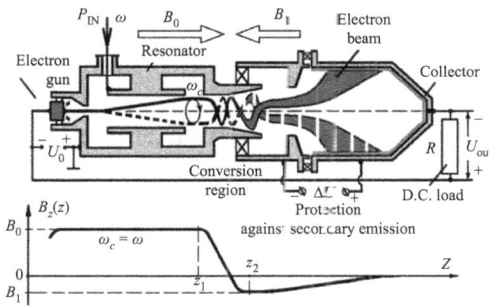

Figure 9.20 Schematic of a cyclotron wave converter

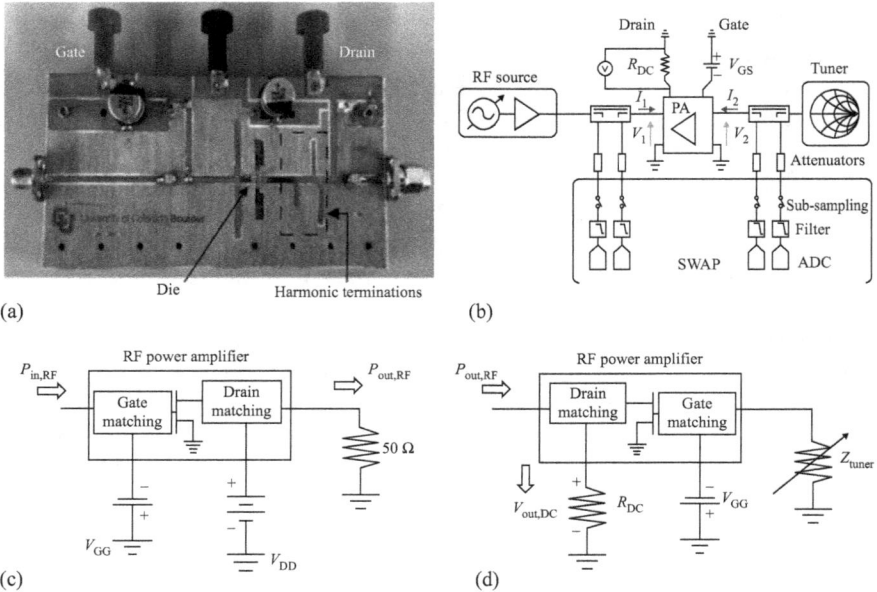

(a)   (b)   (c)   (d)

Figure 9.21   PA rectifier: (a) PA rectifier developed by University of Colorado [35]; (b) measurement block of a PA rectifier [23]; (c) schematic of circuit as an amplifier; and (d) schematic of circuit as a rectifier

9.4   RF amplifier/rectifier with semiconductor

The circuit for an RF amplifier can be applied as a rectifier. Figure 9.21 shows a photograph of the class F^{-1} PA with GaN HEMT [35]. The performance of the amplifier is characterized at 2.14 GHz with the drain voltage bias of 28 V and a bias current of 160 mA. PAE of 84% is achieved with the output power of 37.6 dB m and gain of 15.7 dB under −3 dB compression. When this circuit is used as a rectifier, RF power is input into the drain, which is unbiased and the gate is terminated in a variable impedance and biased close to pinch-off. The measurement of the rectifier performance shows that the RF–DC conversion efficiency reached 85% with the DC output voltage of 36 V and input power at the drain of 42 dB m with $R_{DC} = 98.5\ \Omega$. This is named PA rectifier. The design method of the circuit at the GHz range is the same as that at the kHz–MHz range but with a concept of an additional mode and impedance.

References

[1]   K. Honjo, "High Power Amplifier for MPT (*in Japanese*)", Section 2.1.1, "*Solar Power Satellite* (ed. N. Shinohara)", Ohm Publishing, Tokyo, 2011.

[2]   C. Zheng, T. Yoshida, R. Ishikawa, and K. Honjo, "GaN HEMT Class-F Amplifier Operating at 1.5 GHz (*in Japanese*)", Proc. of IEICE, C-2-27, Mar. 2007.

[3]   M. Kamiyama, R. Ishikawa, and K. Honjo, "C-band High Efficiency AlGaN/ GaN HEMT Power Amplifier by Controlling Phase Angle of Harmonics (*in Japanese*)", Proc. of IEICE, CS-3-1, Sep. 2011.

[4]   S. Mihara, M. Sato, S. Nakamura, *et al.*, "The Result of Ground Experiment of Microwave Wireless Power Transmission", Proc. 66th International Astronautical Congress, IAC-2015-C3.2.1, 2015.

[5]   H. Yanagawa, K. Ijichi, S. Shimazaki, and O. Kashimura, "The Outline and the Up-to-Date Status of the Power Transmission System Development Project towards the Realization of the SSPS", Proc. 30th ISTS & 6th NAST, 2023-q-12, 2023.

[6]   T. Mitani, N. Shinohara, H. Matsumoto, and K. Hashimoto, "Experimental Study on Oscillation Characteristics of Magnetron after Turning Off Filament Current (*in Japanese*)", *IEICE Trans. C*, Vol. J85-C, No. 11, pp. 983–990, 2002.

[7]   W. C. Brown, "The Sophisticated Properties of the Microwave Oven Magnetron", 1989 IEEE MTTS Digest, pp. 871–874, 1989.

[8]   M. C. Hatfield and J. G. Hawkins, "Design of an Electronically-Steerable Phased Array for Wireless Power Transmission Using a Magnetron Directional Amplifier", Proc. of 1999 MTT-S Int. Microwave Symp., pp. 341–344, 1999.

[9]   N. Shinohara, T. Mitani, and H. Matsumoto, "Development of Phase-Controlled Magnetron (*in Japanese*)", *IEICE Trans. C*, Vol. J84-C, No. 3, pp. 199–206, 2001.

[10] C. Liu, H. Huang, Z. Liu, F. Huo, and K. Huang, "Experimental Study on Microwave Power Combining Based on Injection-Locked 15-kW S-Band Continuous-Wave Magnetrons", *IEEE Trans. Plasma Sci.*, Vol. 44, No. 8, pp. 1291–1297, 2016.

[11] R. Adler, "A Study of Locking Phenomena in Oscillators", Proc. of IRE, No. 34, pp. 351–357, Jun. 1946.

[12] N. Shinohara, H. Matsumoto, and K. Hashimoto, "Phase-Controlled Magnetron Development for SPORTS: Space Power Radio Transmission System", *Radio Sci. Bull.*, No. 310, pp. 29–35, 2004.

[13] N. Shinohara, B. Shishkov, H. Matsumoto, K. Hashimoto, and A. K. M. Baki, "New Stochastic Algorithm for Optimization of Both Side Lobes and Grating Lobes in Large Antenna Arrays for MPT", *IEICE Trans. Commun.*, Vol. E91-B, No. 1, pp. 286–296, 2008.

[14] B. Yang, T. Mitani, and N. Shinohara, "Experimental Study on a 5.8 GHz Power-Variable Phase-Controlled Magnetron", *IEICE Trans. Electron.*, Vol. E100-C, No.10, pp. 901–907, 2017.

[15] T. Ohira, "Power Efficiency and Optimum Load Formulas on RF Rectifiers Featuring Flow-Angle Equations", *IEICE Electron. Exp. (ELEX)*, Vol. 10, No. 11, pp. 1–9, 2013.

[16]   J. Guo and X. Zhu, "Class F Rectifier RF–DC Conversion Efficiency Analysis", Proc. of Int. Microwave Symp. (IMS), WE4G-1, 2013.

[17]   K. Hatano, N. Shinohara, T. Seki, and M. Kawashima, "Development of MMIC Rectenna at 24 GHz", Proc. of 2013 IEEE Radio & Wireless Symp. (RWS), pp. 199–201, 2013.

[18]   R. J. Gutmann and J. M. Borrego, "Power Combining in an Array of Microwave Power Rectifiers", *IEEE Trans. MTT*, Vol. 27, No. 12, pp. 958–968, 1979.

[19]   S. Dehghani and T. E. Johnson, "A 2.4 GHz CMOS Class D Synchronous Rectifier", Proc. of Int. Microwave Symp. (IMS), WE1F-1, 2015.

[20]   M. N. Ruiz, R. Marante, and J. A. García, "A Class E Synchronous Rectifier based on an E-pHEMT Device for Wireless Powering Applications", Proc. of Int. Microwave Symp. (IMS), TH1A-3, 2012.

[21]   T. W. Yoo and K. Chang, "Theoretical and Experimental Development of 10 and 35 GHz Rectennas", *IEEE Trans. MTT*, Vol. 40, No. 6, pp. 1259–1266, 1992.

[22]   J. O. McSpadden, L. Fan, and K. Chang, "Design and Experiments of a High-Conversion-Efficiency 5.8-GHz Rectenna", *IEEE Trans. MTT*, Vol. 46, No. 12, pp. 2053–2060, 1998.

[23]   W. C. Brown, "Optimization of the Efficiency and Other Properties of the Rectenna Element", MTT-S Int. Microwave Symp., pp. 142–144, 1976.

[24]   K. Kawai, N. Shinohara, and T. Mitani, "Novel Structure of Single-Shunt Rectifier Circuit with Impedance Matching at Output Filter", *IEICE Trans. C*, Vol. E106-C, No. 2, pp. 50–58, 2023.

[25]   C. Wang, B. Yang, and N. Shinohara, "Study and Design of a 2.45GHz Rectifier Achieving 91% Efficiency at 5-W Input Power", *IEEE Microw. Wirel. Compon. Lett.*, Vol. 31, No. 1, pp. 76–79, 2021.

[26]   K. Kikkawa, T. Saen, N. Sakai, and K. Itoh, "A 2.4-GHz 10-W Class Bridge Rectifier and Its Efficiency Analysis with the Behavioral Model", *IEEE MTTS Trans.*, Vol. 70, No. 3, pp. 1994–2001, 2022.

[27]   N. Sakai, K. Noguchi, and K. Itoh, "A 5.8-GHz Band Highly Efficient 1-W Rectenna with Short-Stub-Connected High-Impedance Dipole Antenna", *IEEE MTTS Trans.*, Vol. 69, No. 7, pp. 3558–3566, 2021.

[28]   K. Itoh and N. Sakai, "Highly efficient High-Power Rectenna Techniques", Proc. of RFIT 2021, 2021.

[29]   K. Hooman, "61.5% Efficiency and 3.6 kW/m^2 Power Handling Rectenna Circuit Demonstration for Radiative Millimeter Wave Wireless Power Transmission", *IEEE Trans. MTT*, Vol. 70, No. 1, pp. 650–659, 2022.

[30]   S. Hemour, C. H. Lorenz, and K. Wu, "Small-Footprint Wideband 94 GHz Rectifier for Swarm Micro-Robotics", Proc. of Int. Microwave Symp. (IMS), WE1F-5, 2015.

[31]   V. A. Vanke, V. M. Lopukhin, V. K. Rosnovsky, V. L. Savvin, and K. I. Sigorin, "Ground-Based Receiving/Converting System for Space Solar Power Systems", *Radiotech. Electron.*, Vol. 27, No. 5, p. 1014, 1982.

[32] V. A. Vanke and V. L. Savvin, "Cyclotron Wave Converter for SPS Energy Transmission System", Proc. of SPS'91, pp. 515–520, 1991.

[33] V. A. Vanke, V. L. Savvin I. A. Boudzinski, and S. V. Bykovski, "Development of Cyclotron Wave Converter", Proc. of WPT'95, p. 3, 1995.

[34] W. C. Brown, "Rectification of Microwave Power", IEEE Spectrum, pp. 92–100, Oct. 1964.

[35] M. Roberg, T. Reveyrand. I. Ramos, E. A. Falkenstein, and Z. Popović, "High-Efficiency Harmonically Terminated Diode and Transistor Rectifiers", *IEEE MTT Trans.,* Vol. 60, No. 12, pp. 4043–4052, 2012.

Chapter 10

Applications of coupling WPT for electric vehicle

Yukio Yokoi[1]

10.1 Introduction

The report of MIT Team in 2007 where they proved wide and variable air-gap power transfer based on resonant coupling WPT was made big impact to realize social implementation of WPT for e-mobility and ERS (Electric Road System). That is called "MIT's revolution or Columbus's egg (2007)" [1]. After the MIT report, many concept EV cars with WPT capability have been demonstrated at various motor shows. At Tokyo Motor Show 2011 Japan, many Original Equipment Manufacturers (OEMs), such as Toyota, Nissan, Mitsubishi motors, Yamaha, and GM, displayed their concept EVs with WPT option [2]. Also at New York International Auto Show 2012, Nissan announced that they will deliver the Infiniti LE Concept with the wireless charging option within 24 months [3]. It was doubtless the world's first announcement to install the WPT system on mass production passenger EV. Unfortunately, the plan was not realized.

Afterward, during the Paris Motor Show of October 11, 2016, Qualcomm Halo announced that "2013 Mercedes-Benz S550e will offer wireless EV charging technology built using Qualcomm Halo breakthroughs" through their news release [4]. They also said that with the facelift of the Mercedes-Benz S550e in 2018, Daimler AG plans to launch a WEVC system, manufactured by a tier 1 power electronics supplier, which has licensed Qualcomm Halo inventions. The commercial application of WEVC is a first among hybrid vehicles. Drivers of the S550e (badged S500 in Europe) equipped with the wireless charging option will simply park atop a special pad and charging will begin—no cables to manage or untangle, just "park it and charge it." Figure 10.1 shows a concept of WPT system for EV at private parking spot.

Standardization of such new function for mass production passenger EV has started from September 2011 as IEC JPT61980 by TC69 and from February 2014 ISO 19363 NWIP has begun. After 10 years deliberation, in May 2023 the IS (International Standard) IEC61980-2; 2023 has released, then IS set for static WPT,

[1]WPT System Technical Committee of JSAE (Society of Automotive Engineers of Japan), Japan

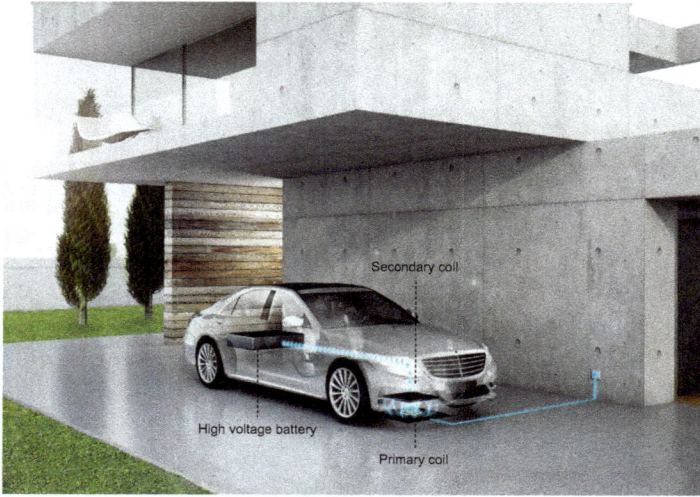

Figure 10.1   Concept of the WPT system for passenger EV [3]

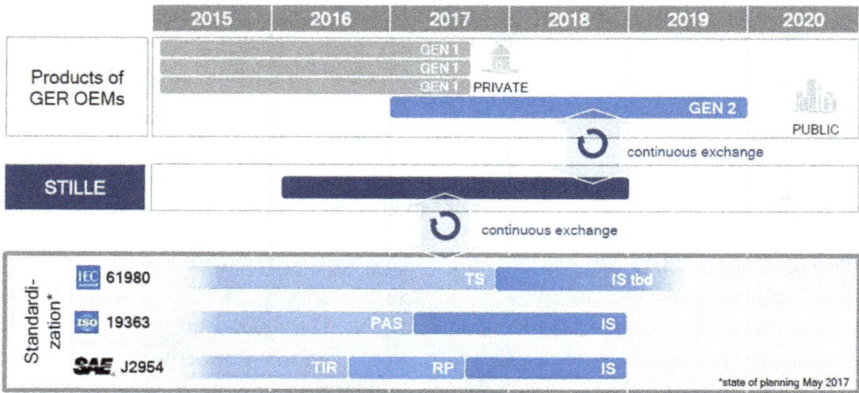

Figure 10.2   Overall timeline for standardization with respect to market introduction [5]

IEC 61980-1 Ed. 2.0:2020 (b), IEC61980-2; 2023 and IEC 61980-3:2022 are completed in one set (Figure 10.3).

It also needs change of domestic and international regulations, International Telecommunication Union (ITU), Comité International Spécial des Perturbations Radioélectriques (CISPR), etc. Figure 10.2 shows the overall timeline for standardization with respect to market introduction prepared by STILL project in 2017 [4]. In Germany, STILL is considering Gen1 production for private WPT charging, which will expected come around the end of 2017, and Gen 2 production for public WPT charging will in the market around the end of 2019.

According to world-wide effort after publishing ISO/IEC international standards (IS), standard of SAE J2954, UL 2750, ITU frequency recommendation and China standard GB/T for static wireless power transfer for EV have released. The progress is shown in Figure 10.3

After ed1, remarkable progress of applications of coupling WPT has been made. After many field evaluation tests, commercialization for mass production of

2010-09 ; IEC JPT61980 NWIP
2013-06 ; ITU-R SG1 CG-WPT
2013-10 ; CISPR Ottawa assembly
2014-02 ; ISO 19363 NWIP
2015-07 ; IEC61980-1 IS released
2016-03 ; Japan New Reg for EV by MIAC
2019-10 ; ITU-R-REC-SM.2110-1-Released
2019-02 ; IEC 63243(61980-5) DWPT NWIP
2020-04 ; ISO19363:2020
2020-10 ; SAE J2954 _202010, Light-Duty
2020-10 ; ISO5474 Unified EV charging STD started
2020-11 ; IEC 61980-1:2020 released
2021-04 ; IEC 63381 (61980-6) DWPT COM NWIP
2022-11 ; IEC 61980-3:2022 released
2022-12 ; SAE J2954/2_202212. Heavy-Duty
2023-05; ; IEC 61980-2:2023 released
2023-05; IEC IEC-61980-5,6 under Development

STD/
Regulation

Figure 10.3   Progress of Standard and Regulation (by May 2023)

WHY WIRELESS CHARGING?

Wireless charging to grow substantially, in premium vehicles

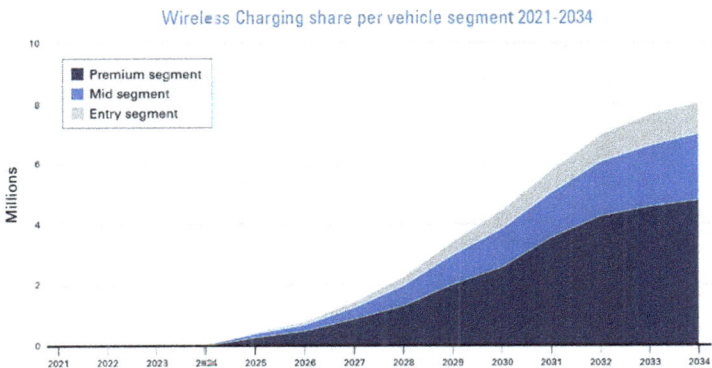

Wireless Charging share per vehicle segment 2021-2034

Figure 10.4   Vehicle-side charging forecast in September 2022 (by S&P Global 2022)

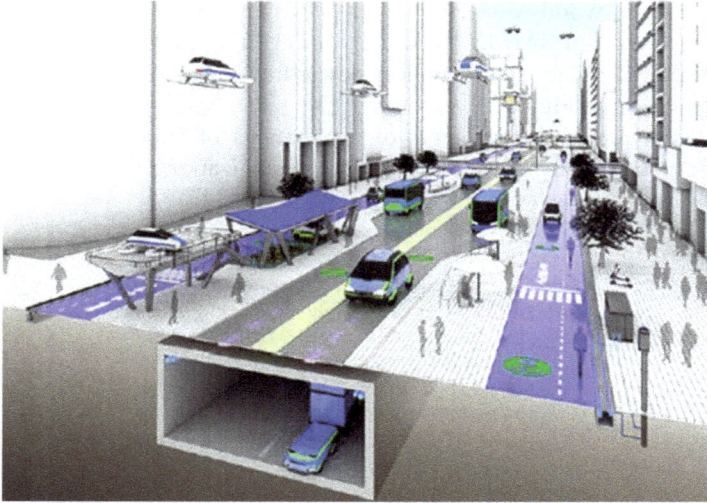

Figure 10.5   Future Urban Road Image "e-MoRoad" (by Obayashi 2022)

passenger EV has started at China and Korea from 2022. S&P Global has released vehicle-side charging forecast in September 2022 as shown in Figure 10.4 [6].

Another point of view, to realize carbon free society on transport sector, for social implementation of dynamic wireless power transfer for ERS (Electric Road System) is now developing at many places such as Europe, the USA, Japan, China, and Korea.

Japanese construction company Obayashi and Denso has proposed "e-MoRoad" in September 2022 for Image of Future Urban Road (Figure 10.5) [7].

10.2   EV and charging

History of EV started in 1888 Flocken Elektrowagen (Germany). At the New York Motor Show held in 1901, 58 gasoline automobiles, 58 steam automobiles, and 28 electric automobiles were at display (Figure 10.6).

Because of the complicity to handle charging cable and plug, and also relatively poor performance of battery, gasoline automobile has got a leading part of mobility during 100 years. By the end of the twentieth century, due to the environmental problem, air pollution became one of the major problems of transportation. EV has considered good solution for reducing CO_2. For energy storage, Li-ion battery, electric double layer capacitor (super capacitor) has become a common use.

On the other hand to charge or transfer energy into the on-vehicle electric storage device, many e-charging methods are developed and standardized. The classification of e-charging methods is shown in Figure 10.7.

To establish standard of e-charging for commercialization, many technical committees (TCs) of IEC are collaborating and keeping liaison with each other, which is shown in Figure 10.8. The scope of TC69 (electric road vehicles and

Figure 10.6 Cable and plug charging for EV at early stage [6]

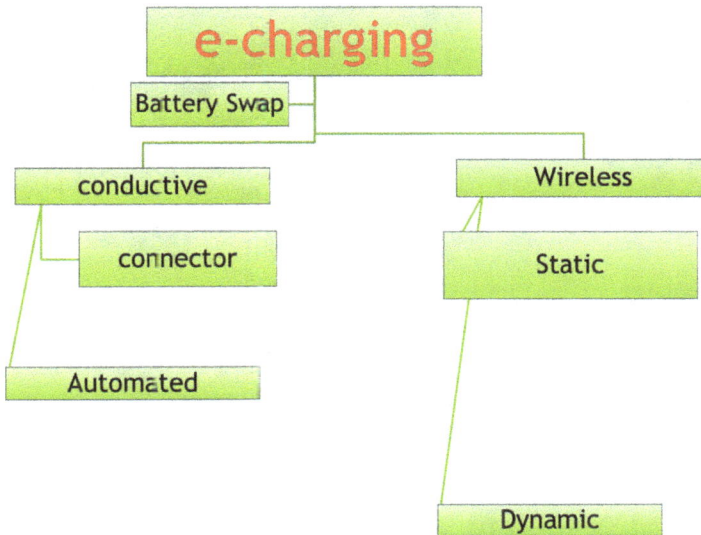

Figure 10.7 Classification of e-charging method

electric industrial trucks) is "to prepare international standards for road vehicles, totally or partly electrically propelled from self-contained power sources, and for electric industrial trucks." In TC69, WG7 for EV WPT systems started their work from 2016.

10.3   Conductive charging

According to Figure 10.8, standardization for conductive charging of EV was developed by many TCs of IEC including cable and connector (IEC15118 series) and communication (command and control, IEC15118 series). Conductive type of charging for EV has become a major method, but it is not the concern of this book. The detail is described in Section 10.4 of *Automotive Engineering Handbook volume 7 (EV/Hybrid) by JSAE* [7].

Conductive AC charging became popular in Europe, and many European cities introduced such charging station. Figure 10.9 shows one case of AC conductive charging station just in front of the NICE central station.

Fast conductive charge station, which enable multiple use of three types, Combo, ChaDeMo of fast DC charging and AC charging, get over the difference of

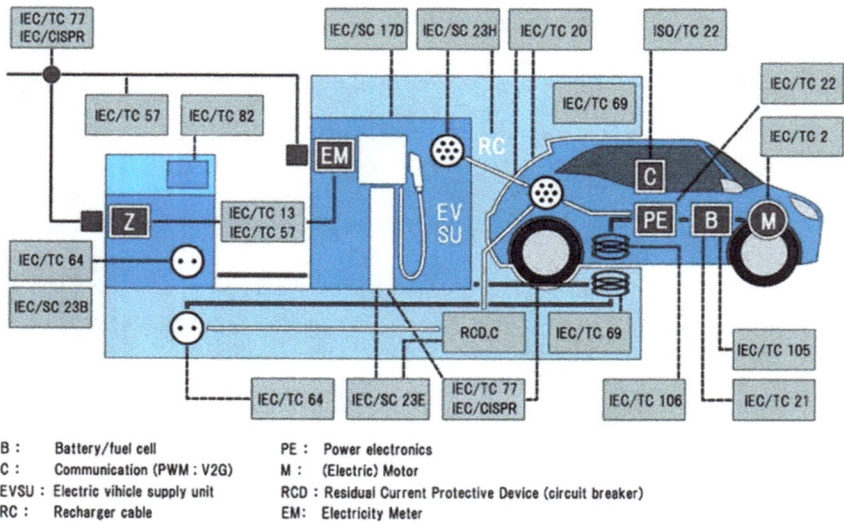

B :   Battery/fuel cell
C :   Communication (PWM : V2G)
EVSU :   Electric vihicle supply unit
RC :   Recharger cable
PE :   Power electronics
M :   (Electric) Motor
RCD :   Residual Current Protective Device (circuit breaker)
EM :   Electricity Meter

Figure 10.8   Technical committees relating e-mobility charging method

Figure 10.9   AC conductive charging spot at NICE central station. Photo by Yokoi

connector standard in Europe, coming popular. Figure 10.10 shows multiple uses of fast conductive charging station at Vattenfall office of Stockholm, Sweden, in April 2016.

These AC and DC charging methods both use connector and cable and need manual handling. At some cities in Europe, EV Bus has deployed automatic high-power (over 100 kW) charging without cable, pantograph, or another method. For one sample at Goteborg, Sweden, August 2015, Volvo bus was already operating automated high-power charging stations at terminal and bus stop. Necessary charging time is only 6 min (Figure 10.11).

Figure 10.10   Fast conductive charging station of Combo, CHaDeMo. and AC at Vattenfall office of Stockholm, Sweden. Photo by Yokoi

Figure 10.11   Automated high-power charging station for Volvo bus of Goteborg, Sweden. Photo by Yokoi

10.4   Wireless charging

Standardization and commercialization of WPT for EV started at California, USA, by the end of 2000. SAE Electric Vehicle Inductively Coupled Charging standard SAEJ1773 was published in 1995.

Charging style of GM EV-1 is shown at Figure 10.12, and Figure 10.13 shows the charging paddle. This paddle type means that the air gap between primary and secondary coil is very close but no connection by wire using inductive charging.

10.4.1   Field evaluation in Europe

After "MIT's revolution or Columbus's egg" (2007), restriction of the air gap problem of WPT for EV was solved. Many feasibility studies and field evaluation have proposed to deploy on passenger EV.

Figure 10.12   GM EV-1 1996

Figure 10.13   Charging paddle for EV-1

One of earliest evaluations was announced on November 18, 2011, by Qualcomm Halo [5].

Their press release [8] said, "Qualcomm Incorporated today announced the first Wireless Electric Vehicle Charging (WEVC) trial for London in what is a UK and industry-leading initiative. Qualcomm is collaborating with the UK Government, as well as the Mayor of London's office and Transport for London (TfL) to deliver the trial." They also announced that "the pre-commercial trial is expected to start in early 2012 and will involve as many as 50 electric vehicles (EVs)." Unfortunately we could not find the report of the pre-commercial trial yet (Figure 10.14).

In Europe, for EV bus operation of public transport has widely evaluated and partially commercial operation in use. Figure 10.15 shows the situation of field evaluation of WPT deployed on EV bus, which is assembled and introduced by Takahashi [9].

Figure 10.14   In London, wireless charging for EVs

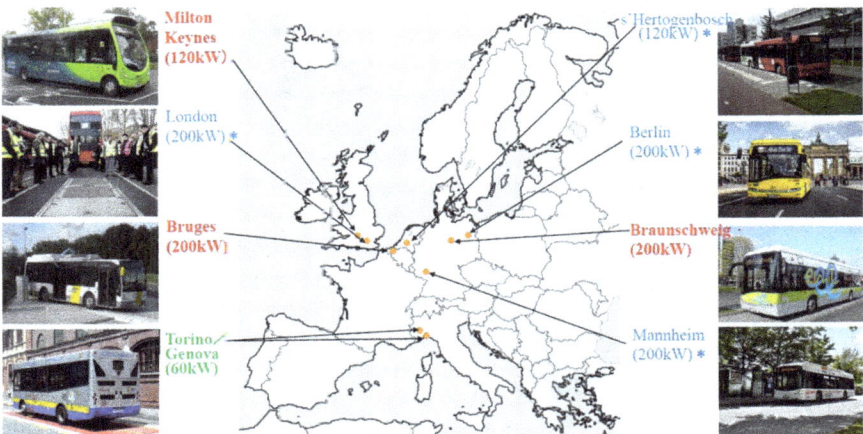

Figure 10.15   High-power wireless charging application for public service in Europe

10.4.2   Field evaluation in Japan

In Japan in February 2012 Mitsui Home and IHI announced to evaluate WPT operating with passenger EV at Smart Home project HEMS at Kashiwanoha, Chiba Prefecture [10]. Field evaluation was done in March 2016 with use of smartphone for operation monitor and control of WPT charging [11] (Figures 10.16–10.21).

OEMs in Japan, Toyota, in 2014, and Honda also in 2014 have field evaluated wireless charging for EV [9].

Toyota evaluates using three cars and residential during 2 years [12]. Honda uses smart home "E-KIZUNA Project" with government of Saitama city.

Toshiba with support of Waseda University and ANA has evaluated wireless charging of both passenger EV and bus during 2 years using funding of the Ministry of the Environment project [13].

Figure 10.16   *EV wireless charging of smart house of Mitsui home and IHI. Smartphone is used for monitoring and control charge operation.*

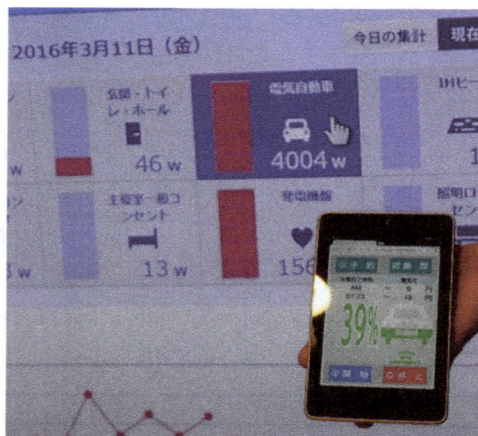

Figure 10.17   *Control with smartphone of EV wireless charging*

Figure 10.18   Wireless charging EV under test by Toyota

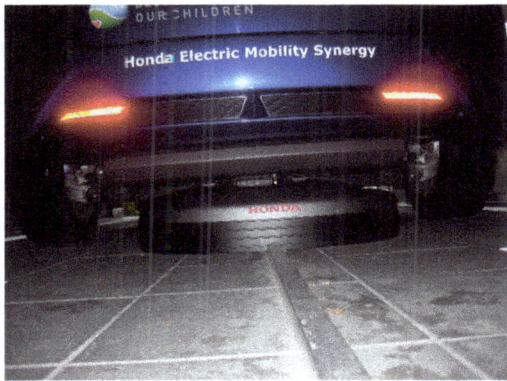

Figure 10.19   Wireless charging coils under test by Honda

Figure 10.20   Wireless charging primary system for EV and bus

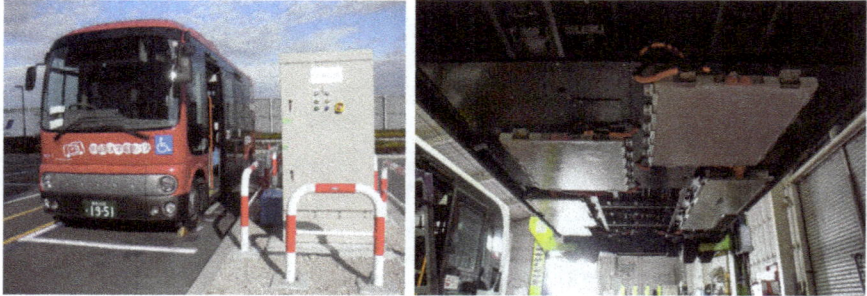

Figure 10.21   Wireless charging bus and secondary coils by Toshiba

Figure 10.22   OLEV at KAIST campus, Daejeon, Korea

10.4.3   Field evaluation in Korea

In Korea, KOREAN Wireless Power Forum (KWPF) was started in 2011; they have alliance with domestic companies and Standards Development Organizations (SDOs). But the forum almost excluded WPT charging of EV [14]. Study and field evaluation on EV bus in Korea are developed by Korean Advanced Institute of Science and Technology (KAIST). They are studying and developing both static and dynamic wireless charging of EV bus named "online electric vehicle (OLEV)." Figures 10.22 and 10.23 show their OLEV operation in their campus at Daejeon, Korea.

They demonstrated their OLEV at their campus during IEEE WoW in June 2015, Daejeon.

10.4.4   Field evaluation in China

China has continuously studied and begun standardization to use WPT for EV bus and passenger EV [15]. Their first report on WPT for EV was presented at UNECE/

Figure 10.23   Primary coil on road and secondary coils on the bus

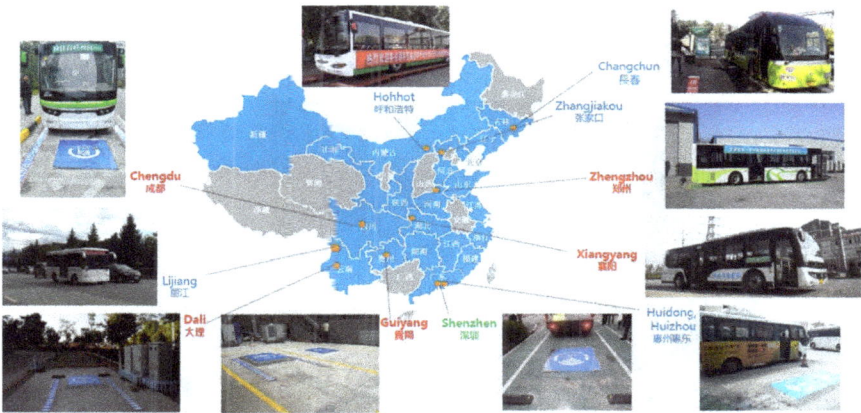

Figure 10.24   Cities of 38 strategical partnership with ZTE and EV bus

WP29, EVE 7th session Beijing, China, October 17–18, 2013 [16]. After 5 years of that report, ZTE developed WPT for bus and EV. They made strategical partnership with 38 cities in mainland China as shown in Figure 10.24.

ZTE China has developed conductive and wireless charging for EV and EV bus. Figures 10.25 and 10.26 are situations of their field trial at a laboratory located at Shenzhen.

According to Article 15, ZTE is actively promoting EV WPT standardization in China. ZTE sponsored Wireless Power Transmission Alliance, an open industry alliance oriented to developing WPT technology, group standards development, and market promotion. They also promote establishment of a joint working group of China Electricity Council (CEC) and National Technical Committee of Automotive Standardization (NTCAS), responsible for EV WPT GB/QC/NB standardization. ZTE promotes local standardization for EV WPT in several provinces and cities.

*Figure 10.25   Wireless parking primary and secondary coil for EV bus at ZTE
           Shenzhen*

*Figure 10.26   Wireless (right) and conductive (left) parking power supply unit at
           ZTE Shenzhen*

10.5   Regulation and standardization for WPT

For deployment and commercialization of WPT for EV, adopting regulation
relating wireless services is required. Both to keep safety for human body and to
keep minimum affect existing wireless service should be important. Domestic and
international regulations managed by local government is required. Ministry of
Internal Affairs and Communications (MIC) in Japan, Federal Communications
Commission (FCC) in the United States, European Telecommunications Standards
Institute (ETSI) and Comité Européen de Normalisation Électrotechnique
(CENELEC) in Europe are concerning regulatory work.

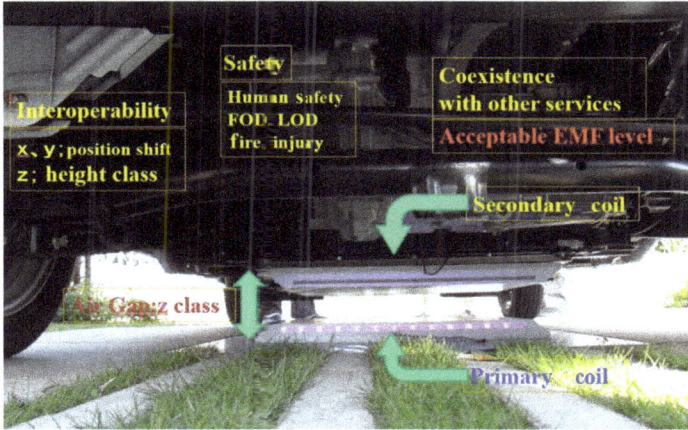

Figure 10.27 Major regulation, standardization aspects for EV WPT

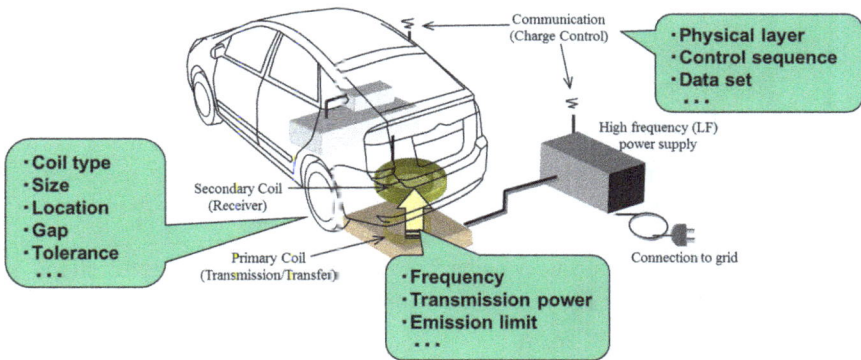

Figure 10.28 Standardization part for EV WPT

Standardization and regulation for passenger EV should be basically considered as shown in Figures 10.27 and 10.28. Current situation has been reviewed in a handbook by JSAE 2016 [17].

- Interoperability:
  Interoperable with different OEMs, power class, height class, and so on.
- Safety:
  Human safety, which is guide lined by International Commission on Non-Ionizing Radiation Protection (ICNIRP) should be considered. Effect to active implantable medical devices (AIMD), according to EU directive (90/385/EEC) should also be considered.

  Foreign object detection to avoid fire due to temperature rise should be important.

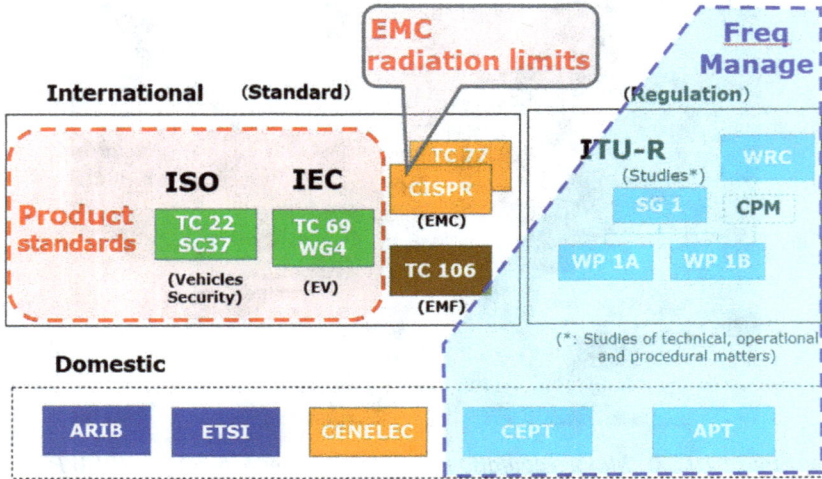

Figure 10.29   WPT relating international and domestic organizations

- Co-existing with other existing service:
  Due to WPT of magnetic coupling method using adequate frequency resource, it should be considered to keep radio act or frequency management rule.

It should also be considered to keep harmonization with existing radio service. WPT system should check emission level of not only the frequency in use for WPT but also harmonizing frequency (co-existing).

International and domestic standardization and regulation organizations are related to each other. The relation is shown in Figure 10.29.

10.5.1   Japan

In Japan, the MIC had continued regulatory modification work on WPT including EV applications from June 2013. The MIC released new rule of Radio Act for EV WPT in March 2016.

The discussion to modify radio act includes the following three WPT systems [18].

1. WPT system for EV/PHEV (79–90 kHz); magnetic coupling
2. WPT system for mobile (6.7 MHz bands); magnetic coupling
3. WPT system for mobile (400 MHz bands); capacitive coupling

The available frequency band was decided as "79–90 kHz" for EV WPT. Power level and emission level for EV WPT system are shown in Table 10.1. Graphed emission level is shown in Figure 10.30.

Standardization for the WPT system in Japan was developed by ARIB as ARIB STD-T113, wireless power transmission systems [19]. It was released in December 2015 as three types of WPT, including 400 kHz for mobile use,

Table 10.1   Power level and emission level for EV WPT system in Japan (2016)

Class	WPT power	Emission level out of frequency in use	Emission level frequency in use
3 kW class (private)	Up to 7.7 kW	68 4 dB μA @ 10 m	CISPR 11 Class B 10 dB mitigation at second to fifth harmonization frequency

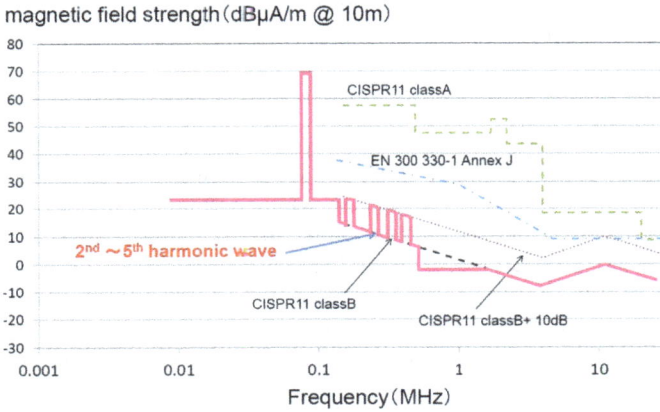

Figure 10.30   Emission level for EV WPT in Japan (2016)

6.78 MHz for mobile use and microwave (2.54 GHz band) surface transfer type of electric magnetic coupling. Magnetic coupling type using 85 kHz band for EV application is under development.

10.5.2   European standards for electricity supply

Based on discussion of IEC TC69 WG4 and ISO TC22 SC37, the European Commission released Commission Implementing Decision M/533 on December 3, 2015. In the document, it stated on p. 7 that "One technical solution for interoperability with a technical specification for wireless recharging for passenger cars and light duty vehicles compatible with the specification contained in IEC/TS 61980-3 Ed. 1.0, or its later edition."

And in p. 12 "APPENDIX I REQUESTED WORK PROGRAMME Publication due date December 31, 2019 where publication means that makes reference to the moment when the relevant ESO makes a standard available for its members or to the public."

According to M/533 request, European OEMs and Tier 1 supply companies have organized STILLE to prepare comment their opinion for IEC TC69 WG7, ISO TC22, SC37, and SAE.

Table 10.2 Finished and plan of WPT standard in China

Document	2015	2016				2017			
	Q3	Q1	Q2	Q3	Q4	Q1	Q2	Q3	Q4
General requirements	◆			◆	◆	CD	◆		
Specific requirements				◆	◆	◆	◆	D	CD
Communication protocols				◆	◆	◆	◆	D	CD
EMF				◆	◆	◆	◆	D	CD

Note: ◆: Finished by Q2 of 2017; D: Planned after Q2 of 2017.

10.5.3 China

China is preparing work for their standards for passenger EV in the long-term, till 2027, planning. Development planning of wireless charging system for passenger car market-driven, cross-start, reasonable planning, industrial services, and improved guidance role is under way [15]. They also work with standardization work with IEC/ISO and SAE.

Their progress and plan of EV WPT in drafts are shown in Table 10.2.

China's opinion for EV WPT standards showed in Table 10.2 includes the following topics.

- General requirements for EV WPT
- CD in Q2 of 2017
- Specific requirements for EV WPT
- Communication protocol between Vehicle Control Unit (VCU) and Power Supply Unit (PSU)
- Limits and test methods of EMF for EV WPT
- CD in Q3 of 2017

Their overall standards of WPT for passenger EV are shown in Table 10.3. They also have long-term time plan as shown in Table 10.4.

Development planning of wireless charging system for passenger car market-driven, cross-start, reasonable planning, industrial services, and improved guidance role!

10.5.4 IEC/ISO and SAE

Major IEC/ISO standards for WPT for EV as of April 2017 are shown in Table 10.5 [19].

The WPT system being standardized (as the first phase) (ISO 19363/IEC 61851-3/SAE J2954) is shown in Figure 10.31.

Discussion on the WPT system for passenger EV is based on the following key factors:

- WPT system intended to be used for passenger cars and light duty vehicles.
- WPT system using magnetic field (MF-WPT).

Table 10.3 Overall standards of WPT for passenger EV in China

系统与设备 System & Device	技术要求 Technical Requirements	电动汽车无线充电系统 通用要求	EV WPT General Requirements
		电动汽车无线充电系统 特殊要求	EV WPT Specific Requirements
	设备要求 Device Requirements	电动汽车无线充电系统 地面设备	EV WPT GA
		电动汽车无线充电系统 车载设备	EV WPT VA
接口 Interface	通信 Communication	电动汽车车载设备与无线充电设备 通信协议 and PSU	Communication Protocol between VCU and PSU
	互操作性 Interoperability	电动汽车无线充电系统 互操作性要求	EV WPT Interoperability Requirements
测试 Test	互操作性测试 Interoperability Test	电动汽车无线充电系统 通信一致性测试	EV WPT Communication Consistence Test
		电动汽车无线充电系统 互操作性测试	EV WPT Interoperability Test
	安全性测试 Safety Test	电动汽车无线充电电磁暴露限值与测试方法 WPT	Limits and Test Methods of EMF for EV WPT
		电动汽车无线充电系统 电磁兼容性	EV WPT EMC
		电动汽车无线充电系统 测试规范要求	EV WPT Test Specification
		电动汽车无线充电系统 地面设备测试规范	EV WPT GA Test Specification
		电动汽车无线充电系统 车载设备测试规范	EV WPT VA Test Specification
施工验收 Installation & Acceptance	充电站 Charging Station	电动汽车无线充电站 设计规范	EV WPT Charging Station Design Specification
		电动汽车无线充电站 工程施工和竣工验收规范	Installation and Acceptance Specification
运行维护 Operation & Maintenance	计量 Metering	电动汽车无线充电系统 电能计量	EV WPT Electricity Power Metering
	充电站 Charging Station	电动汽车无线充电系统 运行维护规范	EV WPT Operation and Maintenance Specification
		电动汽车无线充电系统 检修规范	EV WPT Check and Repair Specification

Table 10.4   Long-term time plan of WPT for passenger EV in China

Step	2015	2016	2017	2018	2019	2020	2021	2022	2023	2024	2025	2026	2027
System and device, interface	♦	♦♦	♦ D	DD	D								
System and device, test		♦	♦♦	♦♦ D	♦ DD	♦ DD	♦ D	D	D				
Engineering and acceptance					♦	D	♦		♦ D	D			
Operation and maintain						♦		D		♦	♦	D	D

Note: ♦: started year; D: estimated finish year.

*Table 10.5   Major IEC/ISO standards for WPT for EV as of April 2017 by Miki
[20]. TS, technical specification; PAS, publicly available
specification; IS, International Standard; TIR, technical information
report; RP, recommended practice*

Title	Estimated publication date	Estimated revision date	Committee
IEC 61980-1 Electric vehicle wireless power transfer (WPT) systems – Part 1: General requirements	2015/7 (IS 1st Ed)	2018/12 (IS 2nd Ed)	IEC/TC69
IEC 61980-2 Electric vehicle wireless power transfer (WPT) systems – Part 2: Specific requirements for communication between electric road vehicle (EV) and infrastructure with respect to wireless power transfer (WPT) systems	2018/08 (TS)	2019/6 (IS)	IEC/TC69
IEC 61980-3 Electric vehicle wireless power transfer (WPT) systems – Part 3: Specific requirements for the magnetic field wireless power transfer systems	2018/08 (TS)	2019/6 (IS)	IEC/TC69
ISO 19363 Electrically propelled road vehicles— Magnetic field wireless power transfer	2017/1 (PAS)	2018/12 (IS)	ISO/TC22/SC37
SAE/TIR J2954 Wireless Power Transfer for Light-Duty Plug-In Electric Vehicles and Positioning Communication	2016/5 (TIR)	2017/12 (RP) 2018/12 (IS)	SAE J2954TF

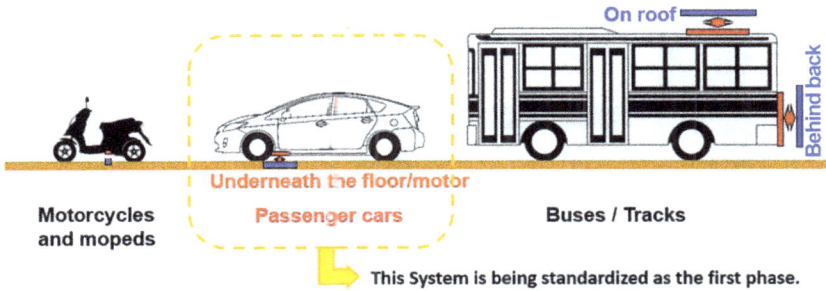

Figure 10.31 System being standardized as first phase

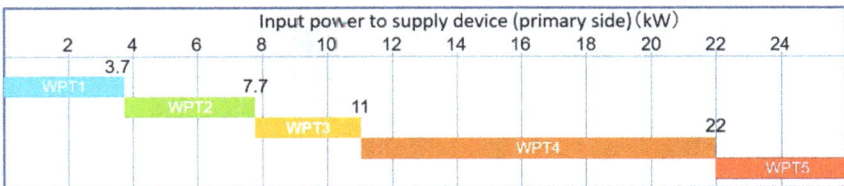

Figure 10.32 Power class for EV

- Primary device (infrastructure side ¼ supply side) is installed on the ground surface.
- Secondary device (vehicle side) is installed underneath the vehicle under body.
- 85 kHz band (79.00–90.00 kHz) has been proposed/standardized as the system frequency range for transferring power in ISO/PAS 19363:2016, SAE/TIR J2954:2016, and IEC 61980-3 Committee Draft.

Publicly available specifications (PAS), technical information report (TIR):

- Requirements for power classes of <3.7 kW, 3.7–7.7 kW, and 7.7–11 kW are being standardized.
- Existing communication technology (Wi-Fi) is used for command and control communication.

Discussion of the WPT system for heavy duty vehicles was started in SAE J2954 TF.

Classification according to input power to supply device (WPT input power class) defined by IEC, ISO, and SAE is shown in Figure 10.32.

To keep interoperability is important item for WPT operation under high efficiency, between different types of primary and secondary (on-vehicle) coil type, circular or Double D-shaped (DD) type, under condition of power classes and Z-class. Interoperability test condition for different coils by STILLE is shown in Figure 10.33.

Product sample of circular and DD coil by Qualcomm is demonstrated in Figure 10.34.

Bombardier proposed secondary (on-vehicle) coil primove is shown in Figure 10.35.

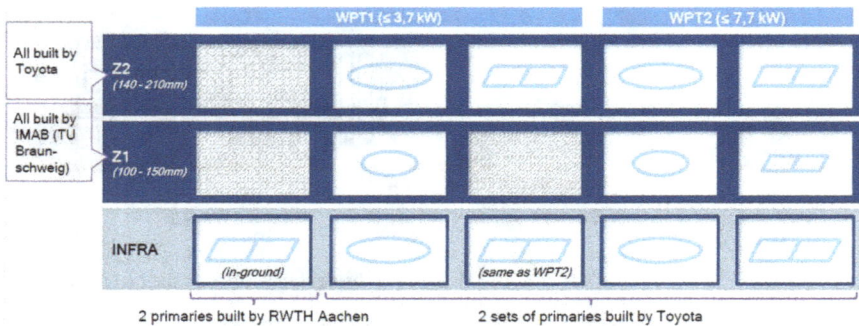

Figure 10.33   Interoperability test condition for different coils by STILLE

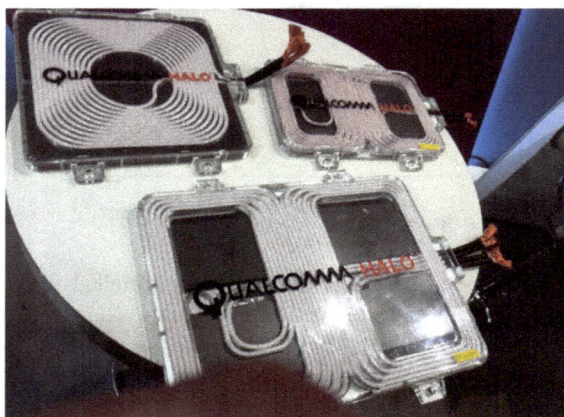

Figure 10.34   Circular and DD coil by Qualcomm

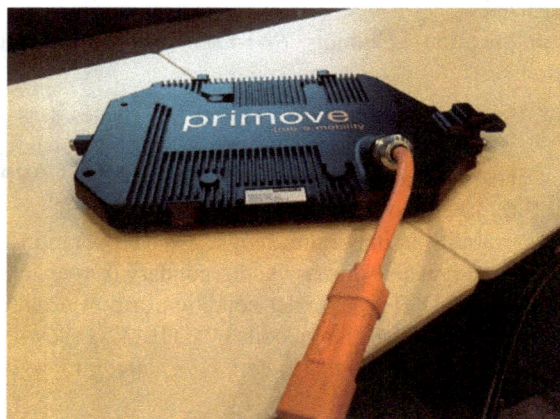

Figure 10.35   Secondary coil by primove

10.6 ITU activity on WPT; frequency allocation

ITU (founded in 1865) cites the purpose for the union such as to promote the extension of the benefits of the new telecommunication technologies to all the world's inhabitants.

ITU-R 210-3/1 Wireless power transmission (June 2015) decides that the following questions should be studied [1,21].

1. Under what category of spectrum use should administrations consider WPT: ISM or other?
2. What radio frequency bands are most suitable for WPT?
3. What steps are required to ensure that radiocommunication services, including the radio astronomy service, are protected from WPT operations?

10.6.1 2014: Approval of non-beam WPT report

According to discussion for Q. ITU-R 210/1 (WPT) in Study Group 1 (SG1)—Working Party 1A (WP1A) in 2014, new Report ITU-R SM.2303—Wireless power transmission using technologies other than radio frequency beam was released.

And in 2015, chairman's report (ITU-R SM.2303-1) and preliminary Draft of New Recommendation was released.

Requirements of new report ITU-R SM.2303 are described in Table 10.6.

According to the discussion of WPT in ITU in June 2017, the results are the following:

Draft New Recommendation ITU-R SM [WPT]

Considering that industrial alliances, consortia, and academia have investigated several frequency bands for WPT technologies, including 19–21 and 59–61 kHz for the shaped magnetic field in resonance EVs, 79–90 kHz for magnetic resonant technology for EVs, 100–300 kHz for magnetic resonant and induction technology for mobile devices, and 6765–6795 kHz for magnetic resonant technology for mobile devices.

Discussion of Draft New Recommendation ITU-R SM [WPT] will continue till 2018 assembly. Final target to modify RR for assign WPT frequency is at World Radio Conference (WRC) on October 19, 2019.

10.7 Coexisting with other wireless service (CISPR)

CISPR is one of the committees of IEC. Their mission is to avoid cross-interference between wireless services and make international harmonization of EMC.

The discussion on WPT started in 2013 plenary held at Ottawa, which is not yet recognized and initiated by Japan. Task Force (TF) has organized at B and F subcommittee.

Plenary meeting at Daejeon, May 2017. The Ad hoc Group 4 (AHG4) agreed to deliver the Committee Draft for Voting (CDV) of Class B Limit of magnetic

Table 10.6   ITU non-beam WPT system in Report SM.2303 (2) in 2017 [1]

	Magnetic resonance and/or induction for electric passenger vehicles	**Magnetic induction for heavy duty vehicles**
Application types	EV charging in parking (static)	Online electric vehicle (OLEV) (EV charging while in motion including stopping/parking)
Technology principle	Magnetic resonance and/or induction	Magnetic induction
Countries under consideration	Japan	Korea
Frequency range	42–48 kHz, 52–58 kHz, 79–90 kHz, and 140.91–148.5 kHz are in study	19–21 kHz, 59–61 kHz
Power range	3.3 kW and 7.7 kW; classes are assumed for passenger vehicle	– Minimum power: 75 kW – Normal power: 100 kW – Maximum power: on developing – Air gap: 20 cm – Time and cost saving
Application types	EV charging in parking (static)	OLEV (EV charging while in motion including stopping/parking)
Technology principle	Magnetic resonance and/or induction	Magnetic induction
Countries under consideration	Japan	Korea
Frequency range	42–48 kHz, 52–58 kHz, 79–90 kHz, and 140.91–148.5 kHz are in study	19–21 kHz, 59–61 kHz
Power range	3.3 kW and 7.7 kW; classes are assumed for passenger vehicle	– Minimum power: 75 kW – Normal power: 100 kW – Maximum power: on developing – Air gap: 20 cm – Time and cost saving

field strength for EV as shown in Figure 10.36. Voting limit of the CDV will be till December 1, 2017.

Some changes and relaxations from Japanese rule (Figure 10.30) are included at the proposed limit shown in Figure 10.36.

This class B limit is under voting and not yet decided at CISPR. After voting and then accepted, each domestic rule will be modified.

10.8   Human safety: IEC TC106 and ICNIRP

Human safety of wireless service is quite important. ICNIRP of WHO has released improved guidelines. MIC Japan has modified technical standard in Japan according to the guideline methods to test compliance to guideline [22].

Japan proposed compliance method with incident field and measure the field intensity in the vicinity of system without human as shown in Figure 10.37 [23].

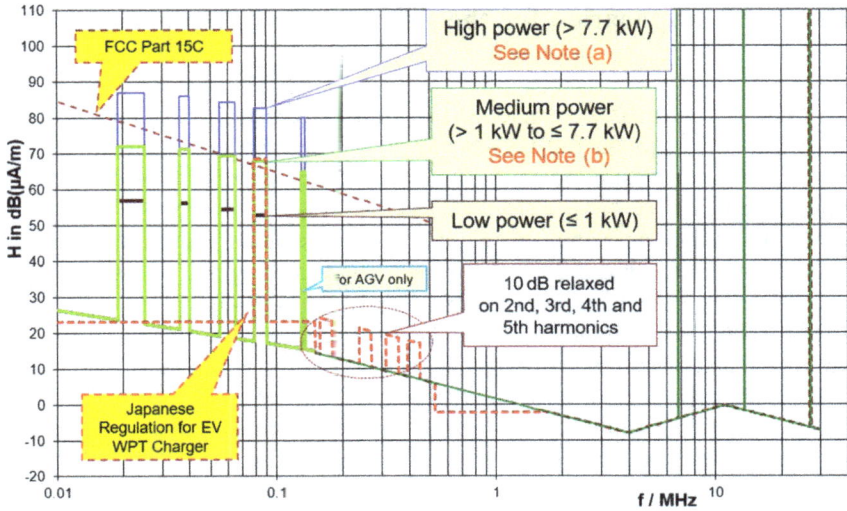

Figure 10.36 Class B limit of magnetic field strength for EV at CDV of CISPR. (a) Limits reduced by 15 dB shall be applied for installations where sensitive equipment in public space is used within a distance of 10 m. (b) For WPT system >3.6 kW the limits may be relaxed by 15 dB when no sensitive equipment is used within a distance of 10 m (to be documented in the manual).

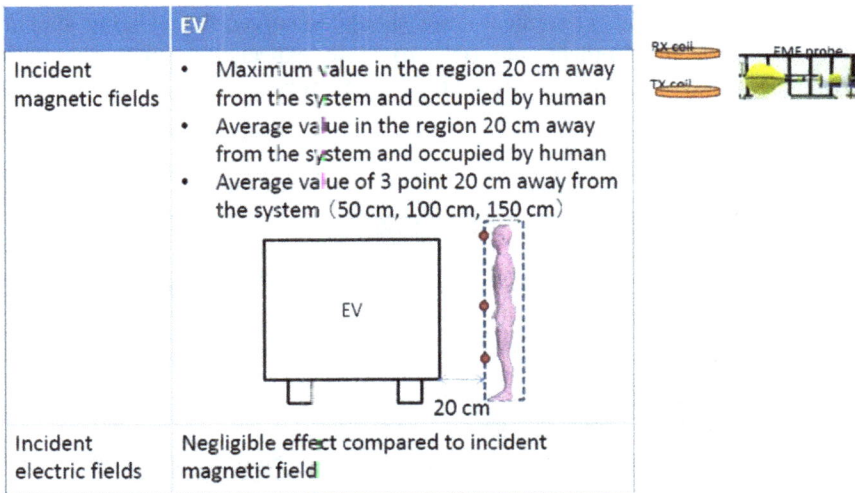

Figure 10.37 Compliance method with incident field for EV

10.9   WPT application for the future: dynamic charging for EV

According to Figure 10.4, vehicle-side charging forecast in September 2022 (by S&P Global 2022), product of EV with WPT function for static charging spot has released mid of 2022 at Korea and China and will be expanded for commercialization phase in 2030s.

Major and important applications of WPT for the public transportation service will be dynamic charging for EV from road. Highway England released report of feasibility study on dynamic power supply for EV at highway [24] (Figure 10.38).

The technology could allow EVs to be driven for longer distances, without the need to stop and charge their batteries.

In Japan, NEXCO is also reported at Japan highway in May 2015 [25].

In Japan, Tokyo University has developed unique wireless in-wheel motor and demonstrated dynamic charging from road [26]. Their EV deployed second-generation wireless in-wheel motor and road-side coil for dynamic WPT as shown in Figures 10.39 and 10.40.

One unique trial using electro capacitive coupling method, Toyohashi University of Technology, has developed battery less EV [27] in the campus at Toyohashi city.

The EV can drive without battery along the road where metal layer is buried to supply energy (Figure 10.41).

In Europe, VEDCOM and Qualcomm announced in June 2017 that they are starting evaluation of dynamic road. Trino Technical University also announced in July 2017 that they are starting "Charge while driving" at highway using the prototype system.

Some technical and economical items should be solved for realizing dynamic road. Technical development and field trial are expected to clear the items. Dynamic road, energy transfer from road to moving EV, will be one of killer application for EV generation.

Figure 10.38   Electric recharging lane planned by the UK

Figure 10.39   Second-generation wireless in-wheel motor

Figure 10.40   Road-side coil for dynamic WPT

Figure 10.41   Battery-less EV at Toyohashi University of Technology

10.10   Progress on standardization after ed1

International and national standardization and regulation status at the moment of May 2023 of wireless power transfer are listed (Table 10.7) [28].

10.10.1   IEC/ISO activity

NWIP (New Work Item Proposal) of IEC61980 was accepted in September 2010. After discussing with TC69 NP (National Party) members, IS of IEC61980-1 ed1 was released in July 2015. Ed 2 was released in November 2020 as IEC61980-2020. IS of IEC61980-3 for Magnetic Power transfer was released in November 2022. IS of IEC61980-2 for Control and Communication was released in May 2023. Then full set of IS for Static Power Transfer has completed after 13 years of first NWIP of IEC 61980 [29].

Title of IEC61980 series released IS are as follows:

IEC61980-1:2020; Electric vehicle wireless power transfer (WPT) systems – Part 1: General requirements.

IEC61980-2:2023; Electric vehicle wireless power transfer (WPT) systems – Part 2: Specific requirements for MF-WPT system communication and activities.

IEC61980-3:2022; Electric vehicle wireless power transfer (WPT) systems – Part 3: Specific requirements for magnetic field wireless power transfer systems.

The extension up to high power wireless power transfer, following PT is under working to prepare new IS.

PT 61980-4; Interoperability and safety of high power wireless power transfer (H-WPT) for electric vehicles.

NWIP for Dynamic Wireless Power Transfer was proposed in February 2019, from Korea and started discussion.

Table 10.7   Standardization and regulation status for EV (in May 2023)

Item	Org	Activity
International Standards (IS) for System	IEC TC69 WG7	IS; IEC61980-1 ed2 October 2020 IS; IEC61980-2 ed1 May 2023 IS; IEC61980-3 ed1 December 2022 WD;IEC61980-4 Under Discussion **WD;IEC61980-5(IEC63243) Under Discussion** **WD;IEC61980-6(IEC63381) will prepare**
IS for Vehicle	ISO TC22 SC37	IS; ISO19363-2020 in October 2020 **CD; ISO5474 series Under Discussion**
National Standard	SAE	J2954 202010 in October 2020 Light Duty J2954/2 in December 2022 Heavy Duty **J2954 for DWPT under WG**
	UL	Outline of Investigation; in March 2020
	China GB	GB/T 38775-1 to -7; by October 2021 GB/T 38775-8 at CDV, -9, -10 under WD
Freq. Manag.	ITU-R SG1	Rec.SM.2110-1 in November 2019 Pending RR at WRC
EMC	CISPR/3/ WG1 AHG4	CD; CISPR11/FRAG1 ed7 under circulate
Human Safety	ICNIRF	Blue book released April 1998 Revised for 1 kHz to 100 kHz at 2010
	IEC TC106	TR62905 (under 10 MHz) in February 2018 PT63184 for Basic Rec under discuss

IEC 63243 ED1; Interoperability and safety of dynamic wireless power transfer (WPT) for electric vehicles.

This project will be renumbered as IEC61980-5 after preparing CD.

Safety monitoring procedure for DWPT is under discussion as shown in Figure 10.42.

One of the configuration of the system proposed by Electreon is as shown in Figure 10.43.

In April 2021, another NWIP for control and communication of Dynamic Wireless Power Transfer was proposed from Korea and started discussion:

IEC 63381 ED1; Communication requirements of dynamic wireless power transfer (D-WPT) for electric vehicles.

Configuration of the system under discussion, which is proposed by Electreon, is as shown in Figure 10.44.

And KAIST also proposed system as shown in Figure 10.45.

NWIP of ISO19363 which defines vehicle side standards was accepted in February 2014 by proposal from Japan. After long discussion along with IEC 61980, It was released in April 2020 as ISO19363-2020.

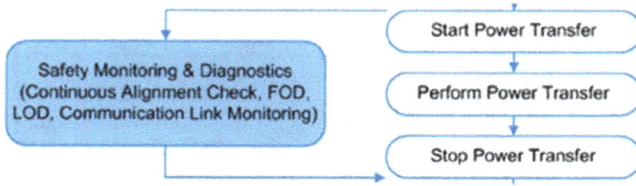

Figure 10.42   Safety monitoring procedure (under discussion)

General overview

The Electreon Dynamic Wireless Charging System (DWCS) comprises the following main components:

- The Management Unit
- The Coil Segments
- The Receiver

IEC-63243

Figure 10.43   Proposed system configuration (proposed by Electreon)

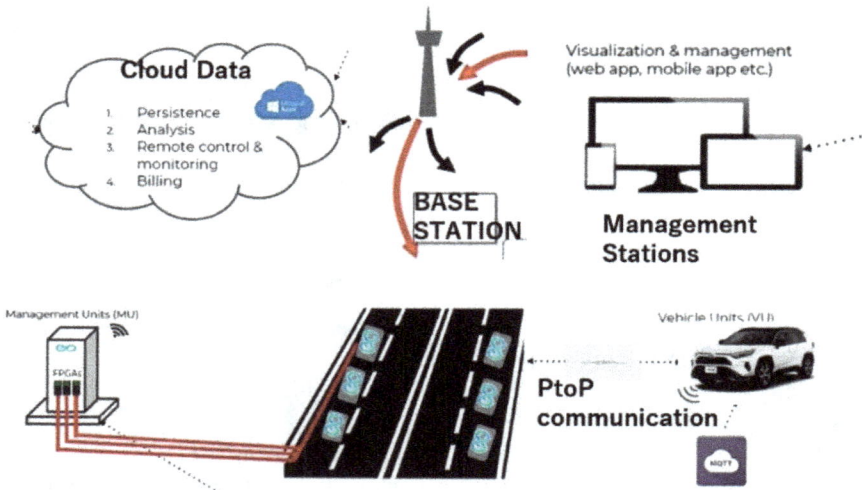

Figure 10.44   Proposed communication system (proposed by Electreon)

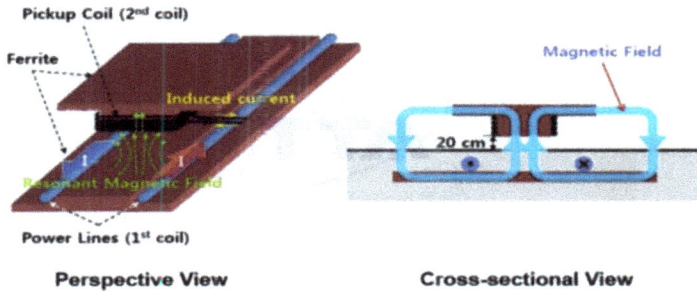

Figure 10.45 Proposed coir configuration (proposed by KAIST)

Title of ISO19363 released IS is as follows:

ISO19363:2020; Electrically propelled road vehicles – Magnetic field wireless power transfer – Safety and interoperability requirements.

After released IS, future edition of this ISO19363 has been moved to ISO 5474-4 which is one of the unified EV charging standard series and titled.

ISO/CD 5474-4; Electrically propelled road vehicles – Functional requirements and safety requirements for power transfer – Part 4: Magnetic field wireless power transfer – Safety and interoperability requirements.

For dynamic wireless power transfer on the vehicle side ISO standard,

ISO/DTS 5474-6; Electrically propelled road vehicles – Interoperability and safety of dynamic wireless power transfer (D-WPT) for electric.

These are under discussion since July 2023.

10.10.2 SAE and UL

SAEJ2954 has been discussed along with ISO/IEC discussion.

In October 2020 the standard for light duty vehicle was released.

J2954_202010; Wireless Power Transfer for Light-Duty Plug-in/Electric Vehicles and Alignment Methodology.

This enables light duty electric vehicles and infrastructure to safely charge up to 11 kW, over an air gap of 10 inches (250 mm) achieving up to 94 percent efficiency.

In December 2022, for wireless charging of heavy-duty EVs, SAE International publishes TIR J2954/2; static and dynamic WPT.

SAE International has published the first Technical Information Report (TIR) that specifies, in a single document, both the electric vehicle- and ground-system requirements for heavy-duty (HD) wireless charging of electric vehicles (EV).

SAE TIR J2954/2 Wireless Power Transfer & Alignment for Heavy Duty Applications helps pave the way for charging HD vehicles without the need for plugging in—widely considered to be a key enabler for accelerating the adoption of EVs and autonomous vehicles. The new guideline builds off the success of the first light-duty SAE J2954 standard published in 2020. The SAE TIR J2954/2

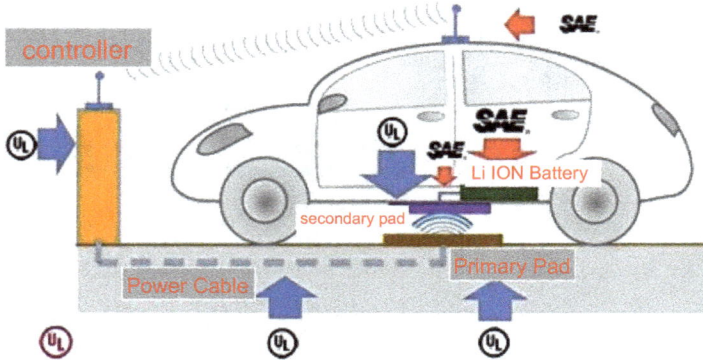

Figure 10.46   Sharing part of SAE and UL in the US

exponentially increases the power level of Wireless Power Transfer (WPT) for heavy duty vehicles to 500 kW.

In US, safety part of standard is prepared by UL. UL2750 covers safety aspect of WPT for EV (Figure 10.46). In March 2023, UL2750 OUTLINE, 2nd Edition was released. UL LLC Outline of Investigation for Wireless Power Transfer Equipment for Electric Vehicles. The first issue was released in May 2020. That is the Outline of Investigation for Wireless Power Transfer Equipment for Electric Vehicles, UL 2750, covers wireless power transfer (WPT) equipment for transferring power to an electric vehicle. WPT equipment consists of two devices, the power source and a ground assembly, as a minimum.

10.10.3   China GB/T

In China National standard GB/T 38775 for WPT covered for this area of techniques as GBT/T 38775-1 to -7 for static WPT was released in October 2021 (Figure 10.47).

Continuing standard such as communications and so on are under progress.

10.10.4   ITU

It was noted that the discussion of Draft New Recommendation ITU-R SM [WPT] will continue till 2018 assembly. Final target to modify RR for assign WPT frequency is at World Radio Conference (WRC) on October 19, 2019. Unfortunately WRC-23 held November, 2023 at UAE (Dubai) has decided to postpone modification of RR for WPT.

After then, assembly of 2019, released the following recommendation [30]:

SM.2110-1 (10/2019); Guidance on frequency ranges for operation of non-beam wireless power transmission for electric vehicles (Table 10.8) (Figure 10.48).

10.10.5   CISPR

At plenary meeting at Daejeon, May 2017, the Ad hoc Group 4 (AHG4) agreed to deliver the Committee Draft for Voting (CDV) of Class B Limit of magnetic field

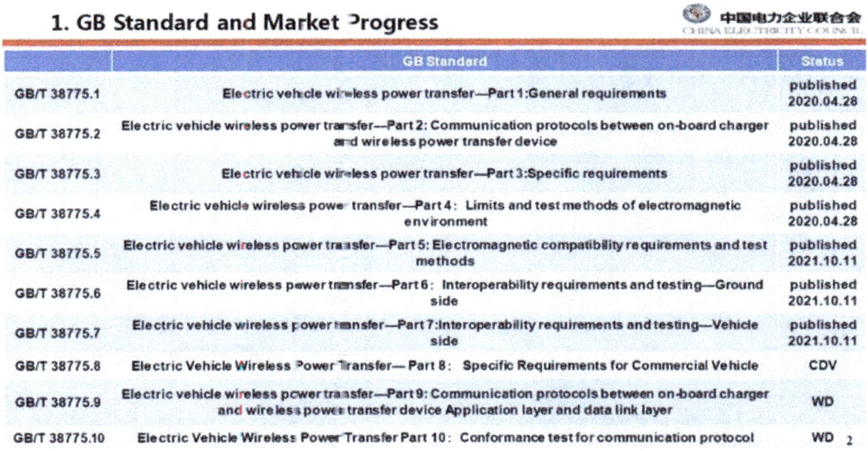

1. GB Standard and Market Progress

中国电力企业联合会
CHINA ELECTRICITY COUNCIL

	GB Standard	Status
GB/T 38775.1	Electric vehicle wireless power transfer—Part 1:General requirements	published 2020.04.28
GB/T 38775.2	Electric vehicle wireless power transfer—Part 2: Communication protocols between on-board charger and wireless power transfer device	published 2020.04.28
GB/T 38775.3	Electric vehicle wireless power transfer—Part 3:Specific requirements	published 2020.04.28
GB/T 38775.4	Electric vehicle wireless power transfer—Part 4: Limits and test methods of electromagnetic environment	published 2020.04.28
GB/T 38775.5	Electric vehicle wireless power transfer—Part 5: Electromagnetic compatibility requirements and test methods	published 2021.10.11
GB/T 38775.6	Electric vehicle wireless power transfer—Part 6: Interoperability requirements and testing—Ground side	published 2021.10.11
GB/T 38775.7	Electric vehicle wireless power transfer—Part 7:Interoperability requirements and testing—Vehicle side	published 2021.10.11
GB/T 38775.8	Electric Vehicle Wireless Power Transfer— Part 8: Specific Requirements for Commercial Vehicle	CDV
GB/T 38775.9	Electric vehicle wireless power transfer—Part 9: Communication protocols between on-board charger and wireless power transfer device Application layer and data link layer	WD
GB/T 38775.10	Electric Vehicle Wireless Power Transfer Part 10: Conformance test for communication protocol	WD

Figure 10.47   China GB/T for WPT

Figure 10.48   ITU-R recommendation SM.2110-1 (10/2019)

Table 10.8 Frequency range of operation of SM.2110-1

Rec. ITU-R SM.2110-1
Table 1
Frequency range for operation of non-beam
WPT systems for electric vehicles

Frequency range	Suitable non-beam WPT-EV
19–21 kHz	Magnetic induction technology or Magnetic resonant technology
55–57 kHz[1]	Magnetic induction technology or Magnetic resonant technology
63–65 kHz[1]	Magnetic induction technology or Magnetic resonant technology
79–90 kHz	Magnetic resonant technology

[1]Not to be used for the fundamental frequency of WPT-EV. Assuming a minimum separation distance of 50 m between WPT-EV and SFTS receivers, the third harmonic must fall within the 64–65 kHz and 55–56 kHz frequency range and the WPT emission be limited to 35 dBµA/m at 10 m. Where a separation distance of greater than 100 m between WPT-EV and SFTS receivers can be guaranteed, the third harmonic may fall within the 63–65 kHz and 55–57 kHz and the WPT emission be limited to 44 dBµA/m at 10 m.

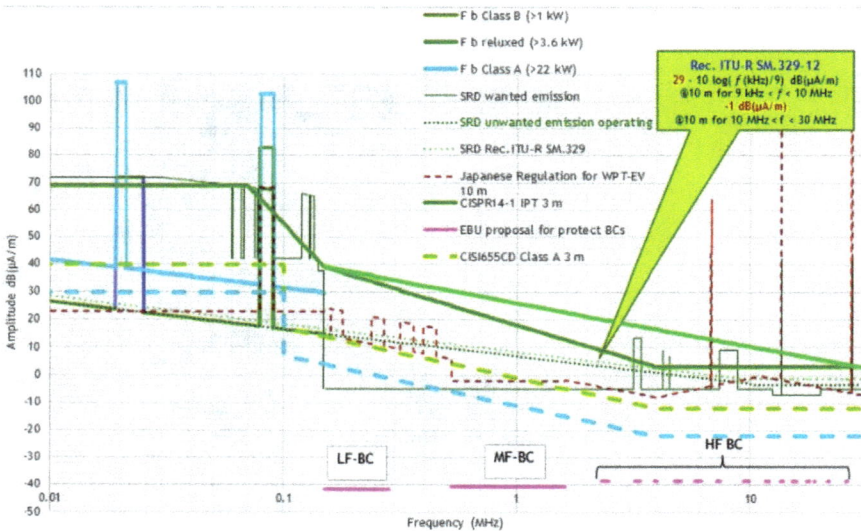

Figure 10.49 Emission requirements I/655/CD (January 2023)

strength for EV as shown in Figure 10.49. Voting limit of the CDV will be till December 1, 2017. Unfortunately the CDV failed to approve mainly because EBU (European Broadcasting Union) and IARU (International Amateur Radio Union) are requested to decrease emission level too low. Then the convener decided the item into fragment as shown in Figure 10.50 and restarted to make international agreement.

CISPR/B/WG1 AHG4 prepared modification and circulated in January 2023 as CIS/I/657A/CC.

Fragment a	Define method: B/763/CDV
Fragment b	Emission level <150 kHz
Fragment c	Introduce electric field strength
Fragment d	Emission level 150 kHz–30 MHz
Fragment e	Conductive level <150 kHz

Figure 10.50   Fragmentation of CISPR/B/WG1 AHG4

Revised Compilation of Comments on CIS/I/655/CD – CISPR 32 ED3: Electromagnetic compatibility of multimedia equipment – Emission requirements [31].

This document is under discussion in July 2023 and has voted but rejected.

After the rejection, the 5 years project limit has expired. But coviener has released CIS/B/828/RR (Review Report on CISPR11/Ed7) at Sept.2023. Then WPT-EVPJ(AHG4) has restarted and now acting to prepare CD.

10.10.6   IEC TC 106

TC106 scope describe that to prepare international standards on measurement and calculation methods to assess human exposure to electric, magnetic, and electromagnetic fields. The tasks include: characterization of the electromagnetic environments with regard to human exposure; measurement methods, instrumentation and procedures; calculation methods; assessment methods for the exposure produced by specific sources (in so far as this task is not carried out by specific product committees); basic standards for other sources; assessment of uncertainties. It covers the whole frequency range from 0 Hz to 300 GHz. It applies to basic restrictions and reference levels.

In February 2018 IEC TR 62905:2018 Exposure assessment methods for wireless power transfer systems has reported. Figure 10.51 shows the protection area of EV WPT.

IEC/TR 62905:2018(E) is a Technical Report. It describes general exposure assessment methods for wireless power transfer (WPT) at frequency up to 10 MHz considering thermal and stimulus effects. Exposure assessment procedures and experimental results are shown as examples such as electric vehicles (EVs) and mobile devices.

In June 2021 CDV (Committee Draft for Voting) of Assessment Methods Frequency Range of 3 kHz to 30 MHz was released; IEC/IEEE 63184 ED1 [32].

Assessment Methods of the Human Exposure to Electric and Magnetic Fields from Wireless Power Transfer Systems – Models, Instrumentation, Measurement

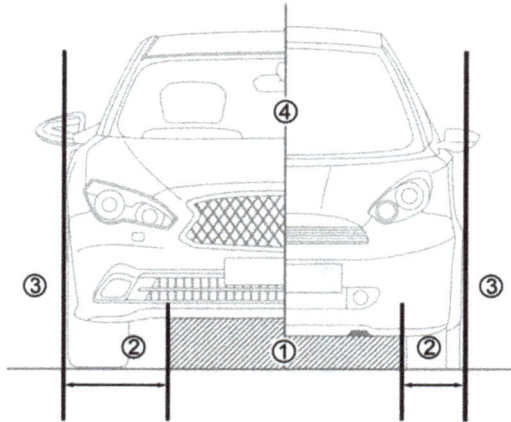

Figure 10.51   Protection area of EV WPT

Figure 10.52   Wireless charging for PHEV S5501e (2016)

and Computational Methods and Procedures (Frequency Range of 3 kHz to 30 MHz).

This is under discussion under JWG 63184 Human exposure to electric and magnetic fields from wireless power transfer systems.

10.11   Commercialization for mass production of passenger EV

Ed1 covers possibility of commercialization till 2016 that was ZTE (China) EV Bus project.

After that, at Paris motor show 2016 Mercedes Benz has announced the first series produced car with wireless charging for PHEV S5501e and will be released in the market 2017 (Figure 10.52). But it does not appear in the market.

In Japan, at Tokyo Motor Snow 2017, BMW has displayed their BMW530e with WPT (Figure 10.53) and showed video that demonstrated how it will be convenient.

BMW started pilot test at Germany from 2018 and also California from 2019 (Figure 10.54).

Unfortunately they are still open for commercial EV to install WPT function in Japan.

For non-commercial car, in 2018, Daihen has installed WPT function for patrol car in the Osaka Castle Park. The car is modified by Tajima Motors (Figure 10.55).

Figure 10.53   WPT by BMW (Tokyo Motor Show 2017)

Figure 10.54   Pilot test by BMW (California 2019)

Figure 10.55   Patrol car with WPT in the Osaka Castle Park (2018)

Figure 10.56   Display by Pues electronics at World Automotive Exhibition (2021)

At World Automotive Exhibition 2021 at Big-Site Tokyo, PUES electronics displayed WPT system with 3.3 kW (Figures 10.56–10.58). Pues is the only manufacture to get WPT type approval of MIC (Ministry of Internal Affairs and Communications) in Japan.

Unfortunately, in Japan commercialization for mass production of passenger EV was not started at the moment of March 2024.

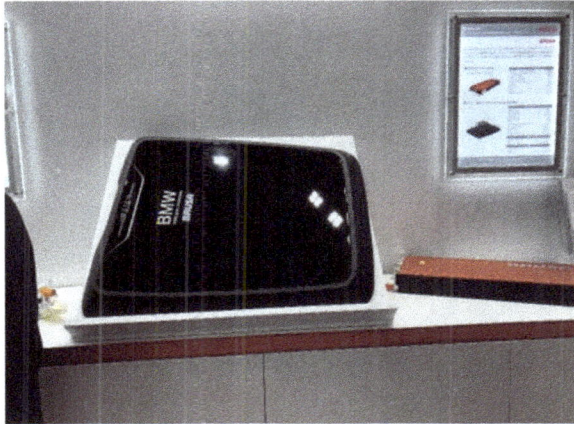

Figure 10.57 Type approved primary coil by BMW

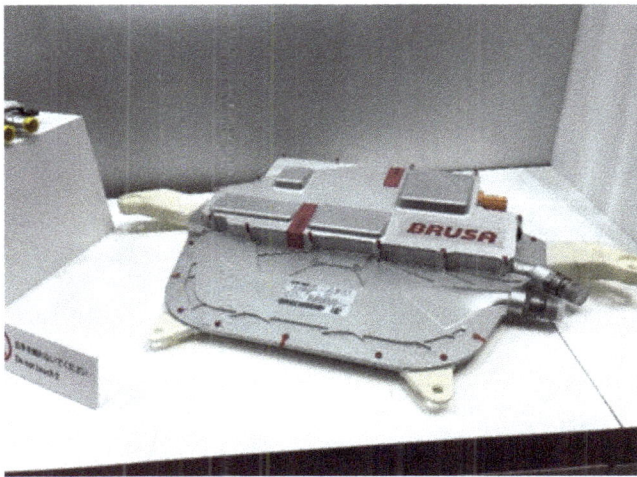

Figure 10.58 Secondary (on-board) coil by Brusa

In UK, for taxi application using multiple primary coils on the taxi pool, quasi dynamic charging was started at Nottingham in 2021 by WiCET (Wireless Charging of Electric Taxis) project [33].

This is the UK's first wireless charging electric taxi demonstration which is taking place in Nottingham. This trailblazing trial will see nine electric taxis retrofitted with wireless charging systems, and five wireless charging pads installed on the road at the Trent Street taxi rank, to demonstrate the application and impacts of wireless charging technology for electric vehicles (Figure 10.59).

WiCET Taxi

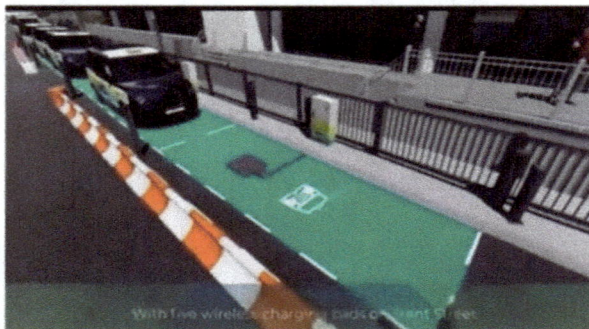

Figure 10.59   Taxi pool of WiCET at Nottingham

Figure 10.60   Genesis GV60 with wireless charging

At Korea in 2021, commercialization for mass production of passenger EV was started. Hyundai Genesis GV60 Gets Factory-Installed Wireless Charging Option (Figure 10.60). The wireless charging system is supplied by WiTricity [34].

China has also started commercialization for mass production of passenger EV from 2021.

FAW Hongqui E-HS9 and SAIC IM L7 with 11 kw WPT system was started in the market in 2021. BAIC Arc Fox and Zhiji L7 were introduced in 2022 (Figure 10.61) [35].

Launching in production

Figure 10.61 Launching production from 2020 of wireless charging

10.12 Field trials of DWPT for EV after 2015

The feasibility study in UK on 12 August 2015, England, to test charge-as-you-drive "electric motorways" "has been committed to is that by 2016 or 2017 they will hold off-road trials." The potential to recharge low emission vehicles on the move offers exciting possibilities (Figure 10.38).

During 2014–2018 EU Horizon support FABRIC (FeAsiBility analysis and development of on-Road charging solutions for future electric vehiCles) project for FS on Dynamic Wireless Transfer at France, Italy.

The project supported Framework: FP7 of EU and started on 1 January 2014 and ended on 31/12/2017 [36].

Figure 10.62 shows FABRIC France Test Course.

At France site, Qualcomm constructed test course installing dynamic wireless Power Transfer system.

Figure 10.63 shows Qualcomm trial course at FABRIC France site.

Primary Coils using Magnetic resonance method are filled in trench at road and it is filled in the trench (Figure 10.64).

DWPT zone is a part of test course and the layout is shown in Figure 10.65.

After FABRIC terminated at France site, from 2020 the project is continued as INCIT-EV project and promoted by VEDCOM a technical developing company of Renault group [37].

Objective and name has changed INCIT-EV. They have demonstrated at Velsilles as high-speed wireless loading on highways. With a power of 90 kW along 80 m, it is expected to be able to charge vehicles circulating up to 120 km/h (Figure 10.66).

Figure 10.62   FABRIC France Test Course

Figure 10.63   Qualcomm trial at FABRIC France site

Another demonstration of INCIT is in the city of Paris with a power of 120 kW over a length of 25 m (Figure 10.67).

Italian site of FABRIC was located at Trino and was terminated in December 2017.

In December 2021, Stellantis opened "Arena del Futuro" ("Arena of the Future") circuit built by A35 in collaboration with Stellantis and other partners, to

Figure 10.64   Primary coils filled in the trench

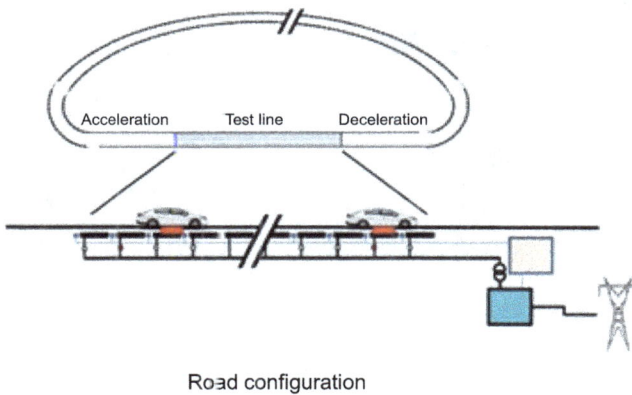

Road configuration

Figure 10.65   DWPT part of the test course

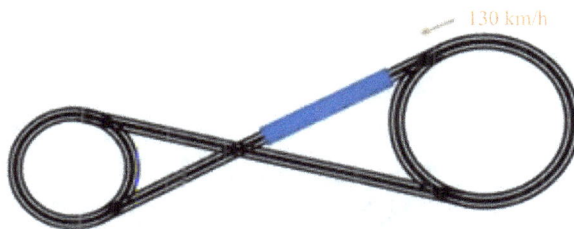

Figure 10.66   High speed DWPT part of the Test Course (INCIT EV)

field test revolutionary electric charging with DWPT (Dynamic Wireless Power Transfer). The 1,050-meter-long circuit is located in a private area of the A35 autostrada, near the Chiari Ovest exit, and is powered with an electrical output of 1 MW (Figures 10.68 and 10.69) [38].

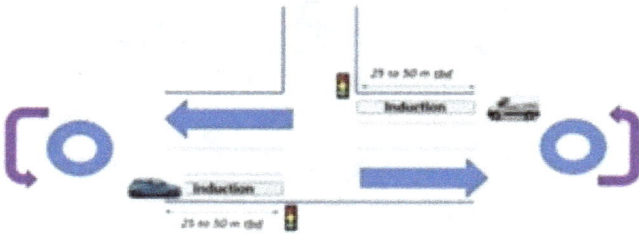

Figure 10.67　Demonstrations of INCIT is in the city of Paris

Figure 10.68　Arena del Futuro by Stellantis in Italy

Figure 10.69　Filling primary coils of wireless charging in trench of rod

In Japan, many trials and evaluation for DWPT and ERS with constructing Test course are undergoing [39].

In 2018, running demonstration of DWPT with WIM was shown at Kashiwanoha campus Chiba, Tokyo (Figure 10.70).

According to the technical base of DWPT with WIM from 2018, Mitsui Fudousan and University of Tokyo deployed for social implementation of DWPT at KOIL Mobility field from 2020 and they announced field test started in July 2023 [40,41].

Figure 10.71 shows KOIL mobility field for DWPT at Kashiwanoha Chiba (2020). Figure 10.72 is the fourth generation car with WIM and DWPT at KOIL mobility field.

Figure 10.70   Field trial of DWPT with WIM at Kashiwanoha 2018

Figure 10.71   KOIL mobility field for DWPT at Kashiwanoha Chiba (2020)

In October 2023, Kashiwa ITS Council and University of Tokyo has started field trial on public road near Kashiwano-Ha Campus station. The trial is supported by MLIT (Ministry of Land, Infrastructure, Transport and Tourism) Japan. [42]. This is the first field trial on public road in Japan [43].

They prepared two test EVs for the evaluation. Figure 10.73 shows one of the test EV for field evaluation.

Figure 10.74 is road side control unit for DWPT at Kashiwano-Ha campus station. Figure 10.75 is power supply zone of DWPT.

TUS (Tokyo University of Science) has deployed DWPT test course at NODA campus. It is designed work with renewable energy PV (Solar energy) [44]. Test EV, lane, and PV for DWPT by TUS are shown in Figure 10.76.

Figure 10.72   Fourth-generation car with WIM and DWPT at KOIL mobility field

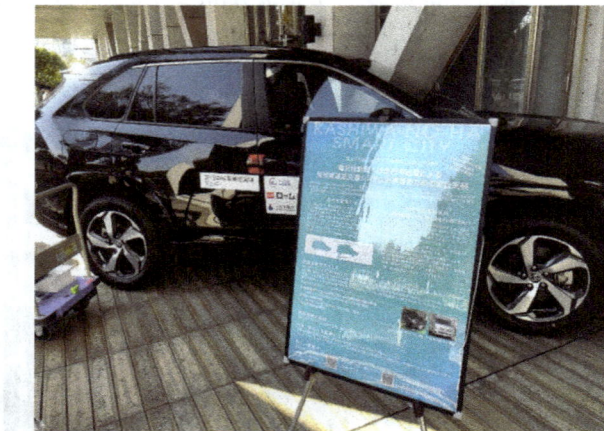

Figure 10.73   Test EV for field evaluation

In June 2022 in Japan, NEDO approved new project Smart Mobility Society Construction under GI (Green Innovation) fund with budget of ¥102bilion. The nearest target of the project is to prepare showcase at Osaka ECPO 2025. The DWT target concept of the project is as shown in Figure 10.77. Energy management

Figure 10.74   Road side control unit for DWPT

Figure 10.75   Power supply zone of DWPT

Figure 10.76   Test EV, lane and PV for DWPT by TUS

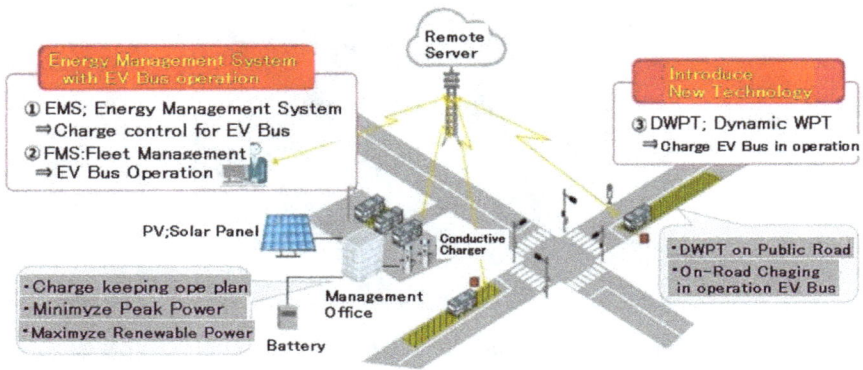

Figure 10.77   Concept of smart mobility society using DWPT

System with EV BUS operation will be developed. Also DWPT system will be introduced to charge under operation of EV BUS [45].

In March 2021, pre-evaluation of DWPT was experienced at Maishima Osaka Bay in the neighboring island of Yumeshima, main exhibition place of Osaka EXPO2025.

Figure 10.78 shows Maishima and Yumeshima at Osaka Bay for Osaka EXPO.

At Osaka EXPO225, DWPT system will operate using around 10 EV Bus prepared by EV Motors Japan. Plan for DWPT at EXPO was released. Figure 10.79 shows DWPT plan at Osaka EXPO225 at Yumeshima.

Another possibility of DWPT utilizing capacitive resonance method, plan of test course for DWPT is undergoing at north of Fuji-san, Fujiyoshida city, Yamanashi from 2021 [46].

Figure 10.78 Maishima and Yumeshima at Osaka Bay

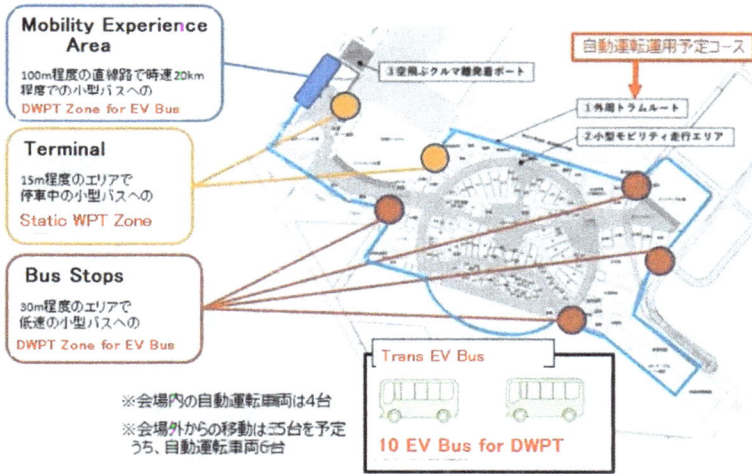

Figure 10.79 DWPT plan at Osaka EXPO225 at Yumeshima

10.13 European project CollERS1 and 2 (electric road system)

DWPT is a technology to supply power to e-mobility (passenger EV and/or EV Bus and Truck). The power transfer system consists of vehicle side on board coil work with primary coil layout under road. The road system for electrification is named "Electric Road System". Different energy transfer technology, conductive or wireless system with above road, on/beside road, in road system (Figure 10.80).

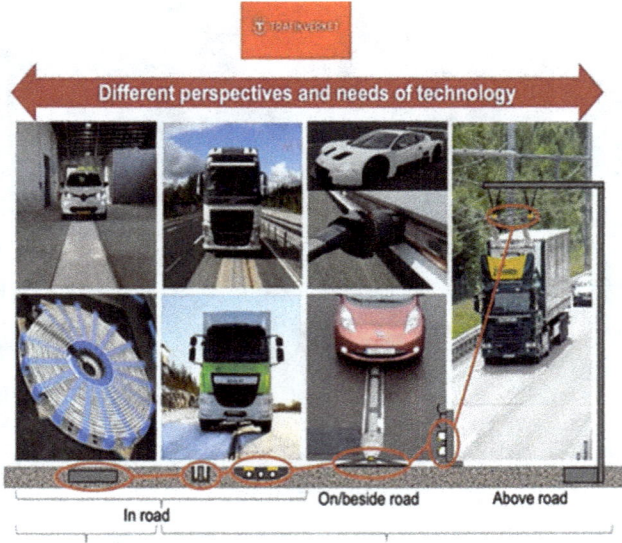

Figure 10.80   Different perspectives and needs of technology

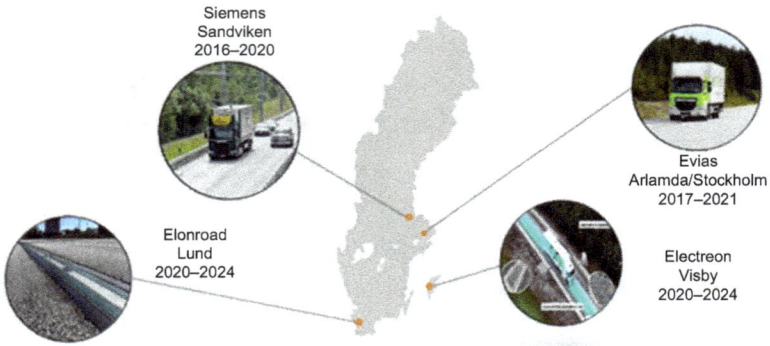

Figure 10.81   Four trial projects in Sweden from 2016 to 2024

Swedish Transport Administration managed four trial in Sweden from 2016 and needs of technology. The four trial projects in Sweden are shown in Figure 10.81 [46];

1. 2016–2020 Above road catenary by Siemens at Sandviken
2. 2017–2021 On road system by Evias at Arianda/Stockholm
3. 2020–2024 In road system by Elonroad at Lund
4. 2020–2024 In road Wireless system by Electreon at Visby Gotoland

During Swedish Transport Administration management of different trials for ERS, European Project has started [47].

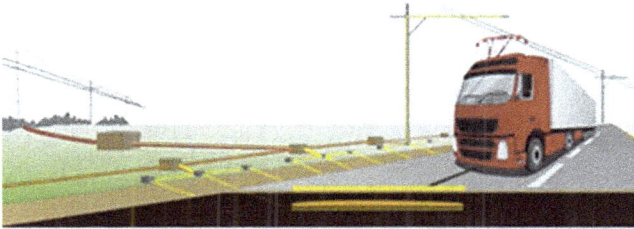

European Electric Road Systems' Symposium in Berlin

Symposium in Berlin, February 14–15, 2023

Figure 10.82 Concept of European road system

Swedish–German Research Collaboration on Electric Road Systems (CollERS) is managed during 2018–2019. The project builds on the official innovation partnership between Germany and Sweden that has been in place since 2017. Figure 10.82 shows the concept of European road system.

That overall goal of the CollERS I project was to increase the common knowledge around ERS by cooperation between Germany and Sweden, and to discuss potential long-term strategies for a successful transnational implementation of ERS in Europe.

The research collaboration "CollERS I" was carried out during the period 2018–2019 and consisted of core members from the Swedish Research and Innovation Platform for Electric Roads and the two national German research projects Roadmap OH-Lkw and StratON.

After the "CollERS I" was terminated the CollERS 2 continued.

The aim of the CollERS2 project is to compile results from field trials in Germany and Sweden on the use of electric road systems (ERS) for the decarbonization of road-based heavy goods transport and the associated accompanying research. These are to be presented and shared with European partners in a European context towards the joint development of long-term strategies for the cross-border implementation of ERS. In addition, the goal is to integrate other European partners into the ERS discussion during the course of the project.

The technical focus of the partners from Germany is on investigating the potential of overhead lines for ERS, as these are already undergoing practical testing in public spaces in the three field trials in Hesse, Schleswig-Holstein, and Baden-Württemberg. In addition, the potentials of conductor rails and inductive technologies will be investigated. The potential of ERS will also be compared to other zero-emission approaches in road freight transport, such as the use of fuel cells or purely battery-electric vehicles with a stationary charging infrastructure, and their synergies will be highlighted. In addition to the technology assessment, regulatory and economic aspects as well as aspects of social acceptance will also be taken into account.

As one result of the project, recommendations for action will be developed for a European expansion of ERS technologies. The starting point for this is a future European ERS network along major TEN-T corridors.

Duration: May 2021–September 2022 (Sweden)/May 2023 (Germany)

Mr. Jan Pettersson Director, Electrification Program, Swedish Transport Administration introduced in May 2023, Permanent ERS in Sweden planned to deploy by 2025. A first permanent ERS will be built on E20 Hallsberg – Örebro. 21 km long taken into use in 2025/26. The government has obtained documentation for regulation. Figure 10.83 shows permanent ERS plan at Sweden in 2025/26.

Figure 10.83   Permanent ERS plan at Sweden in 2025/26

Suggested ERS-roads in France

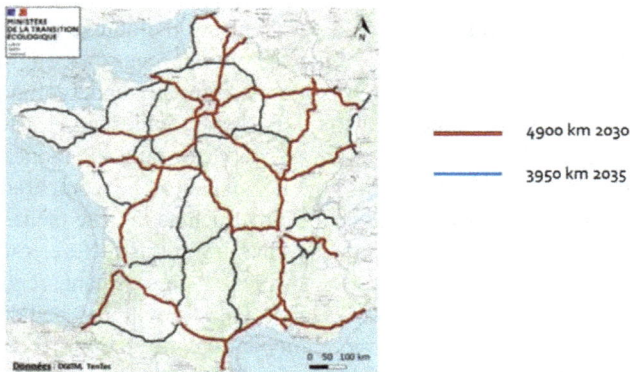

Figure 10.84   ERS-roads plan beyond 2030 and 2035 in France

CollERS 2 expands and cover with France. Sweden, Germany, and France agree to work together to enhance electric road systems technology, identify means for cross-border operability and campaign at European level for the wider spread of this technology.

France suggested ERS-roads plan beyond 2030 and 2035. Figure 10.84 shows ERS-roads plan beyond 2030 and 2035 in France.

10.14 Future possibility of WPT and DWPT for carbon-free society

These 5 years after ed1 was published, technology and standardization of WPT and DWPT have remarkable progress. German "National Platform for e-mobility" was planned at May 2010. At the moment target sale of EV was 1 million per year in Germany by 2020. Sales quantity in the world in 2022 was 10.2 million and 0.83 million in Germany. One of the major problem is still energy transfer from road side into vehicle. Conductive charging has continuously developed AC and DC mode and get high power up to 350 kW class. If when high power charging is achieved but still remain conductive using cable by power.

It is expected that the Wireless Power Transfer technique will lead cableless charge. It will take long time to solve difficulties of safety, coexistence, and mainly cost, and it started in 2022 in Korea and China for static wireless charging capability to mass production of passenger EV.

The roots of DWPT was US patented in 1894 by M. Hutin and M. Leblanc which was 3 kHz inductive WPT for electric train. Recently the request for carbon neutral society, DWPT for ERS has to attract attention to solve.

Many technical breakthrough shall be developed including not only wireless technology but also road construction, energy management, safety management using control and communication.

DWPT will be one of most important way to realize carbon-free society.

Current status and issues of DWPT and ERS in July 2023 in Japan have been reported at IECES2023 Shanghai and will realize good innovative result targeted for carbon neutral in 2050.

NIKKEI BP has published technical roadmap every year. It covers WPT(EV/PHEV) and ERS(DWPT) (48).

References

[1] N. Shinohara, "Current Research and Development of Wireless Power Transfer via Radio Waves and the Application [DML]," presented at WiPoT Symposium 2017, Shinshu Univ, June 2017.

[2] Y. Yokoi, "Evolution for e-mobility Charging in Europe, China and Japan," presented at IET Japan Network 2016; the second IET JN Workshop on ICT, July 15, 2016.

[3] Nissan, Zero-emission luxury car LE concept has displayed at New York international Auto Show (Japanese), https://newsroom.nissan-global.com/releases/120405-03-April 2012-JP [Accessed 17 July 2017].

[4] A. Thomson, 2018 Mercedes-Benz S550e will offer wireless EV charging technology built using Qualcomm Halo breakthroughs, October 2016, https://www.qualcomm.com/news/onq/2016/10/11/2018-mercedes-benz-s550e-will-offer-wireless-ev-charging-technology-built-using [Accessed 11 October 2016].

[5] Volker, "Wireless Charging 2020 – Interoperable and Standardized," Forum proceedings of 17 FORUM-Y16 of JSAE, 20174391, Yokohama, Japan, May 2017, pp. 7–10.

[6] T. Okada, "Wireless Charging Now", WiPoT Symposium, May 2023.

[7] Obayashi press release, "e-MoRoad for road mobility infrastructure in the future", News release September 22, 2022. https://www.obayashi.co.jp/news/detail/news20220926_1.html

[8] Y. Asakura, A. Nojima, T. Miki, and Y. Yokoi, Chap 10. *Automotive Engineering Handbook Volume 7*, Design for EV, Hybrid, JSAE 2016, pp. 535–542.

[9] Qualcomm News Release, First Electric Vehicle Wireless Charging Trial Announced for London, November 10, 2011, https://www.qualcomm.com/news/releases/2011/11/10/first-electric-vehicle-wireless-charging-trial-announced-london [Accessed 29 July 2017].

[10] J. Witkin, in London, Wireless charging for E.V.'s gets a power surge, November 18, 2011, https://wheels.blogs.nytimes.com/2011/11/18/in-london-wireless-charging-for-e-v-s-gets-a-power-surge/ [Accessed 29 July 2017].

[11] S. Takahashi, "Wireless Power Transmission for Public Transportation Service," Forum proceedings of 17 FORUM-Y16 of JSAE, 20174393, Yokohama, Japan, May 2017, pp. 20–26.

[12] N. Kiyomiya, "Smart house MEDIAS with resonant wireless charging by Mitsui Home, car watch, 11 September 2012," http://car.watch.impress.co.jp/docs/news/558803.html [Accessed 30 July 2017].

[13] Y. Yokoi, Roadmap and market trend of WPT for EV, frontiers of research and development of wireless power transfer, pp. 113–126, CMC publication, August 2016.

[14] M. Minakata, "Trend of International standardization for WPT of EV/PHEV," *OHM*, Vol. 102, No. 5, pp. 7–10, May 2015.

[15] S. Obayashi, A. Matsushita, and M. Ishida, "85 kHz band wireless power transfer technologies toward practical realization of contactless charging of electric vehicles and buses," *Toshiba Review*, Vol. 72, No. 3, 2017.

[16] KWPF, "Wireless power conference 2012," Proceedings of KWPF conference, Seoul, November 29, 2012. China Electric Power Research Institute, "Status and Development of China's EV WPT Standards System for SAE J2954 WPT & Alignment Taskforce" presented at SAE meeting, June 2017.

[17]   ZTE, Wireless Power Charging Is Ready for Cars NOW, EVE-07-09e, EVE 7th session, Beijing, China, October 17–18, 2013. https://wiki.unece.org/pages/worddav/preview.action?fileName¼EVE-07-09e.pdf&pageId¼12058681 [Accessed 4 October 2017].

[18]   Y. Yokoi, "Standardization for EV WPT," *Automotive Engineering Handbook Volume 7*, Design for EV, Hybrid, JSAE 2016, pp. 542–546.

[19]   Y. Yokoi, "Regulation improvement and standard for EV-WPT in Japan," presented at IEEE PELS, Workshop on Emerging Technologies: Wireless Power (WoW), Daejeon, Korea, June 5, 2015.

[20]   ARIB Standard STD-T113 version 1.1, Association of Radio Industries and Businesses, December, 2015.

[21]   T. Miki, "Status of standardization activities on wireless power transfer systems for electric vehicles," Forum proceedings of 17 FORUM-Y16 of JSAE, 20174390 Yokohama, Japan, May 2017, pp. 1–6.

[22]   S. Kobayashi, "Status of ITU's study on frequency-related subjects for wireless power transmission," Forum proceedings of 17 FORUM-Y16 of JSAE 20174392, Yokohama, Japan, May 2017, pp. 7–10.

[23]   K. Wake, "Human safety of WPT; technical standard in Japan," Presented at 2015 IEEE WoW, IEEE PELS Workshop on Emerging Technologies: Wireless Power, Daejeon, Korea, June 5–6, 2015.

[24]   GOV.UK, "Off road trials for 'electric highways' technology," 11 August 2015, https://www.gov.uk/government/news/off-road-trials-for-electric-highways-technology [Accessed 30 July 2017].

[25]   M. Sato, "Dynamic power supply for high way Japan," *OHM*, Vol. 102, No. 5, pp. 25–27, May 2015.

[26]   H. Fujimoto, T. Takeuchi, K. Hata, T. Imura, M. Sato, and D. Gunji, "Development second generation wireless in-wheel motor with dynamic wireless transfer," 2017505C, proceedings of 2017 JSAE Annual Congress (spring), May 2017, pp. 277–282.

[27]   T. Ohira, "Battery less EV," Forum proceedings of 17 FORUM-Y16 of JSAE, 20174395 Yokohama, Japan, May 2017, pp. 7–10.

[28]   Y. Yokoi, "World wide Tendency of Energy Transfer for e-mobility–Static and Dynamic Power Transfer– "IEICE Technical Report, WPT 2022-21 (2022-10).

[29]   ITU R, "Guidance on frequency ranges for operation of non-beam wireless power transmission for electric vehicles", Recommendation ITU-R SM. 2110-1 (10/2019)".

[30]   CISPR, "Revised Compilation of Comments on CIS/I/655/CD – CISPR 32 ED3: Electromagnetic compatibility of multimedia equipment – Emission requirements", CIS/I/657A/CC, 2023-01-06.

[31]   IEC TC106 JWG 63184, "Human exposure to electric and magnetic fields from wireless power transfer systems, 2014.

[32]   WiCET, "Wireless Charging of Electric Taxis", https://wicet.co.uk/

[33]   David Schatz, "EV Wireless Charging: Global Update", Plenary speech EVTeC2021, Yokohama, May 2021.

[34] T. Okada, "Wireless Charging is Now", 20231008, EVTeC2023, Yokohama, Japan, May 2023.

[35] FABRIC, "FeAsiBility analysis and development of on-Road chargIng solutions for future electric vehiCles", Grant agreement ID: 605405, January 2014 to June 2018

[36] INCIT EV, "Large demonstratIoN of user CentrIc urban and long-range charging solutions to boosT an engaging deployment of Electric Vehicles in Europe", https://www.incit-ev.eu/about-the-project/

[37] Stellantis press release, "Arena del Futuro", Innovative Dynamic Induction Charging Becomes a Reality", December 2, 2021.

[38] Y. Yokoi, "Current Status and Issues of DWPT and ERS in Japan", 000076, IESES2023, Shanghai July 2023.

[39] Univ. of Tokyo press release, "Start DWPT evaluation at KOIL MOBILITY FIELD for EV", October 8, 2021. https://www.k.u-tokyo.ac.jp/information/category/press/8679.html

[40] Mitsui Fudoan press release, "Deploy DWPT course and start evaluation at KOIL MOBILITY FIELD with Univ. of Tokyo", July 3, 2023 https://www.mitsuifudosan.co.jp/corporate/news/2023/0703/

[41] Univ of Tokyo, "Japan First Public Road Evaluation Program of ERS at Kashiwano-Ha", Press release, October 3, 2023 https://www.k.u-tokyo.ac.jp/information/category/press/10514.html

[42] TUS press release," Succeed real car trial of DWPT with PV", March 24, 2022 https://www.tus.ac.jp/today/archive/20220323_2468.html

[43] Y. Yokoi, "Current Status and Issues of DWPT and ERS in Japan", IESES2023 Shanghai, 000076, July 2023.

[44] NEDO GI fund, "Smart Mobility Society Construction", July 2022. https://green-innovation.nedo.go.jp/en/project/smart-mobility-society/

[45] Y. Yokoi, R. Mizouchi, T. Ohira, *et al.*, "Challenge of Innovative Electric Roadway System Featuring Capacitive Coupling Wireless Power Transfer", 20231061, EVTeC2023, Yokohama, Japan, May 2023.

[46] Jan Pettersson, "Sweden plans for electric Road Systems, ERSE", 20231006_PS6, EVTeC2023, Yokohama, Japan, May 2023.

[47] CollERS, "Project Outline", July 2023 https://electric-road-systems.eu/e-r-systems/project.php

[48] Nikkei BP, "Technical Road Map 2024-2033", 8-8 WPT(EV/PHEV), 8-9 ERS(DWPT) November 2023.

Chapter 11

Applications of long-distance wireless power transfer

Naoki Shinohara[1], Yoshiaki Narusue[2] and Yoshihiro Kawahara[3]

11.1 Introduction

In principle, the distance of wireless power transfer (WPT) via radio waves from a transmitting antenna to a receiving one can be extended if the medium of wave propagation is lossless, e.g., vacuum in space. Technically, the distance limitation of WPT via radio waves is dominated by the power required by users as well as the limitation of system size and efficiency in practical applications. It is easy to develop multiuser WPT systems using WPT via radio waves because they are electromagnetically uncoupled and no interference between circuit parameters at a transmitter occurs even with an increasing number of users. Besides, since the beam efficiency between a transmitting antenna and a receiving one is not high, multiuser WPT systems are also suitable when WPT systems via radio waves are applied in far field.

Ambient radio waves for wireless communication or broadcasting in far field can be gathered and used as electricity using rectennas and rectifying antennas at the user end. It is named the energy harvesting system or energy scavenging system for ambient radio waves (Figure 11.1(a)), which is considered a passive WPT system because the transmitter is not a special one only for WPT but for wireless communication or broadcasting as well. The ambient radio waves for wireless communication or broadcasting are usually modulated to add information.

The modulated radio waves can be used as wireless power as well. When a special transmitter for a WPT system is applied in far field, multiusers can receive high wireless power because the transmitted wireless power can be actively controlled by them. Such WPT system is considered being ubiquitous (Figure 11.1(b)). The term "ubiquitous" means existing or being everywhere, especially at the same time. The so-called ubiquitous information society implies that information can be

[1]Research Institute for Sustainable Humanosphere, Kyoto University, Japan
[2]Graduate School of Engineering, The University of Tokyo, Japan
[3]Graduate School of Information Science and Technology, The University of Tokyo, Japan

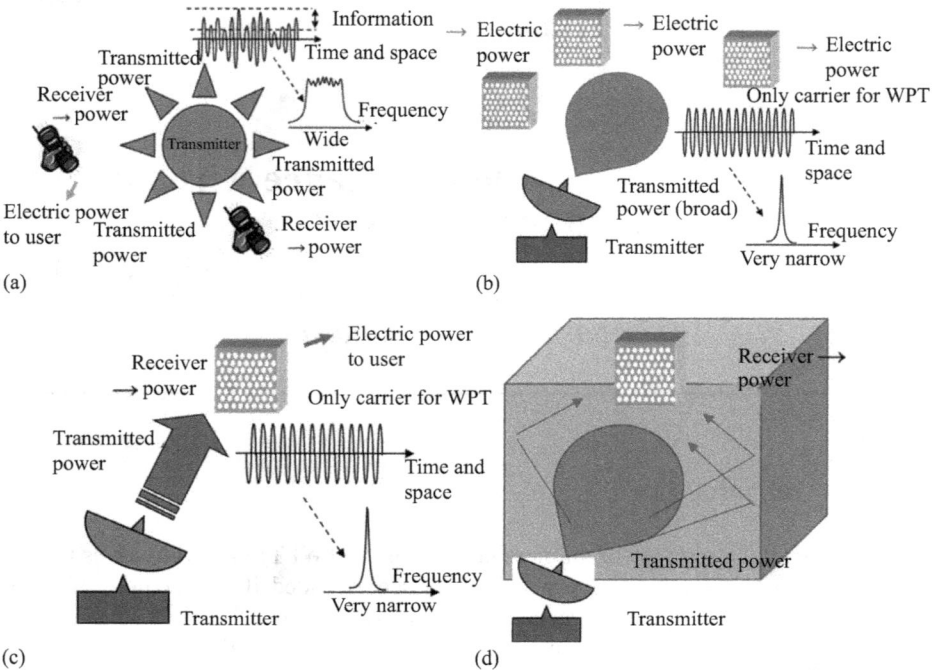

Figure 11.1   Different types of WPT via radio waves: (a) energy harvesting or energy scavenging system; (b) ubiquitous WPT; (c) beam WPT; and (d) WPT in shielded field

used anywhere and anytime. Smartphones and Wi-Fi networks are very close to a ubiquitous information society; however, this is still not enough. Wireless power can also be used anywhere and anytime in ubiquitous WPT systems, which means that power evolution, i.e., electricity, is available unconsciously. The radio frequency identification (RFID) tag is a ubiquitous WPT system where information and power are simultaneously transmitted wirelessly at 920 MHz and 2.45 GHz, respectively. The RFID tag is applied only below 1 W using an antenna below 6 dBi, which is standardized by the International Organization for Standardization/ International Electrotechnical Commission. However, ubiquitous WPT systems are more versatile than the RFID technology.

When high power and efficiency are required for WPT systems via radio waves, instead of wired power lines, a receiver is needed in the radiative near field of a transmitting antenna (Figure 11.1(c)). As described in Chapter 7, when WPT systems via radio waves are in the radiative near field, 100% beam efficiency can be realized theoretically. The beam efficiency depends on the distance between a transmitting antenna and a receiving one, antenna apertures, and frequency. The peak power density in WPT systems in the radiative near field sometimes is high at the center of a transmitting antenna. However, the safety issue of exposure of radio waves to human bodies must be considered. It is easy to consider the coexistence problem between WPT and wireless communication in WPT systems in the radiative near field because the radio wave power is mainly focused at the receiver and the leakage of radio waves is smaller than that in WPT systems in far field. In addition, it is easy to apply WPT systems to moving/flying targets when a phased array antenna, which can electrically control beam direction and form, is applied.

Theoretical requirements for antenna size and distance to realize high beam efficiency can be larger than what we imagine, which depends on radio wave expansion by theory for Maxwell's equations. Sometimes, it is not a practical WPT system. In this case, WPT systems via radio waves should be placed in a shielded field, which can reflect radio waves (Figure 11.1(d)) to increase beam efficiency. The shielded field is very effective to decrease the interference with conventional wireless communication systems and to reduce the risk of radio waves to human bodies.

11.2 Long-distance WPT in far field

11.2.1 Energy harvesting and scavenging

Radio waves can be applied not only for wireless communication and broadcasting but also for WPT as well. Their requirements for wave spectrum and power density of radio waves, however, are different. Broad and narrow wave spectra are needed for wireless communication and WPT systems, respectively. Besides, power density can be low enough to carry information for wireless communication, which must be enhanced for WPT. Figure 11.2 shows the measured wave spectrum and

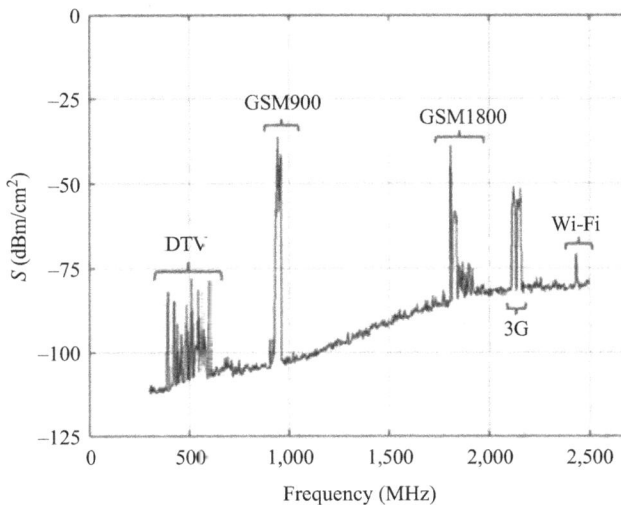

Figure 11.2 Measured wave spectrum and power density of ambient radio waves in London [1]

power density of ambient radio waves in London [1]. A rectifier that is expected to drive only by the harvested radio wave power requires a sufficient voltage under a low-power density circumstance. A charge pump rectifier is frequently applied for rectennas for energy harvesting systems (Figure 11.3). The charge pump, combining with some double voltages, is a good rectifier that can rectify radio waves and boost the voltage simultaneously. RF–DC conversion efficiency, however, is not high enough. Therefore, a good rectenna that can rectify wideband and weak radio waves must be developed to harvest or scavenge the modulated radio waves around us.

A broadband antenna that has been widely applied for wireless communication and broadcasting systems is required at first to develop the wideband rectenna. A dual-circular-polarized spiral antenna array as a rectenna active over a range of 2–18 GHz with single- and multitone incident waves was developed as a broadband antenna for an energy harvesting system [2]. Dual- and triple-band antennas are often developed for a broadband rectenna [3,4].

In the case of ambient energy harvesting from TV broadcasting radios, audio and video streams are sent at digitally modulated closely spaced carrier orthogonal frequency-division multiplexing frequencies referred to as orthogonal frequency-division multiplexing, rather than a single narrow-band frequency. Under integrated services digital broadcasting standard, HDTV channels use 6 MHz wide channels each between frequency bands of 470 and 770 MHz. Each 6 MHz channel has 5,617 carriers at 0.99206 kHz frequency spacing. Conventional energy harvesters especially that are based on RFID-based systems are optimized for high sensitivities at a single tone within the RFID operating bands between 862 and 928 MHz. In most RFID systems, the wireless power is focused within a single 250 or 500 kHz frequency channel. This difference in modulation method significantly affects the efficiency of RF–DC conversion, and thus, the use of a conventional charge pump is not suitable. In order to take advantage of the ambient TV wireless

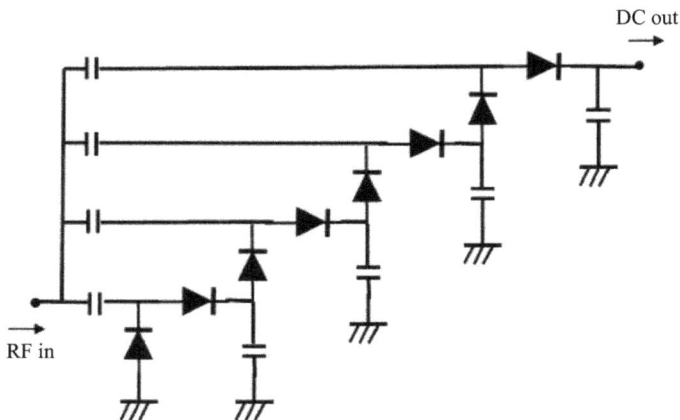

Figure 11.3   Charge pump rectifier

power, the proper design and integration of an RF front-end and embedded software are essential. A six-element log-periodic array with an antenna element scaling factor (τ) of 0.94, a relative element spacing (σ) of 0.18, and dipole lengths of between 24 and 30 cm is determined to yield maximum gain and bandwidth between 500 and 600 MHz for an A-4 size array structure [5]. A log-periodic antenna is used to maximize the gain and return-loss bandwidth based on [6,7] (Figure 11.4).

A broadband rectifier is required as well, which is based on a broadband impedance-matching technique. The related circuit is frequently applied for energy harvesting systems, while a developed one for a broadband rectifier is shown in Figure 11.5 [8]. Besides, the developed broadband rectifier with a broadband matching circuit at an output filter is shown in Figure 11.6 [9]. To achieve the rectification of broad radio waves, broadband resonators of a sector-distributed constant line (open stub) as class F load are put at the output filter, instead of resonators of a normal sector-distributed constant line (open stub).

Sensors for Internet of Things (IoT) and other low-power sensor devices are expected and suitable as energy harvesting applications [10–12]. A thermo-sensor is often applied to demonstrate a developed energy harvesting system [13,14]. The wireless power source is not only from Wi-Fi, mobile phone, Bluetooth, and TV, but also from the leakage of a microwave oven [14]. The ambient power from not only radio waves but also solar light, heating, and vibration is harvested. The term

Figure 11.4   An ambient wireless power harvesting from a TV tower in Tokyo [7,8]

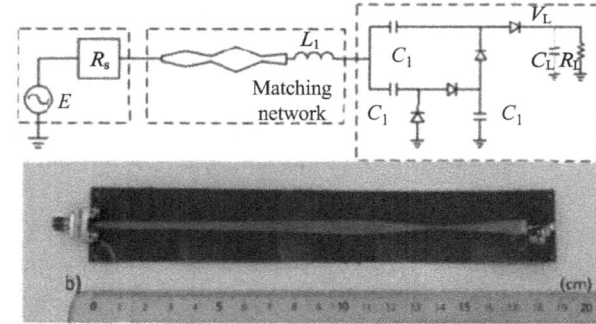

(a)

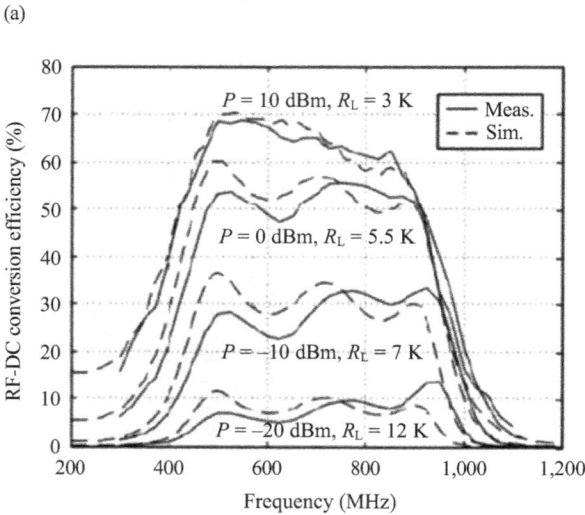

(b)

*Figure 11.5   One octave bandwidth based on nonuniform transmission line for UHF
rectifier: (a) equivalent circuit and developed rectenna; (b) simulated
and measured RF–DC conversion efficiency of rectifier [8]*

"energy harvesting" covers all power generation technologies from distributed
power around us. We can choose a suitable ambient power source or combine
two ambient power sources, e.g., sunlight and radio waves, for energy harvesting
systems [15].

Typical wireless sensor network platforms constitute of a microcontroller,
flash memory, RF transceivers, and an A/D converter for sensors. These typically
consume several dozen milliwatts for operation. Although the power consumption
of each electrical component is low, system-level power consumption amounts to a
few tens of milliwatts. This power level is not easily found in such ambient energy
sources. The technical challenges to harvest sufficient amount of energy out of the
ambient power source are duty cycling of the system. Typical low-power

Sector-type class F load

(a)

(b)

(c)

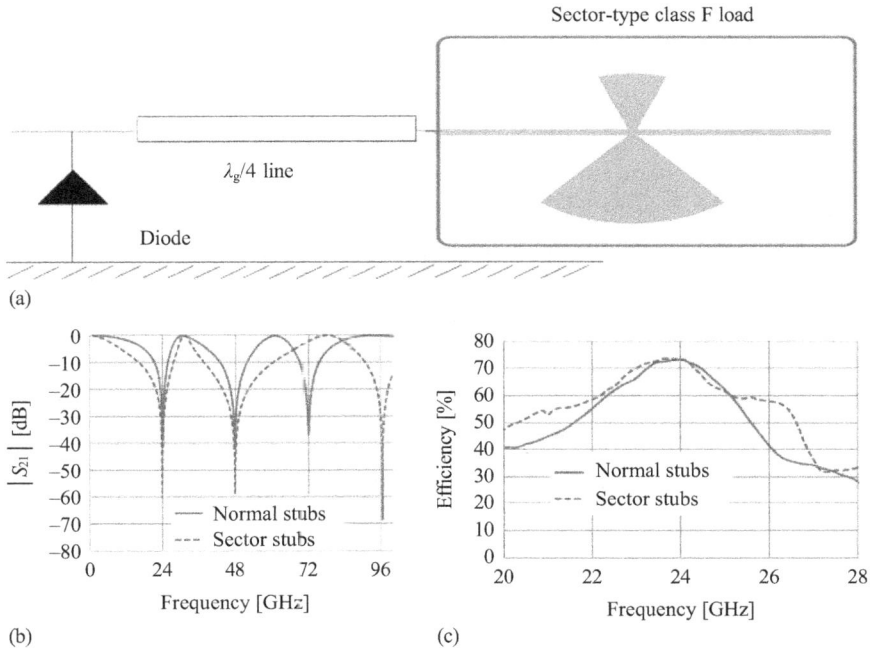

*Figure 11.6 Broadband rectifier with sector open stubs: (a) schematic diagram;
(b) simulated S_{21} of output filter with sector open stubs; and (c)
simulated RF–DC conversion efficiency of rectifier [9]*

microcontrollers support active, sleep, and off-modes. Unlike the off-mode, the
sleep mode consumes a few dozen nA; however, the microcontroller can maintain the status of the register and resume the task with a minimum time and power
overhead. The adaptive duty cycle management plays an important role in bal-
ancing the energy intake and expenditure. Most IoT sensing systems store the
energy into the temporary energy storages such as capacitors. The main advantages of using capacitors over batteries are as follows: (1) A capacitor can be
charged and discharged quickly, (2) a number of recharges are not restricted
(recharges of a Li-ion battery are restricted to a thousand cycles), and (3) the
charging circuit is simpler. However, the main disadvantages are as follows:
(1) Energy density per volume is low, and (2) self-discharge rate is higher. The
self-discharge can cause a trouble for the determination of the optimal duty cycle
in IoT sensors. When a sensor node is set in a duty cycle that is too high compared to harvested energy, the sensor node consumes the stored energy and
finally turns off because of energy shortage. When the sensor is completely off
state, it consumes five times more energy for turning into the active mode compared to wake up from sleep to active. A leakage-aware duty cycle control was
employed to realize the energetically autonomous sensing device that harvest
energy from TV broadcasting signals [16].

The energy harvested from ambient radio waves cannot be applied for high-power applications because the ambient radio wave power is very low. For example, although the battery of a smartphone can persist additional 30% time using the energy harvested from ambient Wi-Fi, LTE, Bluetooth, etc. [17], the smartphone cannot be charged using the energy harvesting technology alone. For charging smartphones using radio wave power, active wireless power transfer (WPT) technologies must be used to increase the wireless power while reducing the power consumption of smartphones or cellphones.

In 2017, a battery-free cellphone operated by the energy harvested from ambient radio waves was developed at the University of Washington, USA, which was driven with 3.48 µW only [18] (Figure 11.7). In addition, the power consumption of 3.48 µm from ambient radio waves was low enough. Key insight to realize a battery-free cellphone was to take advantage of backscatter-based communication systems. Conventional RF energy harvesting systems tried to accumulate the very weak power to the temporary power storage such as a capacitor and operate the active power-hungry microcontrollers intermittently. For the realization of audio-based communication systems, the voice needs to be processed continuously instead of duty cycle-based operation. The most power-consuming device in the cellphone terminal is actually the AD/DA converters for voice communication. Instead of using digital IC for AD/DC conversion, the battery-free cellphone

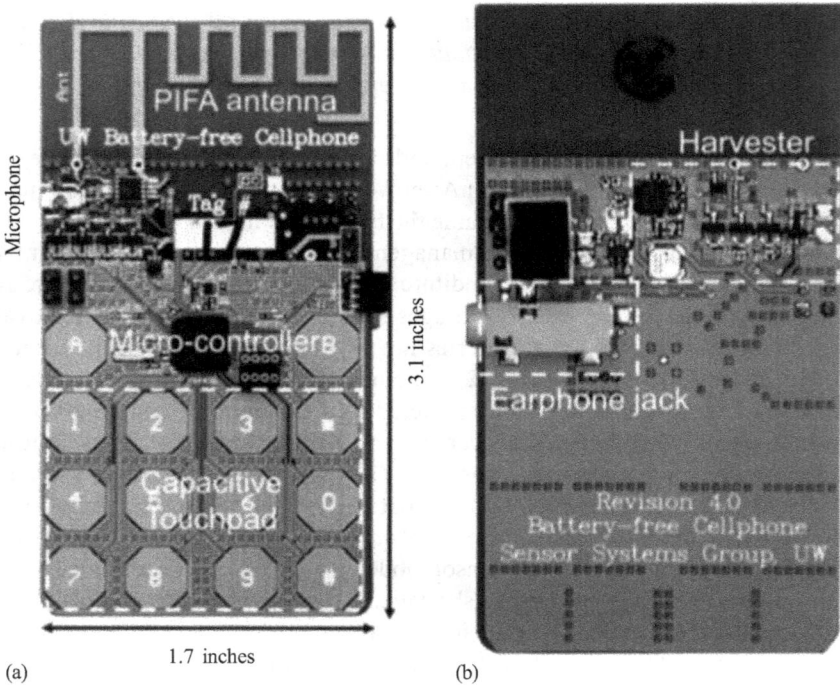

Figure 11.7   Battery-free cellphone: (a) front side; (b) back side [19]

bypasses the conventional computational module between the sensors and communication. Instead of using AD/DA converters, they take advantage of the backscatter module. For transmitting the voice over wireless, the impedance change of an electric microphone connected to an antenna is used to modulate the signal from the transmitter. Similarly, on the downlink, AM (analog) modulation is employed to deliver the continuous voice from the transmitter. This idea is inspired based on the classical crystal radio that can demodulate the voice without battery. The battery-free cellphone can communicate with a base station that is 50 ft (15.2 m) away.

11.2.2   Ubiquitous WPT

If the harvested radio wave power is not high enough to drive a sensor or to charge a smartphone, the WPT technology with an exclusive transmitter should be applied. The multiuser WPT system can be named a ubiquitous WPT system in far field, i.e., wireless power can be provided to everyone anywhere and anytime. There is no difference in antenna and propagation between ubiquitous WPT systems and techniques with a wireless communication system and a broadcasting system. The only difference is at the receiver, which in WPT systems is contributed by a rectenna.

Three issues must be addressed in ubiquitous WPT systems: safety issue of radio waves for human beings, suppression of interference with conventional wireless systems, and practical high system efficiency. For a safe WPT system, wireless power density at receivers should be limited below 1 mW/cm^2, which is the safety limit of microwaves at 2.45 and 5.8 GHz for human beings. The power density below 1 mW/cm^2, which is higher than that of ambient radio waves, is lower than the required power to apply a high-speed wireless charger for smartphones. Radio waves below 1 mW/cm^2 cannot be rectified via a conventional rectenna because of the junction voltage effect of a diode. Therefore, the approach to increase RF–DC conversion efficiency of rectennas in lower power density should be considered for ubiquitous WPT systems as well as energy harvesting systems with a practical high-conversion efficiency. The easiest way to increase RF–DC conversion efficiency of rectennas is to use large antennas. However, a large antenna is not suitable as a mobile application for ubiquitous WPT systems. The best antenna for ubiquitous WPT systems for mobile applications can be a thin super-gain one. Such small antenna is a large-aperture one to receive incident radio waves from wide areas around it. However, such thin super-gain antenna has not been developed yet.

Therefore, the best way to increase RF–DC conversion efficiency is to develop a rectifier with high RF–DC conversion efficiency at the weak radio wave power input. A zero-bias diode can be applied for a low-power rectifier, which unfortunately cannot realize high RF–DC conversion efficiency either. A normal Schottky barrier diode is applied as well, and the circuit should be improved to increase RF–DC conversion efficiency at the low incident radio wave input. It is good to install an optimum impedance matching at an input circuit [19] and an output circuit [20]. As reported previously [19], an active impedance matching circuit was applied to simultaneously increase RF–DC conversion efficiency at low incident radio wave power and at a wide frequency band.

An additional resonator is installed to increase the voltage at a diode to increase RF–DC conversion efficiency at the low incident radio wave power (Figure 11.8) [21,22]. Sufficient voltage at a diode close to a breakdown voltage is required for high RF–DC conversion efficiency, which cannot be added under the weak radio wave power circumstance. The resonator, however, can create a high voltage even at the low incident radio wave power. The rectifier with a short stub resonator can rectify −20 dBm radio wave at 920 MHz with RF–DC conversion efficiency of 40% [21]. A high-impedance antenna was also developed to create a rectenna with a high-impedance circuit, which can create a high voltage [23]. Figure 11.9 shows a developed rectenna with a 1.6 k high impedance folded dipole antenna and a two-stage charge pump rectifier. The RF–DC conversion efficiency of 10% was achieved even at −30 dB input power at 500 MHz.

The extra bias voltage at a diode from an outer power source is often added to increase RF–DC conversion efficiency at conventional detectors. The rectenna in WPT systems, however, is a passive device that cannot be used with an outer extra power source such as a battery. In this case, a self-bias circuit whose power is supplied from a rectifier itself is applied to add the bias voltage at a diode (Figure 11.10) [24]. As shown in Figure 11.10(c), the high RF–DC conversion efficiency curve moves to a lower input power region.

The wireless power below safety level needs to be increased because users require sufficient power for electrical devices. However, the power density of 1 mW/cm^2 is much higher than that for conventional wireless communication systems. The impact of WPT systems on conventional wireless communication

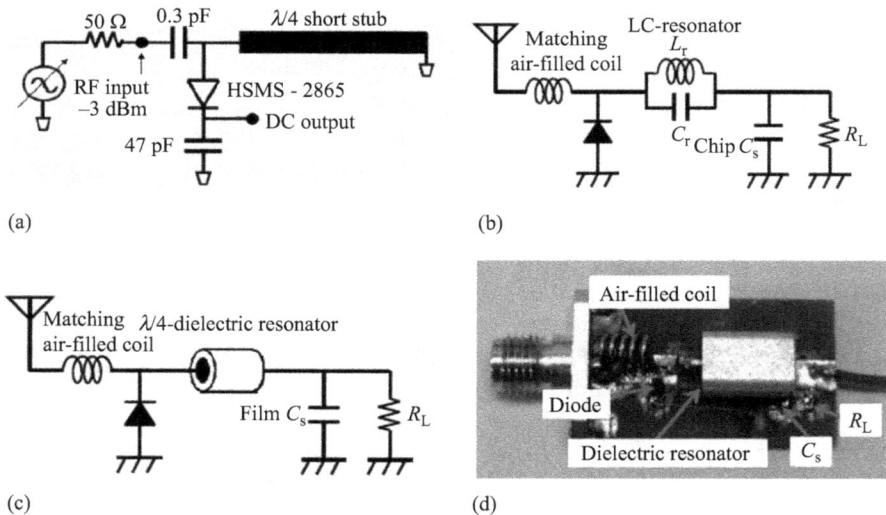

(a)

(b)

(c)

(d)

Figure 11.8   *(a) Rectifier with short stub resonator [21], (b) rectifier with LC resonator [22], (c) rectifier with λ/4 dielectric resonator [22], and (d) developed rectifier with λ/4 dielectric resonator [22]*

(a) (b)

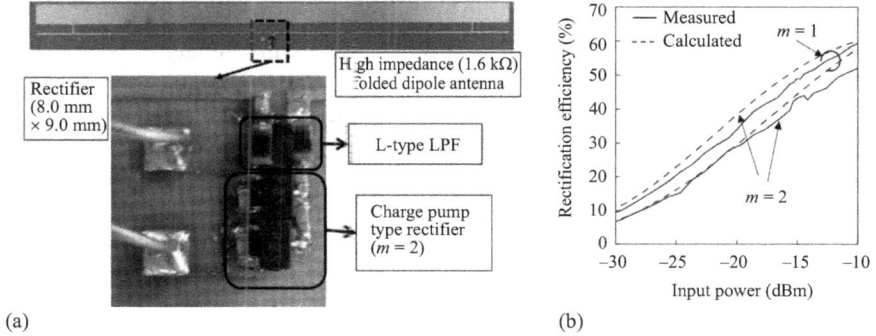

Figure 11.9 (a) Rectenna with high impedance; (b) RF–DC conversion efficiency [23]

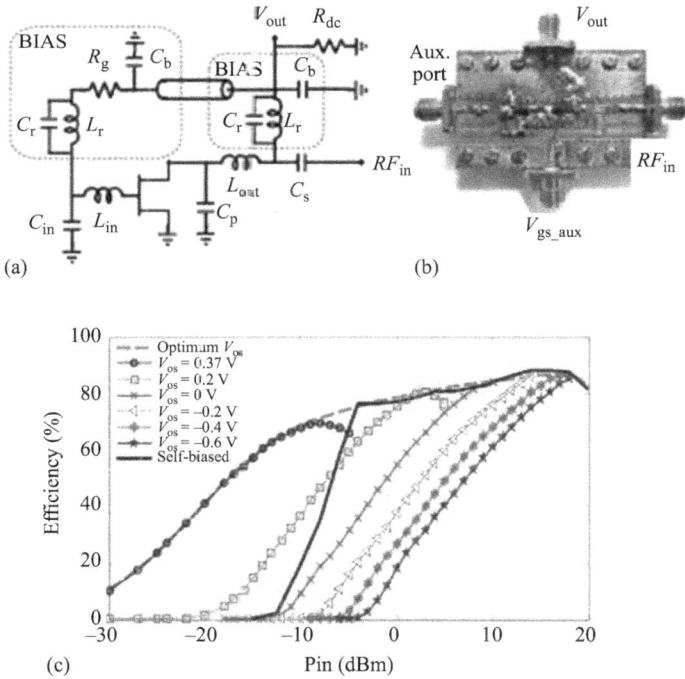

(a) (b)

(c)

*Figure 11.10 E-pHEMT self-biased and self-synchronous class E rectifier:
(a) schematic; (b) developed rectifier; and (c) RF–DC conversion
efficiency at 920 MHz at various bias voltages at the diode [24]*

systems must be considered before the commercialization. Fortunately, the wave
spectrum is very narrow because only non-modulated and continuous radio waves
are usually applied for WPT systems. It is easy to achieve the coexistence of WPT
and other wireless systems by frequency division.

In addition, time division duplex WPT (TDD-WPT) is proposed in the case of a batteryless ZigBee sensor network with WPT at the same frequency band [25]. It is better to use the same frequency band for both wireless communication and WPT to reduce the frequency resource. ZigBee is an IEEE 802.15.4-based specification for a suite of high-level communication protocols used to create a personal area network with small- and low-power digital radios, which is a suitable wireless communication system for WPT because the power consumption is very low. In addition, the interference between high-power WPT and low-ZigBee wireless communication must be considered. Impulse wireless data transmission in ZigBee is adopted to reduce the power consumption. The gaps without any signal exist among information signals (Figure 11.11). So the interference from WPT to ZigBee wireless communication can be suppressed by the TDD-WPT method. With the TDD-WPT system to drive ZigBee, no interference to ZigBee wireless communication under a peak power density of 2.6 mW/cm^2 at the rectenna and ZigBee position existed (Figure 11.12). If a normal WPT system without TDD is applied to

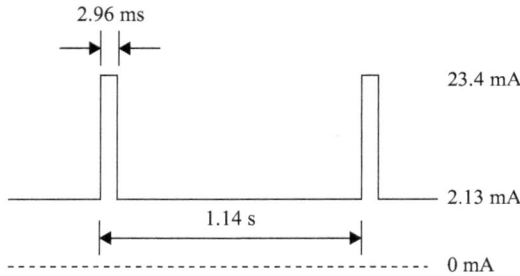

Figure 11.11   Typical ZigBee wireless communication timing

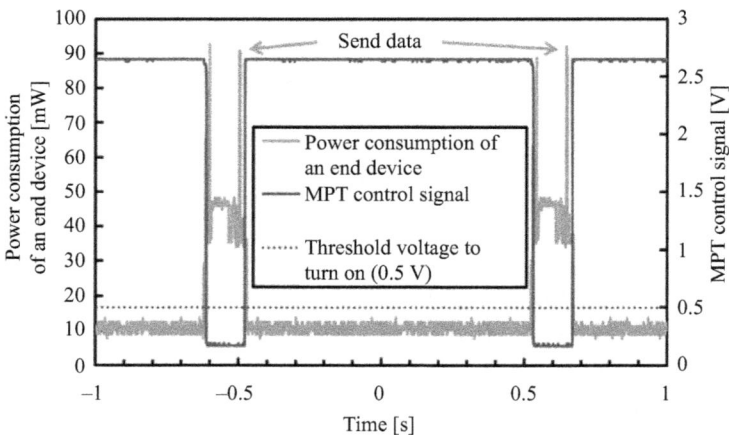

Figure 11.12   Waveforms of power consumption of an end device and microwave power transfer (MPT) control signal of 2 s during the experiment

the batteryless ZigBee system, interference occurs only under the condition of 5 pW/cm^2 power density. The TDD-WPT system is effective in suppressing the interference with other wireless systems.

As described before, RFID is one of the most hopeful applications of ubiquitous WPT systems. The IoT sensor networks and other batteryless sensor applications are expected through not only energy harvesting but also ubiquitous WPT. Besides, a wireless charger for a cell phone is recently expected to be developed. Some venter companies in the United States have been developing wireless chargers for cell phones [26,27]. Frequencies of 920 MHz, 2.45 GHz, and 5.8 GHz at ISM band (Industry Science Medical band) are mainly used to provide the wireless power; 2.45 and 5.8 GHz are at ISM band throughout the world, while 920 MHz is at ISM band in International Telecommunication Union region 2, which includes North and South America. One of the wireless chargers for mobile phones adopted a target-detecting and a beam-forming technology in a multi-pass circumstance like a room, with a phased array antenna to chase a user, keep high wireless power, and decrease power density for human beings [28].

11.3 Long-distance WPT in the radiative near field

Wireless power in far field to multiusers can be provided anywhere and anytime. The receivable wireless power and beam efficiency in far field, however, are low. WPT with high power and efficiency sometimes is needed instead of a wired power line. A receiver should be placed in the radiative near field in this case, which is named "beam-type WPT." The safety issue and suppression of interference with other wireless systems must be considered, which is the same situation as that of a ubiquitous WPT system. The power in beam-type WPT is much higher than that in ubiquitous WPT. A space division technique can be applied to address the safety issue and harmonize with other wireless systems, in addition to the frequency and time division in ubiquitous WPT. In beam-type WPT, the wireless power radiated to a receiver mainly focuses on limited areas. Therefore, power leakage can be suppressed by the beam-forming technology, except at the center of the main beam. In principle, 100% beam efficiency between a transmitting antenna and a receiving one in the radiative near field can be achieved. To increase the beam efficiency, large-aperture antennas should be used. Although high-gain antenna and large-aperture antenna can be used for a radiative near-field WPT system, a phased array antenna, which is composed of many small antennas, is suitable to be used as the transmitting antenna because it is easy to control the beam direction to chase a moving user. On the other hand, an array with small rectennas is suitable to be used as a receiving system. A high-power rectenna is required for high-efficiency and high-power WPT systems, for example, 1 kW WPT system. A diode circuit as a rectifier, however, cannot rectify 1 kW if a large-aperture antenna is used as the receiving antenna. A cyclotron wave converter, which can rectify over 1 kW per antenna, is recommended in the case of one antenna at the user's end in the radiative near field. If the rectenna with a diode is applied, it is better to place many

Figure 11.13   Rectenna array with many rectenna elements

small-power rectennas as an array with a large aperture at the user's end to divide the radio wave power to each rectifier (Figure 11.13).

The optimum array connection of rectennas should be considered. When a rectenna array is developed, the output DC power from each rectifier is combined at the user's end. The mutual coupling between antennas and the impedance matching of antennas and circuits is mainly considered in conventional antenna arrays. The same issue should be considered in a rectenna array as well. A special phenomenon of the rectenna array as an antenna is the breakage of the relationships among an antenna element gain, an array antenna gain, their directivity, and aperture, as described in Chapter 7. The rectenna array is an antenna array with wide directivity, which is the same as that of one antenna element, as well as large aperture, which is sum of apertures of all antenna elements. In addition, a new connection problem occurring with the rectifiers must be considered, especially in radiative near-field WPT systems. In the radiative near field, a power density exists at the receiving array. The power of the incident radio wave is high at the beam center but low at the edge. In such a situation, all rectifiers cannot be operated at the optimum input power for the highest RF–DC conversion efficiency. This means that different DC output powers at different RF–DC conversion efficiencies are combined in the rectenna array in the radiative near field. As a result of the non-linear effect of the diode at the rectifier, the total RF–DC conversion efficiency and the total output DC power theoretically decrease less than those from the sum of individual DC outputs at an individually connected load [29,30]. The connection of rectifiers in series is worse than that in parallel. It depends on the difference between the incident radio wave power and each rectifier. To solve the problem of reduction in efficiency and power, rectenna elements in the radiative near field must be placed as shown in Figure 11.14. In each power density at the receiving

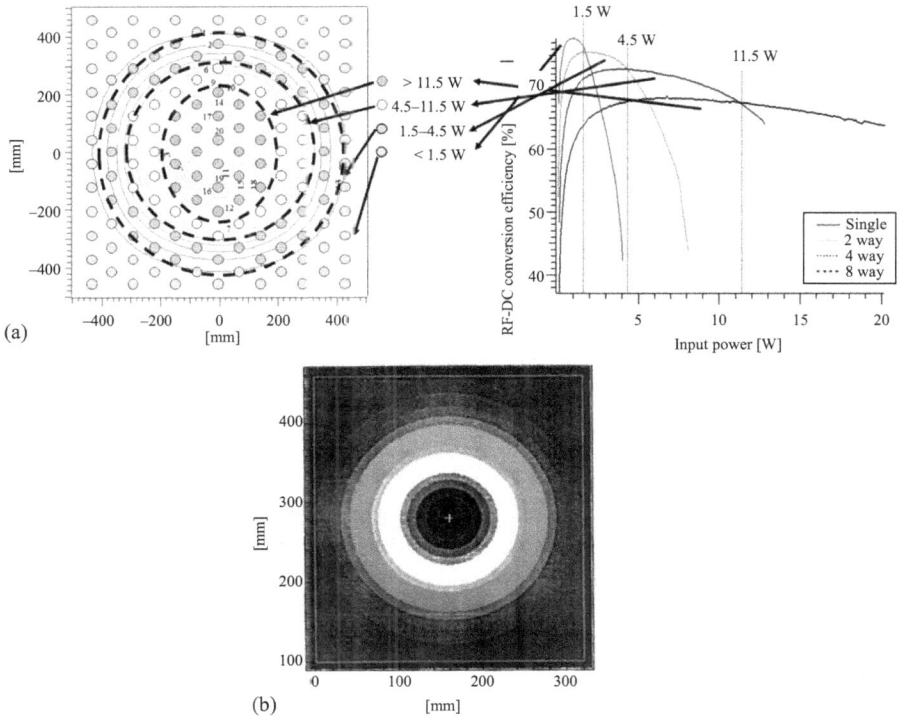

(a)

(b)

Figure 11.14   (a) Optimum rectenna position with different rectenna elements
depends on incident radio wave power in radiative near field;
(b) power density at radiative near field

area, different optimized rectennas should be placed and connected. In the case
shown in Figure 11.14, all rectifiers can receive and rectify their optimum incident
radio waves and all their RF–DC conversion efficiencies reach the maximum.
High-power rectennas are placed at the center, where the incident radio wave
power is high, while low-power rectennas are placed at the edge, where the incident
radio wave power is low in the radiative near field. In this case, the reduction in
power and efficiency would not occur [31]. If all the same rectennas at the rectenna
array are used, the connection of rectifiers is recommended as shown in
Figure 11.15. Balance of the voltage and current of rectifiers should be considered
to maintain high efficiency.

A high-power rectenna is required at the center of the rectenna array. Also, for
other high-power WPT applications, a high-power rectenna is useful. For a low-
power rectifier, how to add voltage at the diode should be considered. To develop a
high-power rectenna, how to decrease the voltage at the diode should be considered
if a conventional Si Schottky barrier diode is used. An increase in the number of
diodes is effective to decrease the voltage at diodes. It is also easy to place a power
divider in front of rectifiers and connect many rectifiers. For example, a high-power

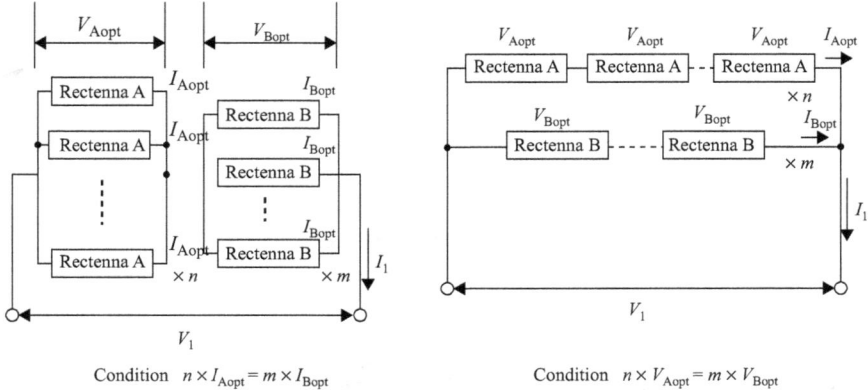

Figure 11.15   Recommended connection of rectifier to maintain high efficiency

rectenna to rectify 100 W at 2.45 GHz with approximately 50% RF–DC conversion efficiency was developed with a 64-way power divider and 256 (= 64 × 4) Si Schottky barrier diodes [32]. However, it was too large to be used as one device.

GaN is one of the hopeful semiconductor materials not only for an amplifier as a field effect transistor or high electron mobility transistor (HEMT) but also for a rectifier as a diode to increase the power at over gigahertz frequency. Figure 11.16 shows a GaN Schottky diode developed by Tokushima University and Kyoto University [33]. The rectenna with the GaN diode can rectify 5 W radio wave at 2.45 GHz with 74.4% RF–DC conversion efficiency (Figure 11.17) [32]. In the future, the rectified radio wave power can be increased theoretically and technically.

The most suitable WPT applied in the radiative near field is the one on moving/flying targets with a phased array antenna [34]. In 1964, W. C. Brown succeeded in performing a WPT-assisted flying drone (small helicopter) experiment whose power generated from a magnetron of 2.45 GHz on ground was approximately 270 WDC. In 1987, the Communications Research Centre group in Canada carried out a WPT-assisted flying airplane experiment using a parabolic antenna from a magnetron. In 1992, a group in Kyoto University, Japan, succeeded in the first WPT-assisted flying airplane experiment with a phased array antenna, which provided 1.25 kW at 2.411 GHz. The airplane received approximately 100 WDC at a 10 m height above the phased array antenna and flew. In 1995, a WPT-assisted airship developed by Kobe University, Japan, flew with a parabolic antenna with the aid of a 10 kW magnetron at 2.45 GHz. Recently, some trials to fly drones with beam-type WPT have been performed throughout the world.

In 2018 in Japan, a large WPT demonstration with a developed new phased array antenna to flying drone was successfully carried out (Figure 11.18). This experiment was carried out by J-Space Systems and was supported by Japanese Ministry of Economy, Trade, and Industry (METI). The transmitted 5.8 GHz microwave power was approximately 1.8 kW from 1.2 m × 1.2 m phased array with developed GaN MMICs. Received Electric Power at the flying drone was approximately 42 W at 30 m

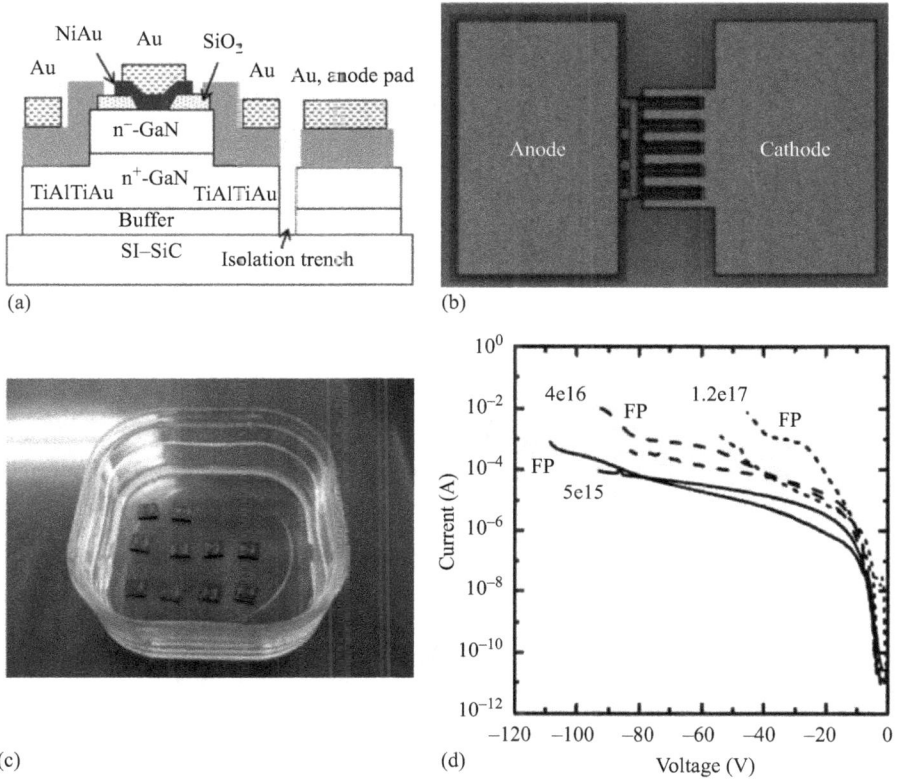

Figure 11.16   *Developed GaN diode: (a) device structure of n-GaN Schottky diode on SI–SiC substrate; (b) photograph of fabricated diode with five anode fingers; (c) photograph of GaN diodes; and (d) breakdown characteristics of diodes under reverse bias [33]*

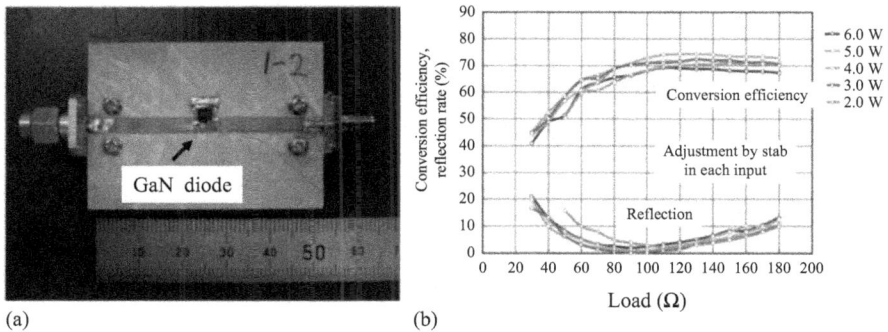

Figure 11.17   *Developed rectenna and measured RF–DC conversion efficiency at 2.45 GHz [32]*

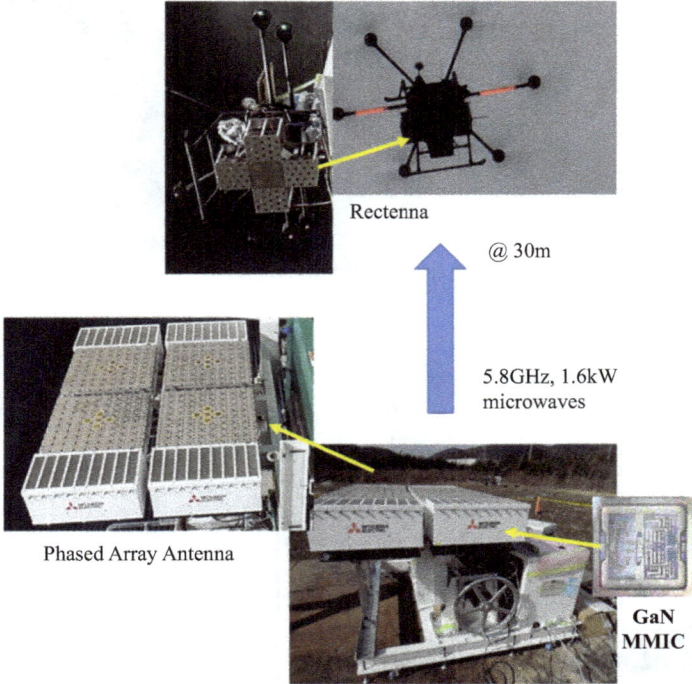

Figure 11.18   WPT-aided drone experiment in 2018 in Japan

above, whose power was used to charge the battery at the drone. In 2021, in Kyoto University in Japan, the WPT experiment to a battery-free micro drone was carried out successfully in laboratory (Figure 11.19) [35]. The transmitted 5.8 GHz microwave power was approximately 134 W to the battery-free micro drone at 80 cm above by designed flat-top beam. The drone could fly only with 27WDC of the rectified microwave in 7 min. The time was restricted only by the motor.

The beam-type WPT can be applied instead of a power line (Figure 11.20) [34]. Especially, the WPT to the isolated venue is one of the several applications when electricity is only required sometimes and the WPT cost can overcome the initial cost of the power lines. In 1975, in Goldstone in the United States, W. C. Brown and R. Dickinson succeeded in the first and largest beam-type WPT field experiment at a distance of 1.6 km between a 24.5 mφ parabolic antenna with 450 kW to 2.45 GHz Klystron and a 3×7 m^2 rectenna array, which received and rectified 30 kW DC at last. In 1995, a group from Kyoto University and Kobe University in Japan carried out a field experiment of the beam-type WPT at 42 m distance using a 5 kW to 2.45 GHz magnetron, which was supported by the Japanese Power Company. Around 2000, a project was intended to develop the beam-type WPT in Re-Union Island near Madagascar, supported by the local government, which has not been realized yet. In 2008, a US TV company supported

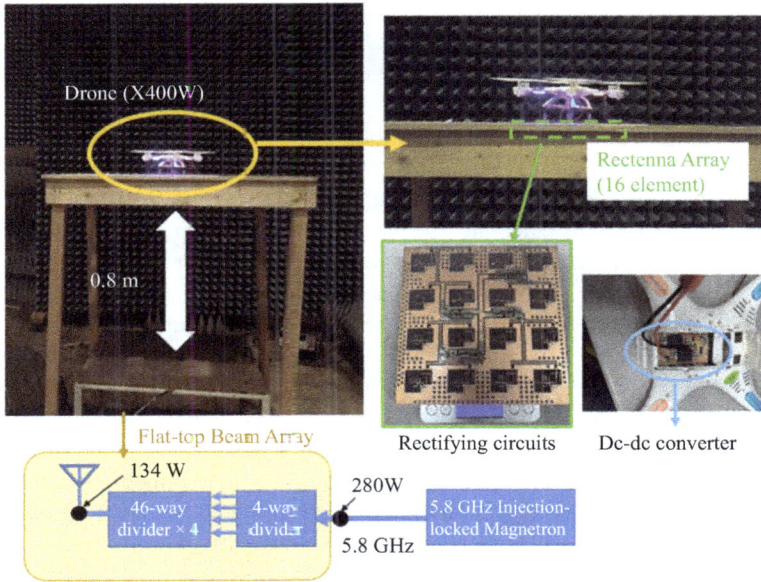

*Figure 11.19   Battery-free micro drone experiment by WPT in 2021 in Kyoto
University*

Figure 11.20   Image of beam-type WPT to isolated place

to carry out the beam-WPT experiment between islands with a distance of
approximately 150 km in Hawaii. In the experiment, Kobe University developed a
phased array antenna. In 2015, in Japan, a large and thin phased array with GaN
HEMT was developed and applied in a field experiment of the beam-type WPT

supported by METI. The transmitted radio wave power was 1.8 kW at 5.8 GHz, and the rectified DC power at a rectenna array, which was 55 m away, was approximately 330 W. The phased array was the same for the drone experiment in 2018. METI simultaneously supported another beam-type WPT experiment in 2015 with a phase-controlled magnetron that had a total power of 10 kW at 2.45 GHz.

X-Band (10.5 GHz) beam WPT experiment was successfully carried out at the US Army Research Field in Blossom Point in 2021 [36]. The microwave power was transmitted from a 5.4 m parabolic antenna and a 91.2 kW pulse magnetron to the rectenna array at 1,046 m away. The received microwave power was 2.27 kW, and the rectified DC power was 1.65 kW. In China, some long-distance beam WPT experiments were carried out in 2020–2022 in series [36]. A startup company, EMROD corp. in New Zealand, carried out the beam WPT demonstration in New Zealand and in Munich in 2022, which were supported by European Space Agency, Airbus corp., etc.

The largest and most hopeful beam-type WPT application in the radiative near field is on a solar power satellite (SPS). The SPS was proposed by P. E. Glaser in 1968 [37] as a new solar power station in space above 36,000 km, which was called a geostationary orbit (Figure 11.21). Sunlight can always irradiate onto the solar cells located in the geostationary orbit. Radio waves, especially microwaves, can propagate without any attenuation by air or ionospheric plasmas. As a result, the total electricity from the SPS was 7–10 times larger than that from solar cells on ground even when 50% of DC–RF–DC conversion efficiency in the beam-type WPT system was considered. In the SPS, instead of the decrease in power transmission efficiency via radio waves, 7–10 times larger electricity can be obtained from CO_2-free solar cells. Based on the Friis transmission equation in the radiative near field, an antenna size of over 2 km is required at 5.8 GHz to realize 90% beam efficiency at a height of 36,000 km. Since the merits of WPT via radio waves for the SPS overcome the demerits, many countries have become recently interested in the SPS [38].

Figure 11.21   Image of solar power satellite

11.4   Long-distance WPT in fielded field

Radio waves can propagate not only in space but also in the fielded field, like a waveguide. The beam efficiency between a transmitting antenna and a receiving one can be improved in the fielded field. Moreover, the leakage to the outside of the fielded field is smaller than that in other WPT systems, which brings less safety issues and interference. A two-dimensional (2D) waveguided WPT system was proposed and developed in Japan, named "surface WPT," which can provide not only wireless power but also wireless information via radio waves propagating through the 2D waveguide (Figure 11.22) [39,40]. The surface WPT was standardized by the Association of Radio Industries and Businesses as STD-T113 [41] in Japan in 2015. The frequency for the surface WPT is allowed at 2.498 GHz $\pm$ 1 MHz, while the power is allowed below 30 W. STD-T113 is the first sterilization of the WPT via radio waves in the world.

Some WPT systems in a one-dimensional (normal) waveguide are proposed and developed in Japan as well. One is named a microwave building with WPT at 2.45 GHz in the waveguides under floors in the building [32]. The advantage of the microwave building is that it reduces the initial building cost. The WPT in the waveguide can also be applied inside a car to provide wireless power to drive sensors in the car [42]. The radio waves can usually be provided in three-dimensional space in a shielded box [43], where a mode of radio waves should be

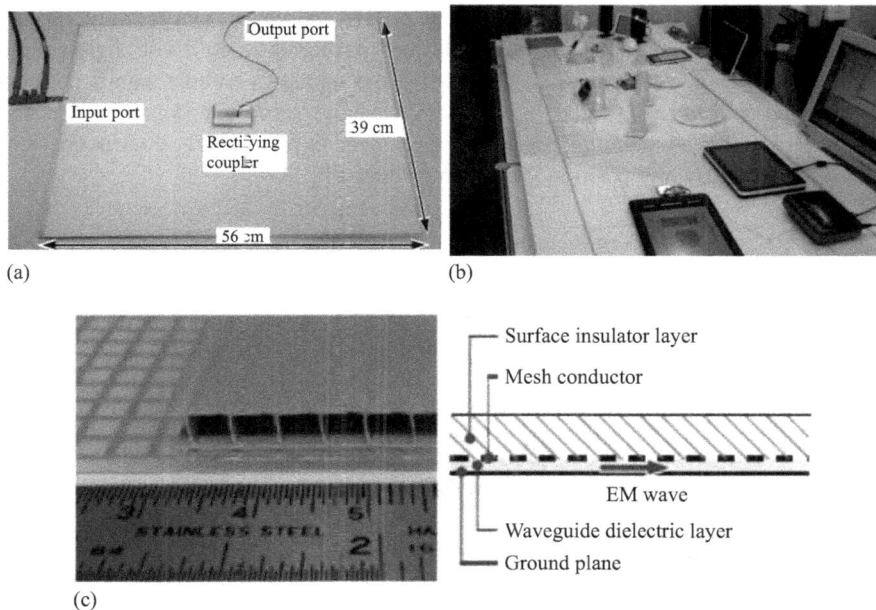

(a)   (b)   (c)

Figure 11.22   (a, b) Photographs of the surface WPT system; (c) inside 2D waveguide [39,40]

considered. In Japan, the WPT in the shield car engine compartment is developed to drive sensors in the compartment [44]. The propagated 2.45 GHz is only considered, while modes of radio waves are not considered.

11.5   Near-field WPT in a cavity resonator

Aside from long-distance wireless power transmission at microwave frequency band, it is also possible to generate strong coupling to coil receivers thousands of times smaller than the size of the room using a resonance-based approach named quasistatic cavity resonance (QSCR) [45]. The main reason of power level of radiative transfer methods is restricted, which is the effect of electric field potentially harmful to human body. Radiative transfer methods have tightly coupled electric and magnetic fields, and those fields are not separable. On the other hand, non-radiative near-field coupling can safely deliver tens to hundreds of watts of power because the magnetic and the electric fields are loosely decoupled. However, such near-field-based system is not suitable for long-distance transmission because transfer efficiency drops rapidly according to the distance of the coils. Moreover, it is not possible to strongly couple coils that are drastically different in size. To solve this problem, QSCR takes advantages of the standing electromagnetic waves that fill the interior of the resonant structure with uniform magnetic fields.

The resonant structure constitutes of metallic walls, a ceiling, floor, and a capacitive loaded vertical pole located in the center of the room. Because power is delivered mainly via magnetic fields in the low megahertz frequency range, daily objects including metal objects such as electrical devices and furniture do not strongly couple to the QSCR. It is also possible to have doors and windows without significant performance drop because of the high Q-factor of the resonator. According to the demonstration in a 54 m^3 room, transmission efficiencies to a small coil receivers were 50%–95% in nearly any position in a room (Figure 11.23).

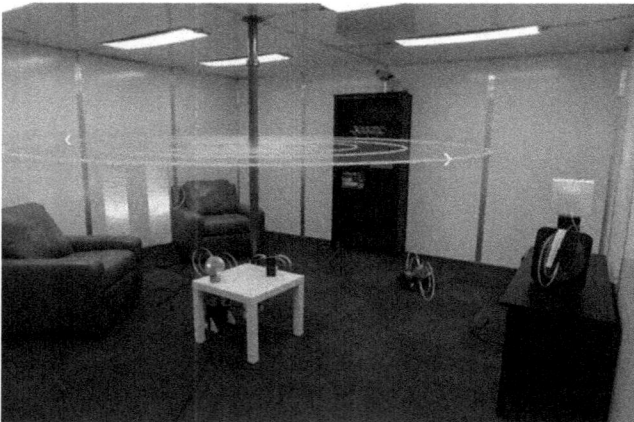

Figure 11.23   Quasistatic cavity resonance demonstrated in a 54 m^3 room

References

[1] M. Pinuela, P. Mitcheson, and S. Lucyszyn, "Ambient RF Energy Harvesting in Urban and Semi-Urban Environments," *IEEE Trans. Microw. Theor. Techn.*, Vol. 61, No. 7, pp. 2715–2726, 2013.

[2] J. A. Hagerty, F. B. Helmbrecht, W. H. McCalpin, R. Zane, and Z. B. Popovic, "Recycling Ambient Microwave Energy with Broad-Band Rectenna Arrays," *IEEE Trans. MTT*, Vol. 52, No. 3, pp. 1014–1024, 2004.

[3] Y. H. Suh and K. Chang, "A High-Efficiency Dual-Frequency Rectenna for 2.45- and 5.8-GHz Wireless Power Transmission," *IEEE Trans. MTT*, Vol. 50, No. 7, pp. 1784–1789, 2002.

[4] B. L. Pham and A.-V. Pham, "Triple Bands Antenna and High Efficiency Rectifier Design for RF Energy Harvesting at 900, 1900 and 2400 MHz," Proceedings of International Microwave Symposium (IMS), WE3G-5, 2013.

[5] R. J. Vyas, B. B. Cook, Y. Kawahara, and M. M. Tentzeris, "E-WEHP: A Batteryless Embedded Sensor-Platform Wirelessly Powered from Ambient Digital-TV Signals," *IEEE Trans. Microw Theor. Tech.*, Vol. 61, No. 6, pp. 2491–2505, 2013.

[6] D. Isbell and R. Duhamel, "Broadband Logarithmically Periodic Antenna Structures," *IRE Int. Conv. Rec.*, pp. 119–128, 1957.

[7] R. Carrel, "Analysis and design of the log-periodic dipole antenna," University of Illinois Electrical Engineering Research Laboratory, Engineering Experiment Station, Antenna Laboratory technical report No. 52, 1961.

[8] B. Ferran, D. Belo, and A. Georgiadis, "A UHF Rectifier with One Octave Bandwidth Based on a Non-Uniform Transmission Line," WE1E-6, Proceedings of the International Microwave Symposium (IMS), 2016.

[9] N. Shinohara, K. Hatano, T. Seki, and M. Kawashima, "Development of Broadband Rectenna at 24 GHz," Proceedings of the 6th Global Symposium on Millimeter-Waves 2013 (GSMM2013), 1569734001.pdf, 2013.

[10] L. Roselli, C. Mariotti, M. Virili, *et al.*, "WPT Related Applications Enabling Internet of Things Evolution," Proceedings of the 2016 10th European Conference on Antennas and Propagation (EuCAP), 2016.

[11] O. Elsayed, M. Abouzied, and E. S-Sinencio, "540 mW RF Wireless Receiver Assisted by RF Blocker Energy Harvesting for IoT Applications with 18 dBm OB-IIP3," Proceedings of the 2016 IEEE Radio Frequency Integrated Circuits Symposium (RFIC), pp. 230–233, 2016.

[12] T. Le, K. Mayaram, and T. Fiez, "Efficient Far-Field Radio Frequency Energy Harvesting for Passively Powered Sensor Networks," *IEEE J. Solid-State Circ.*, Vol. 43, No. 5, pp. 1287–1302, 2008.

[13] S. Kitazawa, M. Hanazawa, S. Ano, H. Kamoda, H. Ban, and K. Kobayashi, "Field Test Results of RF Energy Harvesting from Cellular Base Station," Proceedings of 6th Global Symposium on Millimeter-Waves (GSMM) 2013, T6-6, 2013.

[14] Y. Kawahara, X. Bian, R. Shigeta, R. J. Vyas, M. M. Tentzeris, and T. Asami, "Power Harvesting from Microwave Oven Electromagnetic Leakage," Proceedings of the UbiComp 2013, pp. 373–382, 2013.

[15] J. Bito, R. Bahr, J. G. Hester, S. A. Nauroze, A. Georgiadis, and M. M. Tentzeris, "A Novel Solar and Electromagnetic Energy Harvesting System with a 3-D Printed Package for Energy Efficient Internet-of-Things Wireless Sensors," *IEEE Trans. MTT*, Vol. 65, No. 5, pp. 1831–1842, 2017.

[16] R. Shigeta, T. Sasaki, D. M. Quan, *et al.*, "Ambient RF Energy Harvesting Sensor Device with Capacitor-Leakage-Aware Duty Cycle Control," *IEEE Sens. J.*, Vol. 13, No. 8, pp. 2973–2983, 2013.

[17] Nikola Labs, http://www.nikola.tech/. Accessed on February 27, 2024.

[18] V. Talla, B. Kellogg, S. Gollakota, and J. R. Smith, "Battery-Free Cellphone," Proceedings of the ACM on Interactive, Mobile, Wearable and Ubiquitous Technologies, Vol. 1, No. 2, Article 25 (June 2017), 20 pages. DOI: https://doi.org/10.1145/3090090.

[19] T. W. Barton, J. M. Gordonson, and D. J. Perreault, "Transmission Line Resistance Compression Networks and Applications to Wireless Power Transfer," *IEEE J. Emerg. Sel. Top. Pow. Electron.*, Vol. 3, No. 1, pp. 252–260, 2015.

[20] N. Shinohara, H. Matsumoto, A. Yamamoto, *et al.*, "Development of High Efficiency Rectenna at mW Input (in Japanese)," IEICE Tech. Report, SPS2004-08, pp. 15–20, 2004.

[21] H. Kitayoshi and K. Sawaya, "A Study on Rectenna for Passive RFID-Tag (in Japanese)," Proceedings of IEICE in Spring, CBS1-5, 2006.

[22] T. Yamashita, K. Honda, and K. Ogawa, "High Efficiency MW-Band Rectenna Using a Coaxial Dielectric Resonator and Distributed Capacitors," Proceedings of 2013 International Symposium on Electromagnetic Theory (EMTS2013), pp. 823–826, 2013.

[23] T. Furuta, M. Ito, N. Nambo, K. Itoh, K. Noguchi, and J. Ida, "The 500 MHz Band Low Power Rectenna for DTV in the Tokyo Area," Proceedings of 2016 IEEE Wireless Power Transfer Conference (WPTC), 2016.

[24] M. N. Ruiz and J. A. García, "An E-pHEMT Self-Biased and Self-Synchronous Class E Rectifier," Proceedings of 2014 IEEE MTT-S International Microwave Symposium (IMS2014), TH1C-2, 2014.

[25] N. Shinohara, "Simultaneous WPT and Wireless Communication with TDD Algorithm at Same Frequency Band (Chapter 9)," *Wireless Power Transfer Algorithms, Technologies and Applications in Ad Hoc Communication Networks*, edited by S. Nikoletseas, Y. Yang, and A. Georgiadis, Springer, Berlin, ISBN: 978-3-319-46810-5, 2016, pp. 211–230.

[26] Ossia Inc., http://www.ossia.com/. Accessed on February 27, 2024.

[27] Energous Corp, http://energous.com/. Accessed on February 27, 2024.

[28] Z. Hatem and A. Saghati, "Remote Wireless Power Transmission System 'Cota'", *Frontiers of Research and Development of Wireless Power Transfer (Japanese Book)*, edited by N. Shinohara, CMC Publisher, Japan, ISBN: 978-4-7813-1175-3, 2016, pp. 185–196.

[29] R. J. Gutmann and J. M. Borrego, "Power Combining in an Array of Microwave Power Rectifiers," *IEEE Trans. MTT*, Vol. 27, No. 12, pp. 958–968, 1979.

[30] N. Shinohara and H. Matsumoto, "Experimental Study of Large Rectenna Array for Microwave Energy Transmission," *IEEE Trans. MTT*, Vol. 46, No. 3, pp. 261–268, 1998.

[31] T. Miura, N. Shinohara, and H. Matsumoto, "Experimental Study of Rectenna Connection for Microwave Power Transmission," *Electron. Commun. Jpn., Part 2*, Vol. 84, No. 2, pp. 27–36, 2001.

[32] N. Shinohara, N Niwa, K. Takagi, *et al.*, "Microwave Building as an Application of Wireless Power Transfer", *Wirel Power Transf.*, Vol. 1, No. 1, pp. 1–9, 2014.

[33] K. Takahashi, J.-P. Ao, Y. Ikawa, *et al.*, "GaN Schottky Diodes for Microwave Power Rectification," *Jpn. J. Appl. Phys. (JJAP)*, Vol. 48, No. 4, pp. 04C095-1–04C095-4, 2009.

[34] N. Shinohara, *Wireless Power Transfer via Radiowaves (Wave Series)*, ISTE Ltd. and John Wiley & Sons, Inc., Great Britain and USA, ISBN: 978-1-84821-605-1, 2014.

[35] N. Takabayashi, B. Yang, N. Shinohara, and T. Mitani, "Lightweight and Compact Rectenna Array with 20W-class Output at C-band for Micro-drone Wireless Charging," *IEICE Trans. C.*, Vol. E105-C, No. 10, 2022.

[36] C. T. Rodenbeck, P. I. Jaffe, B. H. Strassner II, *et al.*, "Microwave and Millimeter Wave Power Beaming," *IEEE J. Microw*, Vol. 1, No. 1, pp. 229–259, 2021.

[37] P. E. Glaser, "Power from the Sun: Its future," *Science*, No. 162, pp. 857–886, 1968.

[38] J. C. Mankins, *The Case for Space Solar Power*, The Virginia Edition, Inc., Houston, TX, ISBN: 978-0-9913370-0-2, 2013.

[39] A. Noda and H. Shinoda, "Selective Wireless Power Transmission through High-Q Flat Waveguide-Ring Resonator on 2-D Waveguide Sheet," *IEEE Trans. MTT*, Vol. 59, No. 8, pp. 2158–2167, 2011.

[40] A. Noda and H. Shinoda, "Waveguide-Ring Resonator Coupler with Class-F Rectifier for 2-D Waveguide Power Transmission," Proceedings of 2012 IEEE MTT-S International Microwave Workshop Series on Innovative Wireless Power Transmission: Technologies, Systems, and Applications (IMWS-IWPT2012), pp. 259–262, 2012.

[41] ARIB Standard STD-T113, https://www.arib.or.jp/english/std_tr/telecommunications/desc/std-t113.html. Accessed on April 16, 2018.

[42] S. Ishino, Furuno, Y. Takimoto, *et al.*, "Study on WPT System Using a Radio Wave Hose as a New Transmission Line," Proceedings of IEEE Wireless Power Transfer Conference 2015, P4.1, 2015.

[43] S. Rahimizadeh, S. Korhummel, B. Kaslon, and Z. Popovic, "Scalable Adaptive Wireless Powering of Multiple Electronic Devices in an

Over-Moded Cavity," Proceedings of IEEE Wireless Power Transfer Conference 2013, WEO31-163, 2013.
[44] N. Shinohara, H. Goto, T. Mitani, H. Dosho, and M. Mizuno, "Experimental Study on Sensors in a Car Engine Compartment Driven by Microwave Power Transfer," Proceedings of 2015 9th European Conference on Antennas and Propagation (EuCAP), 2015.
[45] M. J. Chabalko, M. Shahmohammadi, and A. P. Sample, "Quasistatic Cavity Resonance for Ubiquitous Wireless Power Transfer," *PLoS ONE*, Vol. 12, No. 2, pp. e0169045. https://doi.org/10.1371/journal.pone.0169045. Accessed on April 16, 2018.

Chapter 12

Biological issue of electromagnetic fields and waves

Shin Koyama[1]

12.1 Introduction

In the past few decades, the use of electromagnetic fields (EMF) has become pre-valent throughout the world, and we no longer intend to stop using this technology. The widespread use of electromagnetic sources in daily life has induced an inevitable exposure to EMF. Recently, the rapid introduction of EMF has witnessed an upsurge in the use of EMF for telecommunication. Mobile phone use has become ubiquitous. Wireless communication devices emit nonionizing electro-magnetic radiofrequency (RF) fields in the range of 300 MHz to 300 GHz, raising public concern regarding the increasing use of mobile phones and their potential health-related risks. The intermediate frequency (IF) EMF (300 Hz to 10 MHz) is also now widely used for wireless power transmission (WPT) or domestic kitchen appliances such as induction heating (IH) cooking. Unfortunately, although plenty of research has been conducted on RF, little information exists on the potential health effects associated with the exposure to IF magnetic fields. With growing concerns regarding the potential health hazards, it has become necessary to inves-tigate the risks of IF magnetic fields in more detail.

Regarding the response to the risks and anxiety about EMF, the basis of the guidelines is induced by scientific data on the adverse effects of the EMF exposure on human health, which could be differentiated into biological changes and adverse effect on human health. Biological changes signify alterations on the cellular level by the EMF exposure, which might not directly lead to adverse human health, as the homeostasis compensates these biological changes. The International Commission on Non-Ionizing Radiation Protection (ICNIRP) guideline aims to protect from the adverse effects of the EMF exposure and not necessarily avoid all of its biological effects.

This chapter discusses the biological effects of IF and RF exposure. This chapter is divided into the following three parts: epidemiological study, animal

[1]Research Institute for Sustainable Humanosphere, Kyoto University, Japan

study, and cellular study. In the RF field, the frequency of 2.45 GHz is mainly discussed because of the future use of WPT.

12.2 Epidemiological studies

The objective of this epidemiological study is to investigate how often diseases occur in different groups of people and elucidate its causes. In the biological research on EMF, mainly two methods of epidemiological study are used, cohort study and case–control study. The cohort study is an observational study to investigate the relationship between factors and disease outbreak by tracking the population exposed to specific factors and the unexposed population for a given period and comparing the incidence rate of diseases for analysis. The case–control study is an analytical study to investigate the exposure factors for the diseased population. For the unaffected population, which is the control, it is an investigation of the exposure condition to a specific factor. A research method to evaluate the relationship between factors and diseases by comparing the above two groups.

The first report elucidating the adverse effects of the EMF exposure was an epidemiological study. In 1979, Wertheimer and Leeper [1] reported a correlation between extremely low-frequency EMF (ELF-EMF) and pediatric leukemia. Since then, several researchers have examined the effects of such fields, with some studies corroborating the cancer risks with ELF-EMF [2,3]. Although several studies have reported negative results regarding the adverse effect of ELF-EMF [4–6], contradictory results have been published even more recently. In fact, some epidemiological as well as cellular studies are still under discussion (discussed later). However, to date, several cohort studies investigating a relationship between ELF-EMF and cancer have been unable to demonstrate or indicate an increase in this rate.

The WPT system mainly uses IF and RF; however, an epidemiological study about IF has not been established yet. Recently, some researchers have initiated epidemiological studies on IF, but their results will be known in the future.

In contrast, epidemiological studies on high-frequency EMF have been extensively conducted because people who are concerned about the risks of using mobile phones have increased globally. Recently, several studies have already been published about the impact of RF-EMF on human health, of which a large number of studies were related to the frequencies of mobile phones. Of note, several epidemiological studies have suggested the possible risks of RF field exposure [7–9]. Although many studies have denied any adverse correlation between RF exposure and human health, the possibility of risks has not been completely excluded [10–12]. These studies are not necessarily limited to the frequency of WPT; however, most studies denied any adverse effects by exposure to RF-EMF. Based on this literature, the ICNIRP established the guidelines for EMF exposure [13]. In 2011, the international experts organized by the International Agency for Research on Cancer categorized RF/MW radiation as a possible carcinogen group (2B) [14]. The guidelines for exposure to microwaves have been based on thermal effects.

Although these categorizations are usually based on the uncertainty of the obvious effects of EMF exposure, low-level RF/MW radiations at nonthermal intensity are not considered harmful to human health. However, recent research has indicated some adverse effects of RF radiations on human health [15,16].

Perhaps, controversial results remain unclear in epidemiological studies and other research such as animal or cellular studies. Hence, it is essential to clarify the risks of RF/MW radiations.

12.3   Animal studies

Animal responses to EMF have been researched for a long time, particularly low-level and low-intensity ELF magnetic fields because a circadian rhythm or sense of direction in animals is thought to be related to ELF for its utilization. Since the rapid introduction of artificial EMF might affect some biosystems, a large number of studies have been conducted to investigate the relationship between ELF and circadian rhythm or melatonin/serotonin secretion [17–21]. Some studies have suggested that these alterations might lead to a possible etiological factor for the increased risk of certain cancers, depression, and miscarriage [22]. In addition, several studies have indicated a negative impact of ELF exposure on melatonin secretion [23–25]. Although some biological effects of EMF are evident, the results are not consistent, and there are no reported adverse effects of ELF exposure on human health.

Animal studies about IF have not been researched as widely as epidemiological studies. However, a recent widespread introduction of IH cooking hob has induced a concern about human health and IF exposure. Ushiyama *et al.* suggested that an exposure to 21 kHz of sinusoidal IF magnetic field at 3.8 mT for 1 h/day for 14 days did not affect the immune function in juvenile rats [26]. Nishimura *et al.* also found that the acute and sub-chronic exposure to 20 kHz at 0.2 mT or 60 kHz at 0.1 mT of sinusoidal EMFs reveal no toxicity in rats [27]. Furthermore, this group indicated that the exposure to 20 or 60 kHz of IF did not produce any significant teratogenic developmental effects in chick embryos [28]. Likewise, several studies have reported negative results from acute and chronic toxicity studies of 20-kHz magnetic fields at up to 0.03 mT in rats and mice [29–33]. Although these studies are well conducted, the data obtained are insufficient to draw any conclusion on this frequency.

There are few animal studies on IF and many on RF. Since the frequency of 2.45 GHz should be the primary target for WPT, one of the recent animal studies about 2.45 GHz is discussed here. Deshmukh *et al.* [34] exposed groups of six male rats to 2.45 GHz fields at a specific energy absorption rate (SAR) of 0.67 mW/kg for 2 h/day 5 days/week, over a period of 180 days. At the end of the exposure period, the memory of all rats was tested, the brains were removed, and the heat shock protein (HSP) and DNA damage were assessed. Both learning and memory were found to be impaired after the exposure. The results indicated that chronic low-intensity microwave exposure in the frequency range of 2.45 GHz might cause hazardous effects on the brain. Although animal studies focusing on 2.45 GHz are very few, significant positive data have been obtained on other RF frequencies, such as 900 or 1800 MHz, as described earlier. In a rodent brain study, Tang *et al.* [35] exposed 108 male Sprague–Dawley (SD) rats to 900 MHz, 1 mW/cm^2 EMF, or sham (unexposed) for 14 or 28 days (3 h/day). SAR was between 0.016 (whole body) and 2 W/kg (locally in the head). They established that the frequency of crossing platforms and the percentage of time spent in the target quadrant were lower in rats exposed to EMF for 28 days than in those exposed to EMF for 14 days and unexposed rats. Moreover, EMF exposure for 28 days induced cellular edema and neuronal cell organelle degeneration in the exposed rats. In addition, the damaged blood–brain barrier (BBB) permeability, which resulted in albumin and HO-1 extravasation, was observed in the hippocampus and cortex. They found that the EMF exposure for 28 days induced the expression of mkp-1, resulting in ERK dephosphorylation. Overall, these results demonstrated that exposure to 900-MHz EMF radiation for 28 days could significantly impair spatial memory and damage the BBB permeability in rats by activating the mkp-1/ERK pathway. In cancer research on mice, Lerchl *et al.* [36] replicated a positive result by Tillmann *et al.* [37] and observed an increased incidence of bronchoalveolar and hepatic tumors and lymphomas. However, several studies denied the impact of RF on the brain, behavior, genotoxicity, and immunology in animals [38–40].

In summary, most animal studies did not exhibit a significant difference in carcinogenic results between RF-exposed and control or sham-exposed animals. However, several studies specified some adverse effects on the animal brain or oxidative stress, leading to DNA damage. These contradictory results might come from the difference of frequency, animals, or methodology. Thus, it is essential to replicate the positive data in the same way to validate the results.

12.4 Cellular studies

Although the primary target of WPT research is IF, the data for drawing certain conclusions are insufficient. Despite the rapid increase of exposure to IF-EMF from the recent devices, such as IH cooking, anti-theft gate in shops, and WPT systems, only a few in vitro studies have been conducted. To date, several studies on IF have been published. Here, the results of some in vitro experiments, RF frequency used for WPT, the results of RF-exposed cells, and the experimental methods in the RF research useful for drawing results on IF have been discussed.

First, let us discuss some experimental methods. The cellular study can generally be categorized into two types: genotoxic effects and nongenotoxic effects. While genotoxic effects comprise chromosomal or chromatid aberration, DNA strand breaks, micronucleus (MN) formation, and mutation, nongenotoxic effects include cell proliferation, cell cycle distribution, gene or protein expression, immune response, apoptosis, and others. Here, several main studies are introduced and some cellular research results related to IF- or RF-EMF have been simultaneously described.

12.4.1   Genotoxic effects

Research on genotoxicity exposed to EMF is one of the most active areas because this field is strictly related to cancer, which constitutes the most anxious epidemiological study on pediatric leukemia. The genotoxic effects mainly affect DNA, and damaged DNA leads to cancer risks. Several endpoints are described in the following sections.

12.4.1.1   Chromosomal or chromatid aberration

Chromosomal or chromatid aberrations are induced directly by DNA damage and also by other factors in condensed chromosomes in the mitotic phase. For the chromosomal aberration test, cell division is arrested in the metaphase using colcemid when chromosomes are condensed. Then, cells are suspended in a hypotonic solution, centrifuged, and soaked in a fixing solution. Cells are placed on a glass slide, stained with Giemsa solution, and observed under a microscope, revealing several chromosomal aberrations, such as chromosomal break, ring, dicentric chromosome, large fragment, rearrangement, chromatid gap, and chromatid exchange. Apparently, ionizing radiations cut the DNA strand and induce chromosomal aberration. However, nonionizing radiations do not have enough energy to break DNA strands. Although no investigation exists on chromosomal aberrations caused by IF exposure, several studies have reported chromosomal aberrations in RF-exposed cells. Some studies have indicated that exposure to RF-EMF increased chromosomal aberrations [41–43]. In these studies, Mashevich *et al.* [42] insisted that any biological effect (chromosomal aneuploidy and abnormal DNA replication in lymphocytes exposed to 830 MHz) observed was not caused by the increased temperature. Maes *et al.* [43] demonstrated that the chromosomal aberration was counted by exposure to 2.45 GHz with a 50-Hz pulse for 30 or 120 min at a very high SAR of 75 W/kg.

In contrast, several studies have indicated that the exposure to RF-EMF did not cause chromosomal aberrations [44–50]. Kerbacher *et al.* [46] exposed Chinese hamster ovary (CHO)-K1 cells to pulsed 2.45 GHz for 2 h at an SAR of 33.8 W/kg and found that exposure to RF alone had no effect. No differences were observed in the mitotic index (which is defined as the ratio between the number of cells in a population undergoing mitosis to the number of cells in a population not undergoing mitosis) between RF-exposed and temperature-controlled groups. Human blood-derived lymphocytes were exposed to 2.45-GHz RF-EMF for 90 min at an SAR of 12.46 W/kg, and the chromosomal aberration was investigated [47]. No changes were observed in the frequency of chromosomal aberration. Vijayalaxmi [51] reported that the level of damage on a chromosomal aberration in 2.45-GHz RF-exposed and sham-exposed lymphocytes from healthy human volunteers was not significantly different.

In our laboratory, the effects of continuous wave or pulsed wave at 2.45 GHz in mice m5S cells were investigated, and exposure to RF for 2 h at an SAR of 100 W/kg did not induce any chromosomal aberration (Figure 12.1) [52].

Most studies on chromosomal aberration at low SAR demonstrated negative results, particularly 2.45 GHz, the possible usage of which for WPT seems to cause

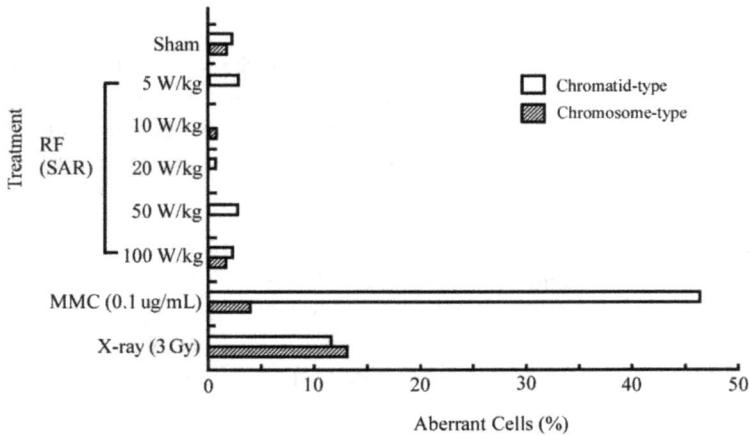

Figure 12.1   *Chromosomal- and chromatid-type aberrations frequency in mice m5S cells exposed to RF, MMC, and X-rays (redrawing data from Ref. [52])*

no adverse effect on chromosomal aberration. Although the chromosomal aberration test is required for cells exposed to IF-EMF, the possible effects to induce chromosomal aberration might be very low.

12.4.1.2   DNA strand breaks

The comet assay has been used in several studies to investigate the effects of EMF exposure. The comet assay can evaluate DNA strand breaks as follows. First, cells are treated with external stimuli, such as EMF, X-rays, or chemical agents. The cell suspension is mixed with agarose and fixed on a glass slide. The glass slide is soaked in the lysis buffer to unfix the cells. Then, electrophoresis is conducted under alkaline (for single- and double-strand breaks) and/or neutral condition (for only double-strand breaks). The cells are fixed with 70% ethanol and stained using SYBR Green. The stained DNAs are observed under a fluorescent microscope, and the images are analyzed by the tail length, tail percentage, and tail moment (Figure 12.2).

Although few studies have investigated the effect of IF exposure on DNA strand breaks, several investigations have been published. Sakurai *et al.* [53] conducted the comet assay to assess DNA strand breaks after exposure to IF magnetic fields. Figure 12.3 shows the effects on the tail length, tail percentage, and tail moment following exposure to an IF-EMF at 6.05 mTrms for 2 h. No statistically significant differences were observed between IF and sham exposures under alkaline or neutral conditions. In comparison to sham exposure, treatment with bleomycin as a positive control significantly elevated the values of all comet parameters. In addition, Shi *et al.* [54] evaluated the effects of IF-EMF generated by WPT based on magnetic resonance on human lens epithelial cells. The cells were exposed to 90 kHz at 93.36 mT for 2 and 4 h, and the comet assay was

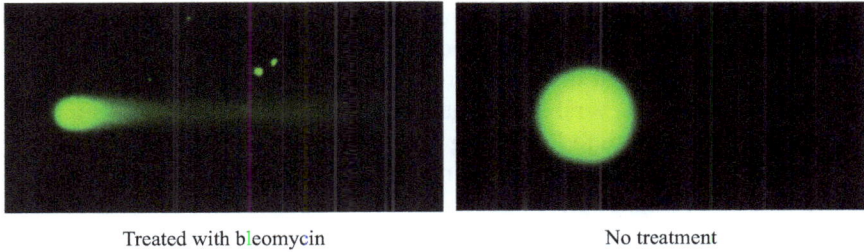

Treated with bleomycin   No treatment

Figure 12.2   *Typical photographs of the comet assay after treatment with bleomycin and without treatment, inducing DNA strand breaks*

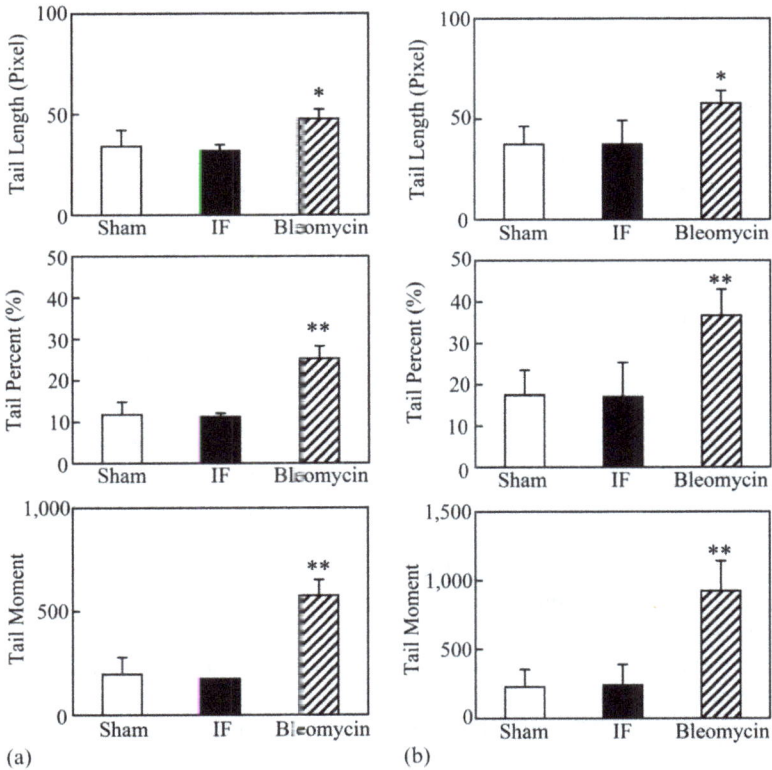

(a)   (b)

Figure 12.3   *Values of the comet parameters (tail length, tail percentage, and tail moment) after electrophoresis under neutral (a) and alkaline (b) conditions. Cells were exposed to sham conditions, IF-EMF at 6.05 mTrms for 2 h or 100 mg/mL bleomycin. *p < 0.05; **p < 0.01 (redrawing data from Ref. [53]).*

conducted. No significant difference was observed in DNA damage between 90 kHz IF field exposure and sham exposure groups. The data about IF on DNA strand breaks are insufficient and more data are required to confirm the effects of IF on DNA damage. Indeed, several studies have demonstrated positive results on DNA strand breaks by exposure to RF. Zhang *et al.* [55] indicated that the low-intensity 2.45-GHz microwave radiation cannot induce DNA and chromosomal damage but can increase DNA damage effect induced by MMC in the comet assay. In addition, Phillips *et al.* [56] indicated that exposure to RF could directly affect DNA and DNA repair mechanisms. However, most papers investigating DNA strand breaks by exposure to RF indicated negative results. No effect of 2.45-GHz RF exposure was found for 2-h exposures at SARs of 25, 70, and 100 W/kg [57]. Similarly, Malyapa *et al.* [58] investigated the chronological effects of 2.45-GHz exposure on cells in the logarithmic growth phase at 0.7 or 1.9 W/kg using the comet assay and indicated no difference in RF-exposed group compared with the sham-exposed group. Several other studies [59–61] have indicated negative results on DNA strand breaks by exposure to RF.

Despite some positive investigations, almost all studies support the consensus that exposure to RF does not break DNA bonds.

12.4.1.3 Micronucleus formation

MN formation is studied in many papers for detecting genotoxicity in cells. This method has been in use for a long time and has recently been considered valuable in detecting cancer [62–65]. For EMF experiments, MN is useful and conducted at many laboratories. A brief protocol is as follows. After cells are exposed to EMF or chemical agents, they are incubated with cytochalasin to arrest the cell division and create binucleated cells. Then, cells are collected and centrifuged on a glass slide by cytospin. Next, cells are fixed with ethanol and stained with propidium iodide and observed under a fluorescent microscope. Of note, 10,000 binucleated cells are examined, and the number of cells including MN is counted (Figure 12.4).

To date, only one study has conducted the MN formation test by exposure to IF [66]. Sakurai *et al.* [53] revealed that MN formation did not increase in cells exposed to the IF field compared with sham-exposed CHO-K1 cells (Figure 12.5). Although the data are insufficient, it does not seem that IF induces MN formation. As the scientific evaluation of potential risks to create MN is scarce, other replicated experiments should be performed to validate the findings.

Conversely, several studies have been conducted on the MN formation test by exposure to RF. However, only a few studies indicated an increase in MN formation by exposure to RF and a majority revealed no increase in MN frequency. Garaj-Vrhovac *et al.* [67] indicated a time-dependent increase in MN formation exposed to 415-MHz RF in human white blood cells. In addition, Zotti-Martelli *et al.* [67] demonstrated the induction of MN formation at 2.45 and 7.7 GHz at a power density of 30 mW/cm^2 for 30 and 60 min. However, many other studies [44,68,69] have indicated negative results for an increase in MN formation following exposure to RF. Koyama *et al.* [70] exposed cells to 2.45-GHz RF for 2 h at a SAR of 5–200 W/kg and investigated the combined effects of RF with chemical

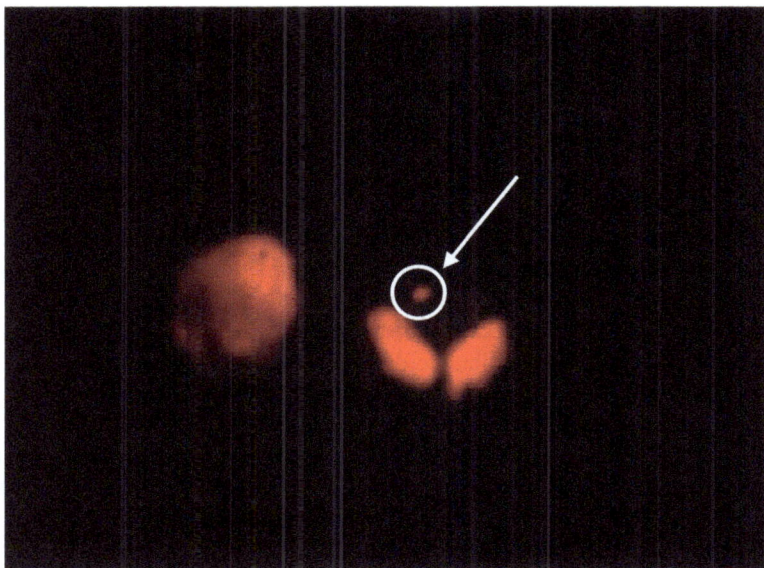

Figure 12.4   A typical photograph of a binucleated cell with MN

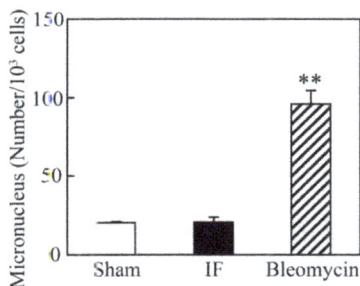

*Figure 12.5   The frequency of induced MN formation in CHO-K1 cells exposed to sham conditions, with IF magnetic field at 6.05 mTrms for 2 h or 10 mg/mL bleomycin. A total of 1,000 binucleate cells were scored. **p < 0.01 (redrawing data from Ref. [53]).*

agent bleomycin. No differences were found between RF alone up to 50 W/kg of SAR; however, the MN formation increased at 100 and 200 W/kg. Regarding combined RF + bleomycin exposure, an increase in MN was observed only at 200 W/kg. Furthermore, at a high temperature (39°C and higher up to 42°C), MN formation increased in a dose-dependent manner. In combined exposure with heat treatment and bleomycin, an increased MN formation was detected only at 42°C. These results suggest that exposure to RF at SAR intensities less than those associated with the normal environments encountered in daily life do not affect the MN formation. MN formation is observed at extremely high SAR associated with rising temperature.

Cells exposed to RF-EMF do not display any increase in MN formation; however, relatively high SAR with elevated temperature may induce MN formation.

12.4.1.4 Mutation

Mutation is the permanent alteration of the DNA nucleotide sequence of the genome in cells. These DNA base sequence changes sometimes lead to harmful results on the cell viability, cancer risks, protein dysfunction, and so on. Although chromosomal aberration, DNA strand breaks, and MN formation indicate genotoxicity, the detection of a mutation is crucial to detect the genotoxicity. The hypoxanthine–guanine phosphoribosyltransferase (HPRT) test has often been used as follows. First, cells are cultured in a medium containing hypoxanthine–aminoprotein–thymidine (HAT medium) to exclude cells with spontaneous HPRT mutations. After transferring to a normal medium, cells are exposed to EMF and incubated for six to ten folds of the doubling time as mutation expression time. After that, cells are incubated in the medium containing 6-thioguanine (6-TG) where only cells with *HPRT* gene mutation can survive until the colony formation is gained. Finally, colonies are stained and counted to calculate the mutation frequency.

Mutations exposed to IF are not adequately investigated. To date, only one study conducted by Sakurai *et al.* [53] has investigated this topic. Figure 12.6 shows the mutation frequencies of the *HPRT* gene following exposure to IF magnetic fields at 6.05 mTrms, sham exposure, or treatment with ethyl methanesulfonate (EMS). After treatment with EMS, as a positive control, the mutation frequency was significantly elevated compared with the sham-exposed cells. In contrast, no difference was noted in the mutation frequency between IF field- and sham-exposed cells.

Despite insufficient data, as mentioned earlier, it does not seem that IF induces mutations.

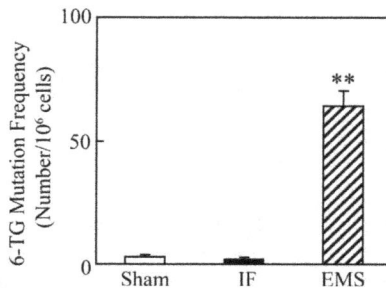

Figure 12.6   6-TG mutation frequency in V-79 cells exposed to the sham conditions, IF magnetic field at 6.05 mTrms or EMS (0.2 mg/mL). **p < 0.01 (redrawing data from Ref. [53]).

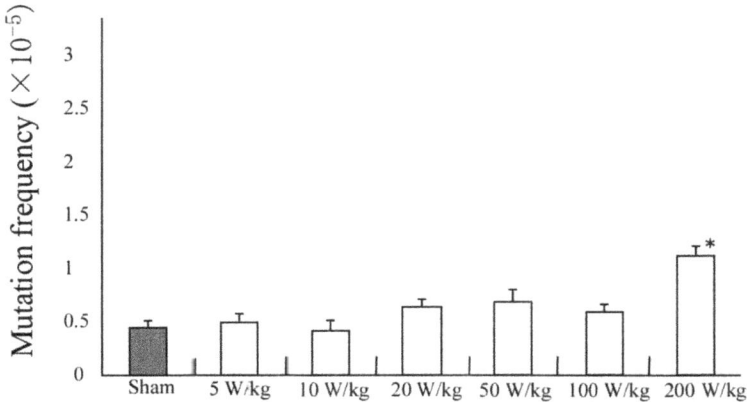

Figure 12.7 Mutation frequency at the HPRT gene in CHO-K1 cells by exposure to 2.45-GHz RF-EMF. *p < 0.05 (redrawing data from Ref. [72]).

Currently, RF-EMF has not been found to induce mutations in cells. Meltz *et al.* [71] used mice leukemia cells to detect mutations of the thymidine kinase gene locus induced by 2.45-GHz RF alone and combined with a concomitant chemical agent. No differences in the frequency of mutation were found between RF-exposed and sham-exposed groups. In addition, we have examined the effects of 2.45-GHz RF on bacterial mutations and the *HPRT* gene mutations in CHO-K1 cells [72]. In the HPRT test, cells were exposed to 2.45-GHz RF field for 2 h at SARs from 5 to 200 W/kg. RF-EMF alone did not induce HPRT mutations up to 100 W/kg. The mutations detected at 200 W/kg were caused by the temperature increase along with high SAR (Figure 12.7).

12.4.2 Nongenotoxic effects

Research on nongenotoxicity exposed to EMF is also one of the key fields because some endpoints indicate positive results by exposure to RF. Although not many studies on IF exist, several papers have been published.

12.4.2.1 Cell proliferation and differentiation

Cell growth and survival are major criteria for assessing the cellular effects of an external factor. Severe damage inhibits or suppresses the cell proliferation and leads to cell death. Changes in the cell proliferation occur because of many phenomena, including apoptosis, necrosis, cell cycle arrest, and malignant transformation. If IF-EMF affects any of these phenomena, they would likely influence the cell proliferation. Sakurai *et al.* [53] described the growth curves of CHO-K1 cells after exposure to sham conditions or IF-EMF at 6.05 mTrms for 2 h or 2-Gy X-ray irradiation (Figure 12.8). No significant difference in the growth rate was observed between sham and IF field exposures. In contrast, the growth of cells was significantly inhibited by X-ray irradiation compared with the sham exposure.

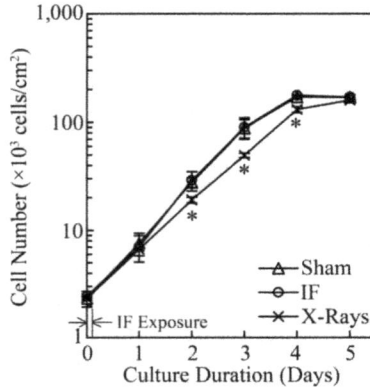

*Figure 12.8   Growth curves of CHO-K1 cells exposed to IF-EMF at 6.05 mTrms for 2 h, sham exposure, or treated with 2-Gy X-rays. *p < 0.05 (redrawing data from Ref. [53]).*

Despite insufficient data on cell proliferation by exposure to IF, several studies have investigated exposure to RF and indicated negative results, except few positive data. Maes *et al.* [43] found no effect of 2.45-GHz RF exposure on the number of cell divisions. In addition, Takashima *et al.* [73] compared the effects of continuous and intermittent exposure of 2.45 GHz at high SARs on the cell growth, survival, and cell cycle distribution. When cells were exposed to a continuous RF at SARs of 0.05–100 W/kg for 2 h, cell growth, survival, and cell cycle distribution were not affected. A recent study [74] indicated that an increase in the cell proliferation after 1-h exposure to 1,800 MHz and a decrease in both the 900- and 1,800 MHz–exposed cells after 4 h was observed. However, in other experimental conditions tested, no effect on the cell proliferation was observed.

Although high temperature decreases the cell proliferation, exposure to RF without an increase in temperature is considered to have no effect.

12.4.2.2   Gene and protein expression

Gene expression is the process by which information from some genes is used in the synthesis of a functional gene product, which are usually proteins. In addition, the protein expression is almost the same but is targeted on protein. The most attractive issue at the cellular level is the effect on HSP, whose expression is induced by various stresses on the gene expression.

Sakurai *et al.* [53] investigated an expression of phosphorylated Hsp27 and immunocytochemical staining of HSP. The phosphorylation of Hsp27 occurs following the stimulation of stress, and the conformation of Hsp27 can change as because of the phosphorylation, which may regulate its activity. They evaluated the expression of the phosphorylated form of Hsp27 by Western blotting (Figure 12.9). The expression of anti-phospho-Hsp27 was not affected by exposure to IF at 6.05 mTrms, although the expression was significantly increased in cells treated with

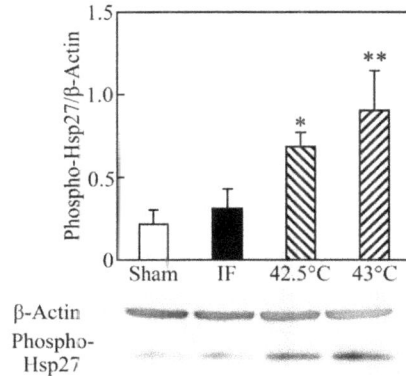

Figure 12.9   *The expression of phosphorylated Hsp27 in cells exposed to the sham conditions, IF magnetic field at 6.05 mTrms for 2 h or heat at 42.5°C or 43°C. The expression of phosphorylated Hsp27 was standardized to that of β-actin. The photograph shows typical results of Western blotting. *p < 0.05; **p < 0.01 (redrawing data from Ref. [53]).*

heat. Some HSPs translocate into the nucleus after treatment with heat. Furthermore, they conducted immunocytochemical staining for Hsp27, 70, and 105 to estimate whether the translocation of Hsp27, 70, or 105 might occur following exposure to IF at 6.05 mTrms. Exposure to IF did not cause translocation of Hsp27, 70, or 105. In contrast, the treatment with heat for 2 h induced nuclear translocation of Hsp27, 70, and 105.

In studies conducted on the RF field, the HSP expression is one of the most active area. Leszczynski *et al.* [75] suggested the possibility that 900-MHz RF exposure had some effects on signaling, particularly on the stress-responding mechanisms of Hsp27 and p38 mitogen-activated protein kinase. In addition, Tian *et al.* [74] and Wang *et al.* [76] investigated the effects of 2.45-GHz RF exposure on HSP expression. Their data indicated possible adverse effects of exposure to RF; however, the effects on high SAR seemed to be related to higher temperatures. This area is still under discussion to clarify the mechanisms of the HSP expression.

12.4.2.3   Immune response

The immune system protects hosts from infections, cancer, or allergens. Several researchers have investigated the immune response exposed to EMF. Despite insufficient data about IF on the immune response, Koyama *et al.* [77] explained some immune response experiments exposed to IF. They investigated the properties of IF-EMF and examined the effect of exposure to a 23-kHz IF-EMF of 2 mT for 2, 4, or 6 h on neutrophil chemotaxis and phagocytosis using differentiated human HL-60 cells. In comparison to sham exposure, exposure to IF magnetic field did not affect neutrophil chemotaxis or phagocytosis (Figures 12.10 and 12.11).

With the rapid increase in exposure to IF, there will be increasing issues related to immune response on exposure to IF.

Migration rate

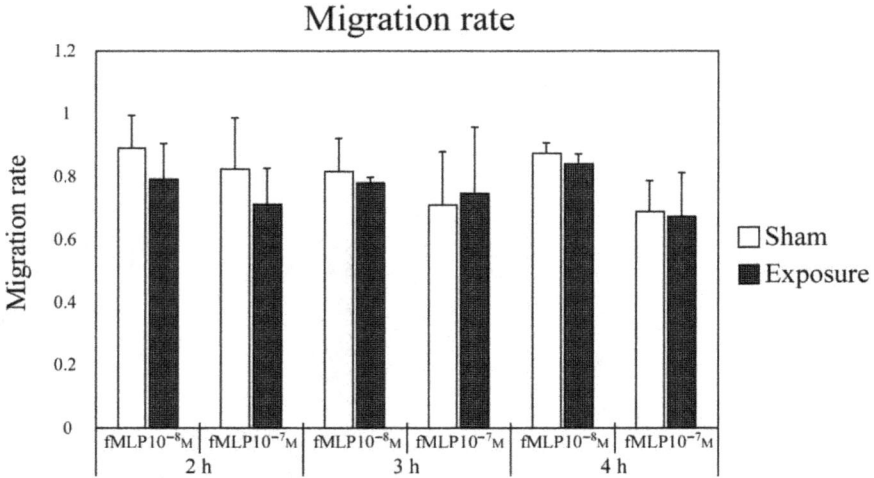

Figure 12.10   Chemotaxis (rate of migration) of differentiated HL-60 cells was assayed using 10 or 100 nM fMLP as the chemoattractant after sham or IF magnetic field exposure for 2, 3, or 4 h (redrawing data from Ref. [77])

Phagocytosis

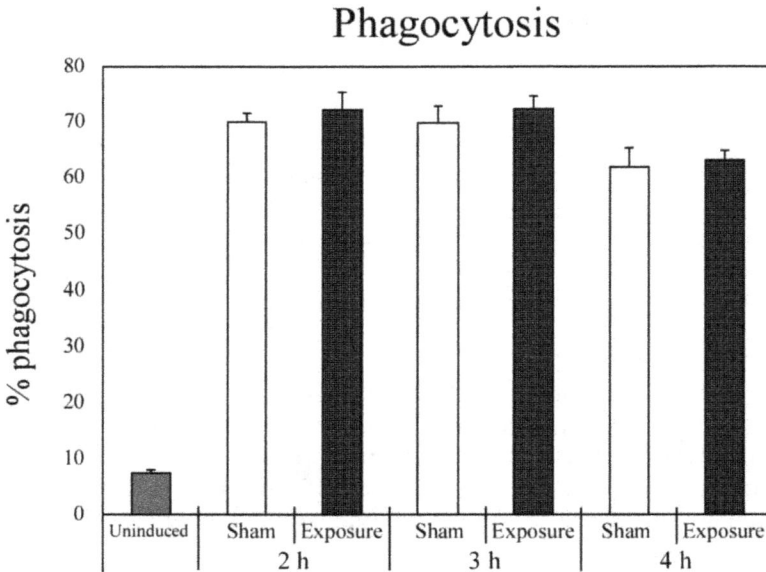

Figure 12.11   The percentage of microspheres phagocytosed by differentiated HL-60 cells after sham or IF magnetic field exposure for 2, 3, or 4 h (redrawing data from Ref. [77])

Furthermore, the immune response research on RF is attractive. Tuschl *et al.* [78] reported the effect of 1950-MHz RF on human immune cells. No statistically significant effects of exposure were established. In addition, Thorlin *et al.* [79] indicated that the study did not provide evidence for any effect of 900-MHz RF used on damage-related factors, such as proinflammatory cytokines IL-6 and TNF-α, in glial cells. However, some groups [80,81] indicated positive effects of exposure to RF on the immune cell activity. Further investigations are required to clarify the effects of exposure to RF on the immune response.

12.4.2.4 Apoptosis

Apoptosis is a process of programmed cell death that occurs in multicellular organisms and a form of cell death that is actively induced by the cell itself to maintain normal status. To date, only one study targeting IF exposure-related apoptosis was conducted by Shi *et al.* [54], as described earlier. The authors did not detect any effect on apoptosis. In the absence of any investigation of apoptosis on the IF exposure, further studies are required.

Thus far, several studies have indicated negative results on RF exposure-related apoptosis [82–85]. Consequently, it is improbable that exposure to RF induces apoptosis; however, several recent studies have also indicated positive results for RF exposure-related apoptosis [86–89]. These results suggest that the mechanisms might be related to the formation of the reactive oxygen species. It is essential to investigate some replicated experiments to validate the mechanisms of positive results.

12.5   Conclusions on IF and RF studies

A handful of cellular studies exist on IF exposure. The usage of IF is now spreading all over the world, and more devices are using this frequency. Although the research on IF is urgent, so far it seems that the low-power intensity of IF might not induce any adverse effects on the cellular response. However, no general conclusions can be drawn on this frequency range.

There are still contradictory results about the effects of exposure to RF on the cellular response. However, RF energy does not directly cleave intercellular DNA since most genotoxicity studies have indicated negative results. Some positive data at high SARs seem to relate mainly thermal effect by exposure to RF. Several interesting results of HSP and apoptosis should be validated shortly. Further studies on RF effects using improved biotechnological methods are warranted for precise conclusions.

References

[1]   Wertheimer N, and Leeper E. (1979) Electrical wiring configurations and childhood cancer. *Am J Epidemiol.* 109(3):273–284.

[2]   Ahlbom A, Day N, Feychting M, *et al.* (2000) A pooled analysis of magnetic fields and childhood leukaemia. *Br J Cancer.* 83(5):692–698.

[3]   Feychting M, and Ahlbom A. (1993) Magnetic fields and cancer in children residing near Swedish high-voltage power lines. *Am J Epidemiol.* 138 (7):467–481.

[4]   McCann J, Dietrich F, Rafferty C, and Martin AO. (1993) A critical review of the genotoxic potential of electric and magnetic fields. *Mutat Res.* 297(1):61–95.

[5]   Stevens RG. (1993) Biologically based epidemiological studies of electric power and cancer. *Environ Health Perspect.* 101(Suppl 4):93–100.

[6]   Murphy JC, Kaden DA, Warren J, and Sivak A. (1993) International commission for protection against environmental mutagens and carcinogens. Power frequency electric and magnetic fields: a review of genetic toxicology. *Mutat Res.* 296(3):221–240.

[7]   Horn Y. (1995) The potential carcinogenic hazard of electromagnetic radiation: a review. *Cancer Detect Prev.* 19:244–249.

[8]   Richter E, Berman T, Ben-Michael E, Laster R, and Westin JB. (2000) Cancer in radar technicians exposed to radiofrequency/microwave radiation: sentinel episodes. *Int J Occup Environ Health.* 6:187–193.

[9]   Santini R, Seigne M, and Bonhomme-Faivre L. (2000) Danger of cellular telephones and their relay stations. *Pathol Biol (Paris).* 48:525–528.

[10]   Jauchem JR. (2008) Effects of low-level radio-frequency (3kHz to 300GHz) energy on human cardiovascular, reproductive, immune, and other systems: a review of the recent literature. *Int J Hyg Environ Health.* 211:1–29.

[11]   Jauchem JR. (2003) A literature review of medical side effects from radio-frequency energy in the human environment: involving cancer, tumors, and problems of the central nervous system. *J Microw Power Electromagn Energy.* 38:103–123.

[12]   Ahlbom A, Green A, Kheifets L, Savitz D, and Swerdlow A. (2004) International commission for non-ionizing radiation protection standing committee on epidemiology. Epidemiology of health effects of radiofrequency exposure. *Environ Health Perspect.* 112:1741–1754.

[13]   International Commission on Non-Ionizing Radiation Protection (ICNIRP). (2010) Guidelines for limiting exposure to time-varying electric and magnetic fields (1 Hz to 100 kHz). *Health Phys.* 99(6):818–836. doi:10.1097/HP. 0b013e3181f06c86.

[14]   IARC. (2013) Monographs on the Evaluation of Carcinogenic Risks to Humans, Volume 102. Non-Ionizing radiation, Part II: Radiofrequency Electromagnetic Fields [includes mobile telephones]. Lyon: IARC.

[15]   Hardell L, Carlberg M, and Mild KH. (2013) Use of mobile phones and cordless phones is associated with increased risk for glioma and acoustic neuroma. *Pathophysiology.* 20:85–110.

[16]   Carlberg M, Hendendahl L, Ahonen M, Koppel T, and Hardell L. (2016) Increasing incidence of thyroid cancer in the Nordic countries with main focus on Swedish data. *BMC Cancer.* 16:426.

[17] Brown FA, and Scow KM. (1977) Magnetic induction of a circadian cycle in hamsters. *J Interdiscipl Cycle Res.* 9:137–145.

[18] Dowse HB, and Palmer JD. (1969) Entrainment of circadian activity rhythms in mice by electrostatic fields. *Nature.* 222(5193):564–566.

[19] Demaine C, and Semm P. (1986) Magnetic fields abolish nychthemeral rhythmicity of responses of Purkinje cells to the pineal hormone melatonin in the pigeon's cerebellum. *Neurosci Lett.* 72(2):158–162.

[20] Lerchl A, Nonaka KO, Stokkan KA, and Reiter RJ. (1990) Marked rapid alterations in nocturnal pineal serotonin metabolism in mice and rats exposed to weak intermittent magnetic fields. *Biochem Biophys Res Commun.* 169 (1):102–108.

[21] Wilson BW, Anderson LE, Hilton DI, and Phillips RD. (1981) Chronic exposure to 60-Hz electric fields: effects on pineal function in the rat. *Bioelectromagnetics.* 2(4):371–380.

[22] Wilson BW, Stevens RG, and Anderson LE. (1989) Neuroendocrine mediated effects of electromagnetic-field exposure: possible role of the pineal gland. *Life Sci.* 45(15):1319–1332.

[23] de Bruyn L, de Jager L, and Kuyl JM. (2001) The influence of long-term exposure of mice to randomly varied power frequency magnetic fields on their nocturnal melatonin secretion patterns. *Environ Res.* 85(2):115–121.

[24] Bakos J, Nagy N, Thuróczy G, and Szabó LD. (2002) One week of exposure to 50 Hz, vertical magnetic field does not reduce urinary 6-sulphatoxymelatonin excretion of male wistar rats. *Bioelectromagnetics.* 23(3):245–248.

[25] Burchard JF, Nguyen DH, and Block E. (1998) Effects of electric and magnetic fields on nocturnal melatonin concentrations in dairy cows. *J Dairy Sci.* 81(3):722–727.

[26] Ushiyama A, Ohtani S, Suzuki Y, Wada K, Kunugita N, and Ohkubo C. (2014) Effects of 21-kHz intermediate frequency magnetic fields on blood properties and immune systems of juvenile rats. *Int J Radiat Biol.* 90 (12):1211–1217. doi:10.3109/09553002.2014.930538.

[27] Nishimura I, Oshima A, Shibuya K, Mitani T, and Negishi T. (2015) Acute and subchronic toxicity of 20 kHz and 60 kHz magnetic fields in rats. *J Appl Toxicol.* 36(2):199–210. doi:10.1002/jat.3161.

[28] Nishimura I, Tanaka K, and Negishi T. (2013) Intermediate frequency magnetic field and chick embryotoxicity. *Congenit Anom (Kyoto).* 53 (3):115–121. doi:10.1111/cga.12018.

[29] Kim SH, Lee HJ, Choi SY, *et al.* (2006) Toxicity bioassay in Sprague-Dawley rats exposed to 20 kHz triangular magnetic field for 90 days. *Bioelectromagnetics.* 27(2):105–111.

[30] Lee HJ, Kim SH, Choi SY, *et al.* (2006) Long-term exposure of Sprague Dawley rats to 20 kHz triangular magnetic fields. *Int J Radiat Biol.* 82 (4):285–291.

[31] Lee HJ, Gimm YM, Choi HD, Kim N, Kim SH, and Lee YS. (2010) Chronic exposure of Sprague-Dawley rats to 20 kHz triangular magnetic fields. *Int J Radiat Biol.* 86(5):384–389. doi:10.3109/09553000903567920.

[32]   Svedenstål BM, and Johanson KJ. (1998) Effects of exposure to 50Hz or 20kHz magnetic fields on weights of body and some organs of CBA mice. *In Vivo.* 12(3):293–298.

[33]   Svedenstål BM, and Johanson KJ. (1998) Leukocytes and micronucleated erythrocytes in peripheral blood from mice exposed to 50 Hz or 20 kHz magnetic fields. *Electro Magnetobiol.* 17(2):127–143.

[34]   Deshmukh PS, Nasare N, Megha K, *et al.* (2015) Cognitive impairment and neurogenotoxic effects in rats exposed to low-intensity microwave radiation. *Int J Toxicol.* 34(3):284–290. doi:10.1177/1091581815574348.

[35]   Tang J, Zhang Y, Yang L, *et al.* (2015) Exposure to 900 MHz electromagnetic fields activates the mkp-1/ERK pathway and causes blood-brain barrier damage and cognitive impairment in rats. *Brain Res.* 1601:92–101. doi:10.1016/j.brainres.2015.01.019.

[36]   Lerchl A, Klose M, Grote K, *et al.* (2015) Tumor promotion by exposure to radiofrequency electromagnetic fields below exposure limits for humans. *Biochem Biophys Res Commun.* 459(4):585–590. doi:10.1016/j.bbrc.2015.02.151.

[37]   Tillmann T, Ernst H, Streckert J, *et al.* (2010) Indication of cocarcinogenic potential of chronic UMTS-modulated radiofrequency exposure in an ethylnitrosourea mouse model. *Int J Radiat Biol.* 86(7):529–541. doi:10.3109/09553001003734501.

[38]   Masuda H, Hirota S, Ushiyama A, *et al.* (2015) No dynamic changes in blood-brain barrier permeability occur in developing rats during local cortex exposure to microwaves. *In Vivo.* 29(3):351–357.

[39]   Kumar G, McIntosh RL, Anderson V, McKenzie RJ, and Wood AW. (2015) A genotoxic analysis of the hematopoietic system after mobile phone type radiation exposure in rats. *Int J Radiat Biol.* 91(8):664–672. doi:10.3109/09553002.2015.1047988.

[40]   Ohtani S, Ushiyama A, Maeda M, *et al.* (2015) The effects of radiofrequency electromagnetic fields on T cell function during development. *J Radiat Res.* 56(3):467–474. doi:10.1093/jrr/rru126.

[41]   Garaj-Vrhovac V, Horvat D, and Koren Z. (1991) The relationship between colony-forming ability, chromosome aberrations and incidence of micronuclei in V79 Chinese hamster cells exposed to microwave radiation. *Mutat Res.* 263(3):143–149.

[42]   Mashevich M, Folkman D, Kesar A, *et al.* (2003) Exposure of human peripheral blood lymphocytes to electromagnetic fields associated with cellular phones leads to chromosomal instability. *Bioelectromagnetics.* 24(2):82–90.

[43]   Maes A, Verschaeve L, Arroyo A, De Wagter C, and Vercruyssen L. (1993) In vitro cytogenetic effects of 2450 MHz waves on human peripheral blood lymphocytes. *Bioelectromagnetics.* 14(6):495–501.

[44]   Vijayalaxmi, Leal BZ, Meltz ML, *et al.* (2001) Cytogenetic studies in human blood lymphocytes exposed in vitro to radiofrequency radiation at a cellular telephone frequency (835.62 MHz, FDMA). *Radiat Res.* 155(1 Pt 1):113–121.

[45] Vijayalaxmi, Bisht KS, Pickard WF, Meltz ML, Roti Roti JL, and Moros EG. (2001) Chromosome damage and micronucleus formation in human blood lymphocytes exposed in vitro to radiofrequency radiation at a cellular telephone frequency (847.74 MHz, CDMA). *Radiat Res.* 156(4):430–432.

[46] Kerbacher JJ, Meltz ML, and Erwin DN. (1990) Influence of radiofrequency radiation on chromosome aberrations in CHO cells and its interaction with DNA-damaging agents. *Radiat Res.* 123(3):311–319.

[47] Vijayalaxmi, Mohan N, Meltz ML, and Wittler MA. (1997) Proliferation and cytogenetic studies in human blood lymphocytes exposed in vitro to 2450 MHz radiofrequency radiation. *Int J Radiat Biol.* 72(6):751–757.

[48] Maes A, Collier M, Slaets D, and Verschaeve L. (1996) 954 MHz microwaves enhance the mutagenic properties of mitomycin C. *Environ Mol Mutagen.* 28(1):26–30.

[49] Maes A, Collier M, Van Gorp U, Vandoninck S, and Verschaeve L. (1997) Cytogenetic effects of 935.2-MHz (GSM) microwaves alone and in combination with mitomycin C. *Mutat Res.* 18(393(1–2)):151–156.

[50] Zeni O, Romanò M, Perrotta A, *et al.* (2005) Evaluation of genotoxic effects in human peripheral blood leukocytes following an acute in vitro exposure to 900 MHz radiofrequency fields. *Bioelectromagnetics.* 26(4):258–265.

[51] Vijayalaxmi. (2006) Cytogenetic studies in human blood lymphocytes exposed in vitro to 2.45 GHz or 8.2 GHz radiofrequency radiation. *Radiat Res.* 166(3):532–538.

[52] Komatsubara Y, Hirose H, Sakurai T, *et al.* (2005) Effect of high-frequency electromagnetic fields with a wide range of SARs on chromosomal aberrations in murine m5S cells. *Mutat Res.* 587(1–2):114–119.

[53] Sakurai T, Kiyokawa T, Kikuchi K, and Miyakoshi J. (2009) Intermediate frequency magnetic fields generated by an induction heating (IH) cooktop do not affect genotoxicities and expression of heat shock proteins. *Int J Radiat Biol.* 85(10):883–890.

[54] Shi D, Zhu C, Lu R, Mao S, and Qi Y. (2014) Intermediate frequency magnetic field generated by a wireless power transmission device does not cause genotoxicity in vitro. *Bioelectromagnetics.* 35(7):512–518. doi:10.1002/bem.21872.

[55] Zhang MB, He JL, Jin LF, and Lu DQ. (2002) Study of low-intensity 2450-MHz microwave exposure enhancing the genotoxic effects of mitomycin C using micronucleus test and comet assay in vitro. *Biomed Environ Sci.* 15(4):283–290.

[56] Phillips JL, Ivaschuk O, Ishida-Jones T, Jones RA, Campbell-Beachler M, and Haggren W. (1998) DNA damage in Molt-4 T-lymphoblastoid cells exposed to cellular telephone radiofrequency fields in vitro. *Bioelectrochem. Bioener.* 45:103–110.

[57] Miyakoshi J, Yoshida M, Tarusawa Y, Nojima T, Wake K, and Taki M. (2002) Effects of high-frequency electromagnetic fields on DNA strand breaks using comet assay method. *Electr Eng Jpn.* 141:9–15.

[58] Malyapa RS, Ahern EW, Straube WL, Moros EG, Pickard WF, and Roti Roti JL. (1997) Measurement of DNA damage after exposure to 2450 MHz electromagnetic radiation. *Radiat Res.* 148(6):608–617.

[59]   Lagroye I, Hook GJ, Wettring BA, *et al.* (2004) Measurements of alkali-labile DNA damage and protein-DNA crosslinks after 2450 MHz microwave and low-dose gamma irradiation in vitro. *Radiat Res.* 161(2):201–214.

[60]   McNamee JP, Bellier PV, Gajda GB, *et al.* (2002) DNA damage in human leukocytes after acute in vitro exposure to a 1.9 GHz pulse-modulated radiofrequency field. *Radiat Res.* 158(4):534–537.

[61]   Speit G, Schütz P, and Hoffmann H. (2007) Genotoxic effects of exposure to radiofrequency electromagnetic fields (RF-EMF) in cultured mammalian cells are not independently reproducible. *Mutat Res.* 626(1–2):42–47.

[62]   Stephens PJ, Greenman CD, Fu B, *et al.* (2011) Massive genomic rearrangement acquired in a single catastrophic event during cancer development. *Cell.* 144:27–40.

[63]   Terradas M, Martín M, Tusell L, and Genescà A. (2010) Genetic activities in micronuclei: is the DNA entrapped in micronuclei lost for the cell? *Mutat Res.* 705:60–67.

[64]   Vargas JD, Hatch EM, Anderson DJ, and Hetzer MW. (2012) Transient nuclear envelope rupturing during interphase in human cancer cells. *Nucleus.* 3:88–100.

[65]   Vral A, Fenech M, and Thierens H. (2011) The micronucleus assay as a biological dosimeter of in vivo ionising radiation exposure. *Mutagenesis.* 26:11–17.

[66]   Garaj-Vrhovac V, Fucic A, Kubelka D, and Vojvodic S. (1996) Effects of 415 MHz frequency on human lymphocyte genome. *IRPA9: International Congress on Radiation Protection. Proceedings.* Volume 3:604–606.

[67]   Zotti-Martelli L, Peccatori M, Scarpato R, and Migliore L. (2007) Induction of micronuclei in human lymphocytes exposed in vitro to microwave radiation. *Mutat Res.* 472(1–2):51–58.

[68]   Bisht KS, Moros EG, Straube WL, Baty JD, and Roti Roti JL. (2002) The effect of 835.62 MHz FDMA or 847.74 MHz CDMA modulated radiofrequency radiation on the induction of micronuclei in C3H 10T(1/2) cells. *Radiat Res.* 157(5):506–515.

[69]   McNamee JP, Bellier PV, Gajda GB, *et al.* (2003) No evidence for genotoxic effects from 24 h exposure of human leukocytes to 1.9 GHz radiofrequency fields. *Radiat Res.* 159(5):693–697.

[70]   Zeni O, Chiavoni AS, Sannino A, *et al.* (2003) Lack of genotoxic effects (micronucleus induction) in human lymphocytes exposed in vitro to 900 MHz electromagnetic fields. *Radiat Res.* 160(2):152–158.

[71]   Meltz ML, Eagan P, and Erwin DN. (1990) Proflavin and microwave radiation: absence of a mutagenic interaction. *Bioelectromagnetics.* 11(2):149–157.

[72]   Koyama S, Takashima Y, Sakurai T, Suzuki Y, Taki M, and Miyakoshi J. (2007) Effects of 2.45 GHz electromagnetic fields with a wide range of SARs on bacterial and HPRT gene mutations. *J Radiat Res.* 48(1):69–75.

[73]   Takashima Y, Hirose H, Koyama S, Suzuki Y, Taki M, and Miyakoshi J. (2006) Effects of continuous and intermittent exposure to RF fields with a wide range of SARs on cell growth, survival, and cell cycle distribution. *Bioelectromagnetics.* 27(5):392–400.

[74] Tian F, Nakahara T, Wake K, Taki M, and Miyakoshi J. (2002) Exposure to 2.45 GHz electromagnetic fields induces hsp70 at a high SAR of more than 20 W/kg but not at 5W/kg in human glioma MO54 cells. *Int J Radiat Biol.* 78 (5):433–440.

[75] Leszczynski D, Joenväärä S, Reivinen J, and Kuokka R. (2002) Non-thermal activation of the hsp27/p38MAPK stress pathway by mobile phone radiation in human endothelial cells: molecular mechanism for cancer- and blood-brain barrier-related effects. *Differentiation.* 70(2–3):120–129.

[76] Wang J, Koyama S, Komatsubara Y, Suzuki Y, Taki M, and Miyakoshi J. (2006) Effects of a 2450 MHz high-frequency electromagnetic field with a wide range of SARs on the induction of heat-shock proteins in A172 cells. *Bioelectromagnetics.* 27(6):479–486.

[77] Koyama S, Narita E, Shinohara N, and Miyakoshi J. (2014) Effect of an intermediate-frequency magnetic field of 23 kHz at 2 mT on chemotaxis and phagocytosis in neutrophil-like differentiated human HL-60 cells. *Int J Environ Res Public Health.* 11(9):9649–9659. doi:10.3390/ijerph110909649.

[78] Tuschl H, Novak W, and Molla-Djafari H. (2006) In vitro effects of GSM modulated radiofrequency fields on human immune cells. *Bioelectromagnetics.* 27(3):188–196.

[79] Thorlin T, Rouquette JM, Hamnerius Y, *et al.* (2006) Exposure of cultured astroglial and microglial brain cells to 900 MHz microwave radiation. *Radiat Res.* 166(2):409–421.

[80] Dabrowski MP, Stankiewicz W, Kubacki R, Sobiczewska E, and Szmigielski S. (2003) Immunotropic effects in cultured human blood mononuclear cells pre-exposed to low-level 1300 MHz pulse-modulated microwave field. *Electromagn Biol Med.* 22:1–13.

[81] Stankiewicz W, Dabrowski MP, Kubacki R, Sobiczewska E, and Szmigielski S. (2006) Immunotropic influence of 900 MHz microwave GSM signal on human blood immune cells activated in vitro. *Electromagn Biol Med.* 25(1):45–51.

[82] Hirose H, Sakuma N, Kaji N, *et al.* (2006) Phosphorylation and gene expression of p53 are not affected in human cells exposed to 2.1425 GHz band CW or W-CDMA modulated radiation allocated to mobile radio base stations. *Bioelectromagnetics.* 27(6):494–504.

[83] Lantow M, Viergutz T, Weiss DG, and Simkó M. (2006) Comparative study of cell cycle kinetics and induction of apoptosis or necrosis after exposure of human Mono Mac 6 cells to radiofrequency radiation. *Radiat Res.* 166 (3):539–543.

[84] Joubert V, Leveque P, Cueille M, Bourthoumieu S, and Yardin C. (2007) No apoptosis is induced in rat cortical neurons exposed to GSM phone fields. *Bioelectromagnetics.* 28(2):115–121.

[85] Merola P, Marino C, Lovisolo GA, Pinto R, Laconi C, and Negroni A. (2006) Proliferation and apoptosis in a neuroblastoma cell line exposed to 900 MHz modulated radiofrequency field. *Bioelectromagnetics.* 27(3):164–171.

[86] Xing F, Zhan Q, He Y, Cui J, He S, and Wang G. (2016) 1800MHz microwave induces p53 and p53-mediated caspase-3 activation leading to cell apoptosis in vitro. *PLoS One*. 11(9):e0163935. doi:10.1371/journal.pone.0163935.

[87] Canseven AG, Esmekaya MA, Kayhan H, Tuysuz MZ, and Seyhan N. (2015) Effects of microwave exposure and Gemcitabine treatment on apoptotic activity in Burkitt's lymphoma (Raji) cells. *Electromagn Biol Med*. 34 (4):322–326. doi:10.3109/15368378.2014.919591.

[88] Hou Q, Wang M, Wu S, *et al.* (2015) Oxidative changes and apoptosis induced by 1800-MHz electromagnetic radiation in NIH/3T3 cells. *Electromagn Biol Med*. 34(1):85–92. doi:10.3109/15368378.2014.900507.

[89] Koyama S, Isozumi Y, Suzuki Y, Taki M, and Miyakoshi J. (2004) Effects of 2.45-GHz electromagnetic fields with a wide range of SARs on micronucleus formation in CHO-K1 cells. *ScientificWorldJournal*. 4(Suppl 2):29–40.

Chapter 13

Impact of electromagnetic interference arising from wireless power transfer upon implantable medical device

Takashi Hikage[1]

Wireless power transfer systems (WPTSs) using nonradiative resonant coupling have attracted considerable research attention and are expected to achieve novel wireless charging and power supply functions for low- and high-power applications, such as home appliances, electric vehicles (EVs), and other electric systems. Because these systems can generate high-strength reactive electromagnetic fields (EMFs), electromagnetic interference (EMI) issues, and human safety must be assessed.

There are various international organizations dealing with WPTS standardization [1,2]. The considered WPTSs have numerous applications, power ranges, and frequency ranges. EMI with other electronic equipment, which affects the progress of applications and commercialization of WPTSs, is a very important issue. To assess the potential for interference, we can refer to international standards. For example, WPTSs using nonradiative resonant coupling are required to conform to the free-space electric- and magnetic field limits cited by the CISPR 11 standard for wanted emissions at the operating frequency and its harmonics and also for unwanted (spurious) emissions up to 18 GHz [3]. The CISPR/B/WG1/AHG4 (Ad hoc Group of WPT for EV) has been continuing activities to add the limit level of radiated/conducted disturbances and the measurement method of WPT for electric vehicle (EV) to the current CISPR11.

Human safety is a significant concern. RF exposure limits in human tissue are provided by international bodies such as the International Commission on Non-Ionizing Radiation Protection [4–6] and the IEEE [7–9]. Compliance with basic restrictions on specific absorption rate (SAR), induced current, and electric field has been investigated [10–12], and WPTS must comply with electromagnetic compatibility (EMC) standards and regulations to guarantee interoperability and safety [13,14].

The International Electrotechnical Commission (IEC) TC106 formed a Working Group (WG9) in 2015 intended to address the following: (a) identification of current gaps for WPT (including implanted medical devices) related to human exposures to electric, magnetic, and EMFs; (b) determinations of requirements for WPT related to exposures assessment.—both stimulation-based frequency ranges

---

[1]Faculty of Information Science and Technology, Hokkaido University, Japan

(<10 MHz, for example) and heat-based frequency ranges (>100 kHz, for example) should be considered; (c) summarization of possible assessment methods for WPT to be applied; (d) determinations whether new IEC standard is required or existing standard can be adapted; (e) preparation of a suitable document addressing this matter—technical report or other deliverables will be prepared; (f) consider potential liaisons with other WG/PT/MTs within IEC TC 106 and other TCs. In addition, the IEEE SA Standards Board authorized a project to develop new standard for the assessment of human exposure to EMFs from radiative WPTS: Measurement and Computational Methods (frequency range of 30 MHz to 300 GHz) in 2023. However, a concern still exists that, when an implantable medical device patient is near a WPTS, the EMFs may be sufficiently strong to cause device malfunction. The EMF distribution generated by resonant coupling-based WPTSs is complicated and varies with the coupling condition (e.g., frequency, air-gap, etc.). Hence, active implantable medical devices (AIMDs) such as cardiac pacemakers and implantable cardioverter defibrillators (ICDs), as well as the EMI risk posed by a WPTS, must be assessed using reliable measurement techniques. In addition, assessment of RF-induced heating of metal implants is important.

13.1   EMI studies on active implantable medical devices

Currently, several reports indicate that various electric or electronic devices may occasionally cause malfunctions in AIMDs via electromagnetic coupling (interference) [15–49]. One well-known example is the EMI created by cellular phones [16–32]. In addition, some reports have confirmed that some radio frequency identification (RFID) interrogators, particularly those operating in the low frequency (LF), high frequency, and microwave frequency bands, can trigger an AIMD malfunction [33–39]. The observed characteristics of the EMI were shown to depend upon the transmission radio wave specifications, AIMD type, and AIMD operation mode setting.

The WPTSs generate a time-varying magnetic field that can induce current and voltage on the pacing leads, in turn generating interference in the pacemaker circuitry. The main coupling mechanism between the pacemaker and the magnetic field generated by the WPTS using nonradiative resonant coupling is due to the loops formed by the leads through the human tissues. The standard methodology for evaluating the effects of EMI from WPTSs on AIMDs can basically be covered by the general EMC test protocol shown in ANSI/AAMI PC69, which covers various EMF-transmitting devices [44].

13.1.1   In vitro EMI measurement system for WPTSs

The EMI measurement system is expected to be applied to various types of WPTSs and AIMDs in practical use and capable of estimating the EMI effects and the maximum interference distance (MID) characteristics. Accordingly, the system comprises a saline-tank torso simulator (referred to as a torso phantom hereafter) with an AIMD, a WPTS as the device under testing (DUT), and electronic measurement devices.

13.1.1.1   Measurement system configuration

The measurement system comprises a torso phantom into which the AIMD can be inserted, a simulated ECG-signal generator/AIMD monitor, a chart recorder, an oscilloscope, a measurement platform, and the WPTS to be tested. In addition, an appropriate measurement protocol is used so that precise and reliable EMI estimations can be performed in an efficient way.

An example of the measurement system configuration with a vertical-type phantom is presented. The basic configuration of the system, as well as an illustration of the system in practice, is shown in Figures 13.1 and 13.2, respectively.

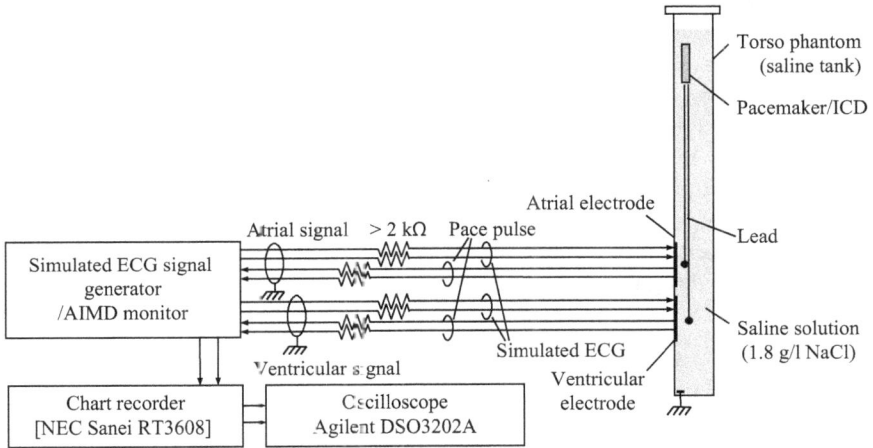

Figure 13.1   Basic configuration of the EMI test phantom

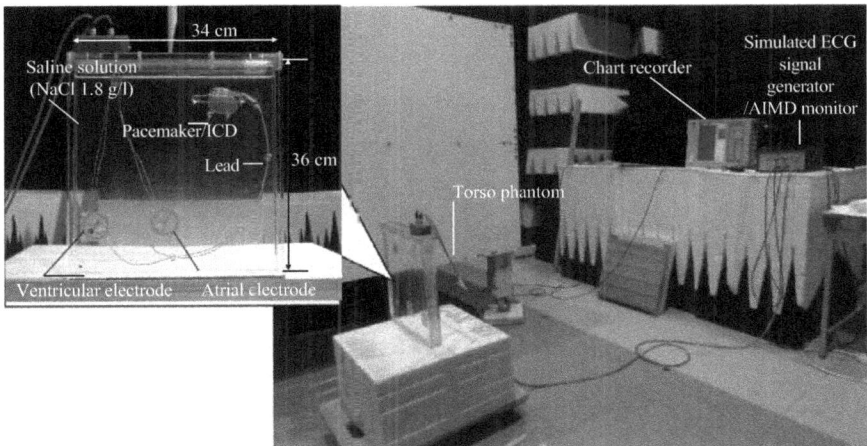

Figure 13.2   Overview of the EMI measurement system

This basic configuration mirrors those for other RF device tests, such as cellular phone handsets [23–27,29,31], base stations [32], and radio transceivers [15]. In Figure 13.1, the simulated ECG signal generator/AIMD monitor supplies the signal to pacemakers and ICDs via the electrodes and leads. Pacemakers and ICDs need to sense the simulated ECG signal to properly operate in the human body. The oscilloscope and chart recorder record the outputs of the pacemaker, or ICD, allowing the occurrence of EMI to be assessed. The distance between the DUT of WPTS and the torso phantom's front surface can be varied during testing; the MID is measured in centimeters.

13.1.1.2 Torso phantom

The torso phantom simulates a standard human chest where an AIMD has been implanted. It is filled with saline solution, and its electronic properties are equivalent to those of a human in terms of simulated ECG signal transmission at a LF. Regarding high-frequency EMI, it has been confirmed that conservative measurement results may be obtained even if the saline solution has a low-frequency tissue-equivalent property [32]. Similarly, a flat-shaped torso phantom is preferred over a realistically shaped one because the former yields conservative estimates [15,23,32]. A vertical-type phantom is preferable for dealing with the WPTS. Examples of the design parameters of the torso phantoms are shown in Figures 13.3 and 13.4.

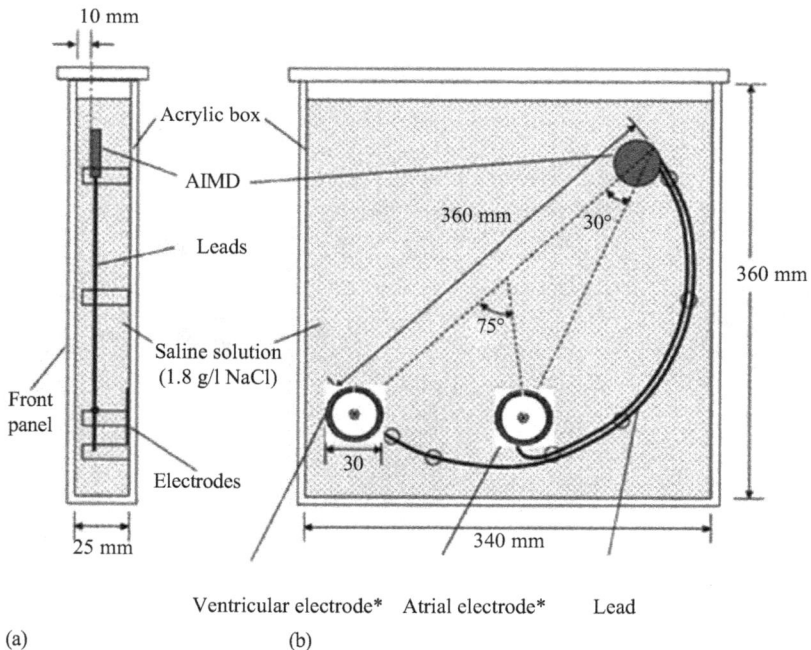

Figure 13.3   Example of a vertical torso phantom: (a) side view and (b) front view

Differential detection-type electrodes are used to apply the simulated ECG signals and monitor the pacing signals through the saline. As shown in Figure 13.4, the electrodes are installed inside the saline tank as dual-chamber pacemakers equipped with atrium and ventricle lead wires or as single-chamber pacemakers. Figure 13.5 shows an example of the interface-circuit construction between the electrodes and the external electronic devices. The fundamental specifications are similar to those described in references [42,44,48]. The electrodes are constructed

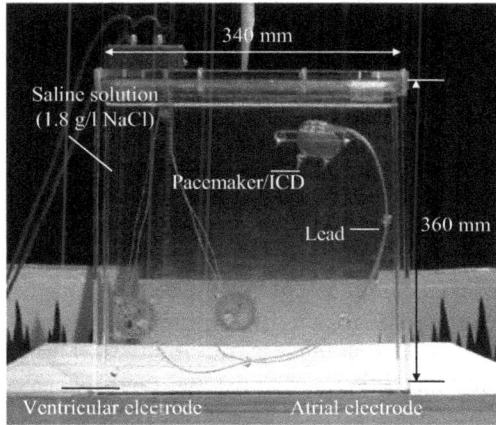

Figure 13.4   Picture of the torso phantom

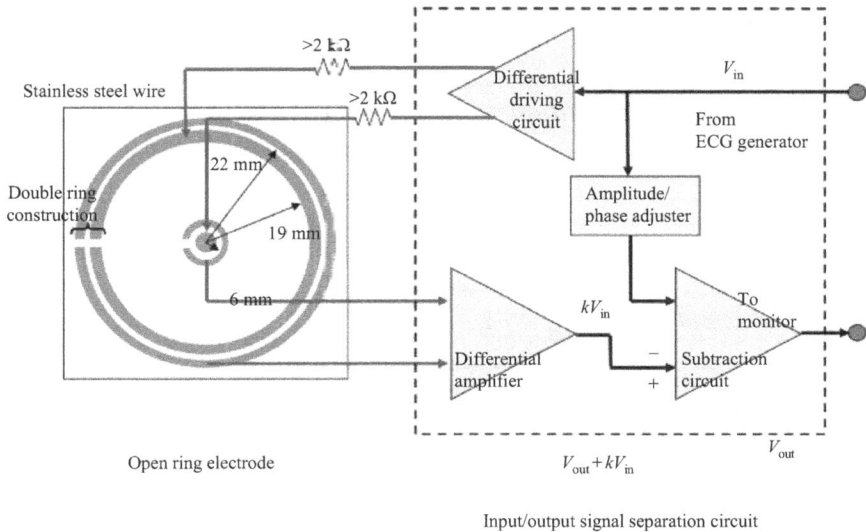

Input/output signal separation circuit

Figure 13.5   Electrode and interface circuit

from coaxial stainless-steel rings and center patch. The electrodes are positioned near the end of the pacemaker leads but without being directly in contact with them. Thus, all types of leads can be easily installed inside the phantom.

Signal flow for the inhibition test
The pacing pulses generated by the pacemaker are received at each electrode. These pulses are then input into the differential amplifier shown in Figure 13.5. This construction detects the differential voltage between the outer rings and the inner ring and patch; the leakage signal from the other end of the lead wire generates approximately the same voltage on these ring and patch, and hence, the atrium and ventricle electrodes can be electrically highly isolated from each other. Thus, the dual-chamber test in one saline tank is performed with high separation performance between the two chamber-output signals.

Signal flow for the asynchronous test
The signals from the ECG generator are fed into the differential-driving circuit and then applied to each electrode. The ECG signals that leak into the monitor circuit can be canceled by the bypass-ECG signals through the amplitude/phase-adjuster circuit. Using this coaxial electrode configuration, the separation of the simulated ECG signals between the lead wires for the atrium and ventricle can be maintained at a level greater than 20 dB, enabling high-sensitivity EMI tests to be conducted on dual-chamber pacemakers [32]. In addition, the double-ring construction allows high immunity against external electromagnetic disturbances to be achieved by adopting differential detection.

13.1.1.3   Electronic devices
The required devices include a simulated ECG signal generator/AIMD monitor, chart recorder, an oscilloscope, and a spectrum analyzer to monitor the interrogator EMF, as well as an AIMD programmer.

13.1.2   Operation conditions of the AIMD
Table 13.1 shows the initial settings of the AIMDs for the EMI experiments. To obtain conservative EMI estimation results, the AIMD is set to its highest sensitivity and shortest refractory period. In addition, its pacing rate is set to 60 pulses per minute (ppm). The experiments are conducted under the atrium-pacing,

Table 13.1   Initial settings of the pacemaker

Parameter	Value
Stimulation mode	AAI or VVI
Heart rate	60 ppm
Pacing and sensing polarity	Unipolar or bipolar
Pulse amplitude and duration	Nominal values (approximately 3.5 V and 0.4 ms)
Sensitivity	Maximum (most sensitive)
Refectory period	Minimum

atrium-sensing, and inhibition-of-response-to-sensing (AAI) mode, as well as the ventricle-pacing; ventricle-sensing, and inhibition-of-response-to-sensing (VVI) mode for single-chamber AIMDs. When the type of DUT is a dual chamber, it should be tested in two single-chamber modes, i.e., the atrial and ventricular chambers separately, to simplify the judgment of EMI occurrence. Moreover, the experiments should be conducted with both bipolar and unipolar sensing and pacing polarities. To detect the ECG signal and pace the heart, an AIMD uses two electrodes: a different electrode and an indifferent electrode. Most AIMDs have two operating modes, which depend upon the configuration of the electrodes. One mode, in which the AIMD uses its metal housing as the indifferent electrode, and the tip electrode at the end of the lead wire as the different electrode, is called "unipolar." The other, in which the tip electrode is the difference electrode, and the ring electrode is the indifferent electrode, is called "bipolar." The ring electrode is located approximately 1 cm away from the end of the pacemaker lead. Details are obtained from references [43–48].

13.1.3   Fundamental test procedure

To assess the EMIs of pacemakers and ICDs, "inhibition tests" and "asynchronous tests" are conducted. The inhibition test examines the dropping of pacing pulses generated by the AIMD. No injection of the simulated pulse occurs in inhibition tests. An example of an inhibition test result is shown in Figure 13.6(a). The required pacing pulse is inhibited because of the EMI occurrence. The asynchronous test examines the generation of fixed-rate asynchronous pulses. The simulated cardiac pulses are injected in asynchronous tests. An example of an asynchronous test result is shown in Figure 13.6(b). In the absence of EMI, the AIMD senses the simulated pulse, and the pacing pulse is inhibited. However, an asynchronous pulse is generated when the AIMD suffers noise and switches into noise-reversion mode.

Inappropriate tachyarrhythmia detection and delivery of therapy or shocks in ICD experiments are also investigated.

To obtain conservative EMI estimation results, the sensitivity and refractory period of the AIMDs are set to their maximum (most sensitive) and minimum values, respectively. The test procedure for the mobile/portable WPTS is identical to that proposed for RFID interrogators [49]. On the other hand, the procedure shown in Figure 13.7 is needed for EV-WPTS to suit the test scenarios [50].

The common points of the tests are as follows:

1. First, set the sensitivity and refractory period of the pacemakers and ICDs to the maximum sensitivity and the minimum time, respectively.
2. Record the ECG signal for each mode. In the absence of interference, the distance between the body of the WPTS and the human torso phantom's front surface decreases. In this case, the MID (distance at which EMI appears) is determined and recorded in centimeters, and the reaction level is also determined based on classification tables [18] (Figure 13.8) for the tested pacemaker/ICD.

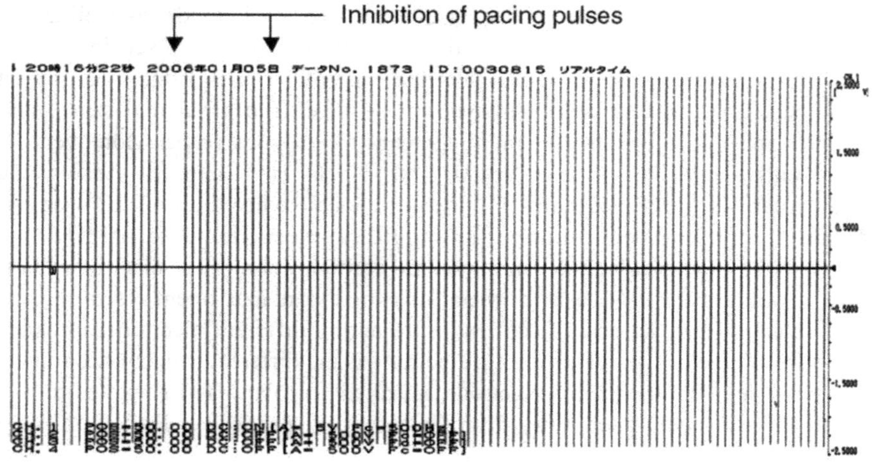

Inhibition of pacing pulses

(a)

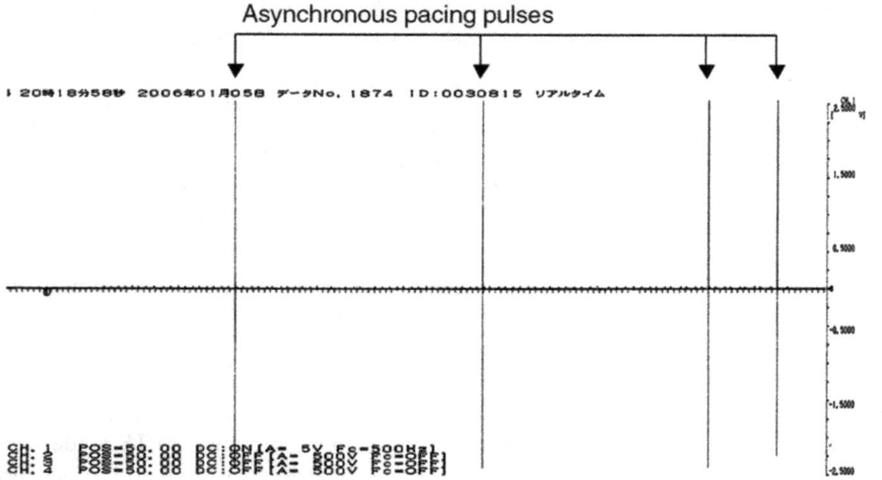

Asynchronous pacing pulses

(b)

Figure 13.6   Examples of (a) inhibition and (b) asynchronous test result

3. Reduce the sensitivity of the pacemakers and ICDs in five levels (maximum, 1.0, 2.4, 5.6 mV, and minimum) and record the MID.
4. Carry out experiments for all scenario and AIMD combinations. The operating modes of pacemakers and ICDs include unipolar and bipolar modes, as well as AAI and VVI modes.

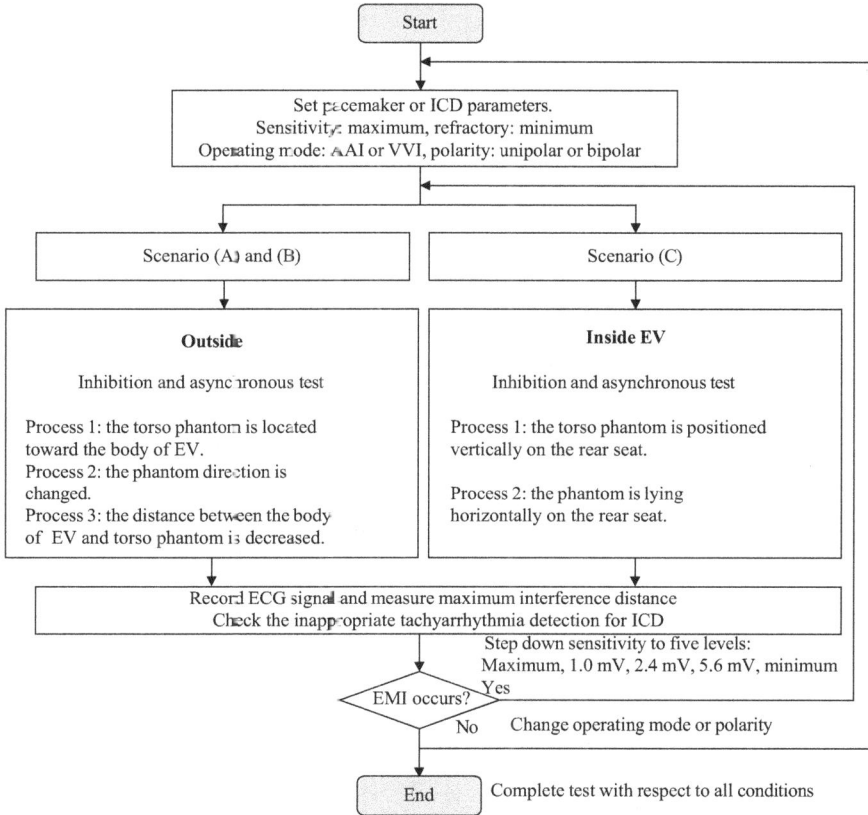

Figure 13.7 The procedure for the experiments

13.1.4 Measurement results for WPTS examples

Fourteen types of WPTSs, including some for mobile/portable applications (12 devices, ≤ 50 W) and EV charging (two devices), were tested [50]. The operational frequency of the tested WPTSs ranged from 70 kHz to 6.78 MHz. The upper transmission power level was set to 3 kW. The tested WPTSs for mobile/portable applications included some commercially available Qi certification devices.

Figure 13.9 defines the position of the mobile/portable WPTS relative to the torso phantom in the test. For the mobile/portable WPTS, we tested two different transmitting conditions. One assumed a power transfer mode in which the receiver device (Rx device) was mounted on the transmitting device (Tx device). The other was a standby mode in which the Rx device was absent. Some of the tested Tx devices emitted intermittent signals in standby mode for Rx device detection.

Figures 13.10 and 13.11 respectively show the basic configuration and definition of the position of the torso phantom relative to the EV-WPTS in the tests.

	Normal operation	Transient response	Permanent response		
			Re-programm able	Surgical procedure required	Biological damage
No reaction (normal operation)	**Level 0**				
Pacing/sensing violation for 1 period or less (recoverable in 2 s or less)		**Level 1**			
Pacing/sensing violation for more than 1 period (2 s)		**Level 2**			
Reset/permanent change in programmed settings of pacemaker			**Level 3**		
Transient pacemaker failure (function stop)			**Level 5**		
Permanent pacemaker failure (function stop)				**Level 5**	
Heating/induced currents in pacemaker lead					**Level 5**

(a)

	Normal operation	Transient response	Permanent response		
			Re-programm able	Surgical procedure required	Biological damage
No reaction (normal operation)	**Level 0**				
Pacing/sensing violation for 1 period or less (recoverable in 2 s or less)		**Level 1**			
Pacing/sensing violation for more than 1 period (2 s)		**Level 2**			
Inhibition ICD shock		**Level 3**			
Inappropriate ICD shock		**Level 4**			
Reset/permanent change in programmed settings of pacemaker			**Level 4**		
Transient pacemaker failure (function stop)			**Level 5**		
Permanent pacemaker failure (function stop)				**Level 5**	
Heating/induced currents in pacemaker lead					**Level 5**

(b)

Figure 13.8　EMI reaction level: (a) for the pacemaker test and (b) for the ICD test

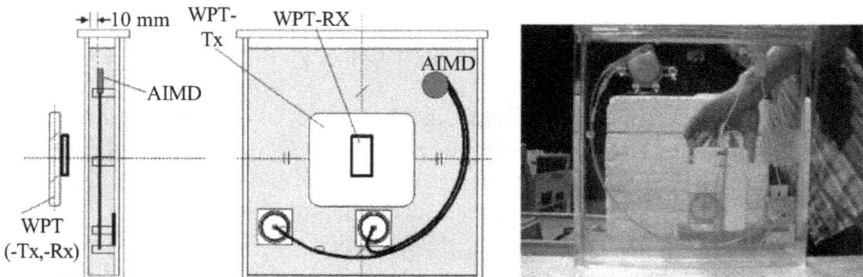

Figure 13.9　Definition of the position of the mobile/portable WPTS relative to the torso phantom in the AIMD-EMI test

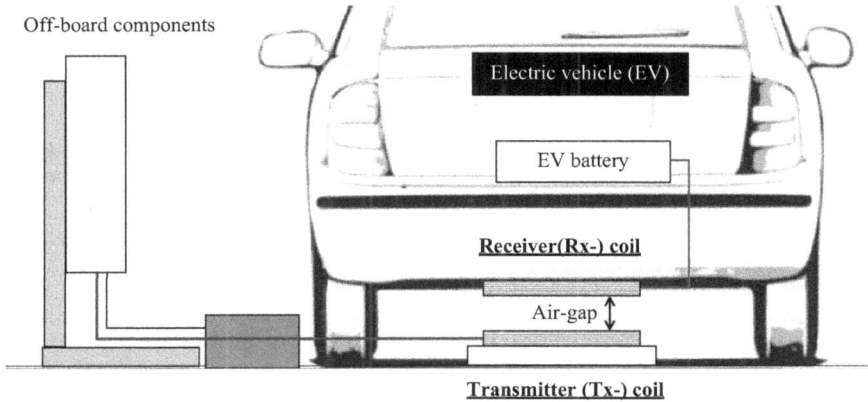

Figure 13.10 Basic configuration of the EV-WPTS

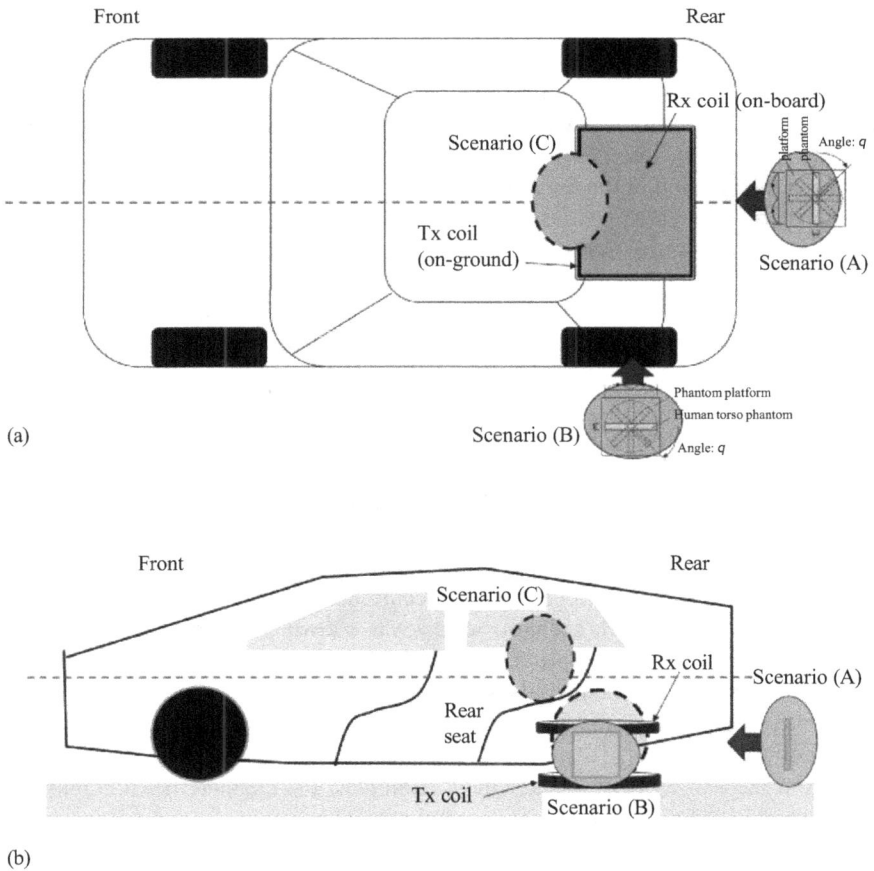

(a)

(b)

Figure 13.11 Definition of the position of the torso phantom relative to the EV-WPTS: (a) top view and (b) side view

Table 13.2   Parameters of the tested WPTSs for EVs

Type	WPT system			EV	
	Frequency (kHz)	Transferring power (kW)	Air gap (mm)	Class	Body type
A	85	2	150	Mid-size car	Five-door hatchback
B	85	3	165	Compact car	Five-door hatchback

Two types of EV-WPTS mounted on EVs currently in production were tested. These systems have different coil structures.

A WPTS transfers electrical energy from a power source to an electrical load via electric and/or magnetic fields or waves between a transmitting coil and receiving coil (Rx coil). The Rx coil was mounted beneath the rear seat of the EV for both types of WPTS tested. The systems tested in this study emitted continuous waves (CWs) with an operational frequency of 85 kHz. Typical characteristics of the EV-WPTSs tested are summarized in Table 13.2. Three different practical scenarios (A, B, and C) were designed for conservative EMI assessments. Scenarios A and B were designed to address the effects of different field distributions and the position of the human body relative to the Tx and Rx coils close to the EV. Scenario C covered the case wherein a human being is sitting upright or lying horizontally on the rear seat of the EV.

The in vitro EMI experiments were conducted using the test system with the procedure and assumed scenarios described earlier. The measured results are summarized in Table 13.3.

For five mobile/portable WPTS combinations, reactions were observed in pacemaker and ICD tests. All the observed EMI events occurred in standby mode (i.e., the tested Tx device emitted an intermittent signal). The MIDs ranged from 1 to 2 cm. The reactions observed included pacing inhibition and inappropriate pacing (the maximum reaction level; pacemaker: level 2, ICD: level 1). All the EMI events observed were transient responses. Once the RF source was turned off, the PMs/ICDs returned to normal operation within a few seconds and no permanent variation in the programmed parameters occurred.

For EV-WPTSs, each scenario tested was shown to yield normal operation during EMI testing. No inappropriate reaction, such as missing of pacing pulses, generation of asynchronous pulses for pacemaker functions, or inappropriate tachyarrhythmia detection and delivery of therapy for ICDs, was observed under EMI testing.

From the entire set of measurement examples, the characteristics common to all types of WPTSs are summarized as follows:

1. The observed EMI for pacemakers is either one or more missing pulses or the undesirable generation of asynchronous pulses. The duration of the EMI

Table 13.3   AIMD-EMI test results for WPTSs

WPT device	Frequency		Trans. power	AIMD-EMI test result
Mobile, portable application	A	70 kHz	0.2 W	
	B	100–200 kHz	5 W (max)	
	C	100–200 kHz	5 W (max)	• Observed electromagnetic interference (EMI) at maximum distance of pacemaker: ≤2 cm ICD: ≤1 cm
	D	100–200 kHz	5 W × 2	• Reaction level (see Figure 13.8) pacemaker: 2 ICD: 1
	E	100–200 kHz	5 W	
	F	110–183 kHz	5–15 W	
	G	110–210 kHz	1–15 W	
	H	134.5 kHz	≒1.4 W	
	I	200 kHz	13 W	
	J	400 kHz	≒0.4 W	
	K	460 kHz	40 W (max)	
	L	6.78 MHz	18.2 W	
EV charging	M	85 kHz	2–3 kW	No reaction occurred
	N	85 kHz	3 kW	

Note: Bold indicates Qi certification device.

occurrences varies from a single pulse to the complete inhibition of continuous-pulse generation during the WPTS operation. The occurrence and characteristics of the EMI depend upon the combination of the device and the operational mode. In the basic EMI mechanism, the EMF penetration components are detected by the nonlinear responses of the internal circuits of the AIMD (envelope detection); when the detected signal is similar to any of the targeted physiological signals (frequency of a few hertz to several hundred hertz) and exceeds the threshold, AIMD malfunctions can occur.

2.   For the EV-WPTS and AIMD combinations examined, no inappropriate generation of asynchronous pulses for pacemaker functions, inappropriate tachyarrhythmia detection, or delivery of therapy for ICDs were observed. Because the transfer signal of the EV-WPTS tested was a CW, AIMD malfunctions did not occur, even when the transfer power level of the WPTS was of a 3-kW class. If the AIMDs are exposed to changing fields, such as pulse-modulated waves or changing magnetic fields (in particular, when the pulsed signals have a repetition time close to the physiological heart rhythm), the probability of the EMI occurrence becomes significant.

13.1.5   Interference voltage measurement for beam-type wireless power transfer

Some voltage sensors for AIMD-EMI interference assessment have been developed. Interference voltage measurements from a pacemaker in the RF frequency band for the beam-type WPTS (RF-WPT) using directly modulated electrical to optical (EO) converter have been reported [51]. Interference voltage measurements due to the 900 MHz-band RF-WPT base-station were conducted using the developed sensor, which was placed inside a torso phantom. Figures 13.12 and 13.13 depict the constructed measurement setup and the measured distance dependence of interference voltage, respectively. The interference voltages were measured when the RF-WPT operated at 915 MHz, both in vertical and horizontal polarizations, under near-field exposure conditions. The developed sensor could accurately measure interference voltage. As indicated in Figure 13.13, vertically polarized waves produce interference voltages that exceed the EMI threshold level at short distances from the antenna. As shown here, based on the AIMD EMI test data collected using actual pacemakers and ICDs, a simple prediction of the relationship

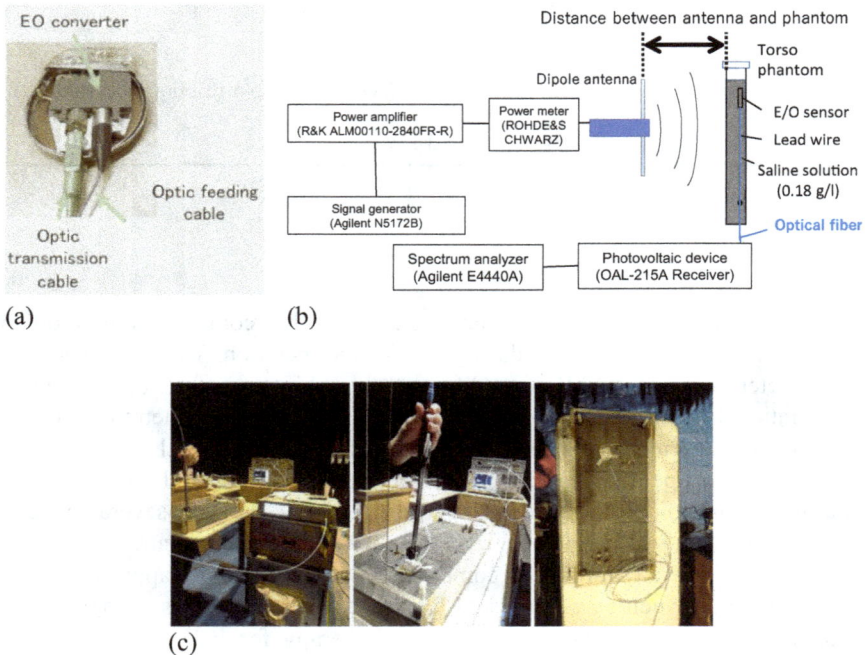

(a)   (b)

(c)

Figure 13.12   Electrical to optical (EO) converter and developed interference voltage sensor: (a) overview of developed interference voltage sensor for AIMD-EMI; (b) block diagram of measurement set-up for interference voltage due to a RF frequency band for beam type WPTS; and (c) overview of measurements of interference voltage

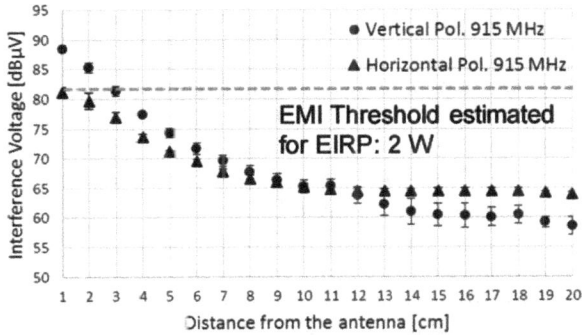

Figure 13.13 Measured distance dependence of interference voltage at RF-WPT frequency (915 MHz)

between maximum exposure level and the potential risk of AIMD EMI occurrence might be possible.

13.2 RF-induced heating of metal implants

Radio radiation protection guidelines [4–9] for human exposure to EMF have been formulated considering the EMF emitted from wireless communication devices. These guidelines do not quantitatively discuss their relevance to humans with metallic objects implanted in their bodies because precisely evaluating local SAR around such implants is difficult. However, given the progress in biomedical technologies, the number of users of technologies such as active implantable pacemakers and medical metal plates, upper-limb prostheses, and prosthetic legs continues to increase. Estimating the amount of exposure that users with metallic implants will experience is necessary. Because of the recent techno-logical advancements in numerical simulation, these difficult issues can be pre-cisely treated. Some studies concerning the interaction of RF EMF with passive metallic implants have been conducted [52–56]. Some experimental estimates of local SAR enhancement for human tissue with metal implants exposed to RF EMF have been conducted. Various types of medical implant have been recently intro-duced into clinical use. In some cases, the temperature increase due to the metallic implants was clearly observed [57,58]. Regarding WTPSs from the human safety standpoint, the assessment of RF-induced heating of metal implants is also important.

Acknowledgment

The author thanks the members of the Japan Arrhythmia Device Industry Association and Broadband Wireless Forum of Japan for their cooperation and support.

References

[1]   APT: "APT report on wireless power transmission (WPT)," No. APT/AWG/REP-62(Rev.1), Feb. 2016.

[2]   https://wirelesspower.ieee.org/technology-trend/standards

[3]   CISPR 11: "Industrial, scientific and medical equipment – radio-frequency disturbance characteristics – limits and methods of measurement, International Special Committee on Radio Interference CISPR Std. CISPR 11:2015 +A1:2016+A2: 2019," 2019.

[4]   ICNIRP: "Guidelines for limiting exposure to time-varying electric, magnetic and electromagnetic fields (up to 300 GHz)," *Health Phys.*, vol. 74, no. 4, pp. 494–522, 1998.

[5]   ICNIRP: "Guidelines for limiting exposure to time-varying electric and magnetic fields (1 Hz to 100 kHz)," *Health Phys.*, vol. 99, no. 6, pp. 818–836, 2010.

[6]   ICNIRP: "Guidelines for limiting exposure to time-varying electric and magnetic fields (100 kHz to 300 GHz)," *Health Phys.*, vol. 118 no. 5, pp. 483–524, 2020.

[7]   IEEE: "IEEE standard for safety levels with respect to human exposure to radio frequency electromagnetic fields, 3 kHz to 300 GHz, IEEE Std. C95.1," 1992.

[8]   IEEE: "IEEE standard for safety levels with respect to human exposure to radio frequency electromagnetic fields, 3 kHz to 300 GHz, IEEE Std. C95.1," 2005.

[9]   IEEE: "IEEE standard for safety levels with respect to human exposure to electric, magnetic, and electromagnetic fields, 0 Hz to 300 GHz, IEEE Std. C95.1," 2019.

[10]   Hirata A., Ito F. and Laakso I.: "Confirmation of quasi-static approximation in SAR evaluation for a wireless power transfer system," *Phys. Med. Biol.*, vol. 58, pp. N241–N249, 2013.

[11]   Christ A., Douglas M.G., Roman J.M, *et al.*: "Evaluation of wireless resonant power transfer systems with human electromagnetic exposure limits," *IEEE Trans. Electromagn. Compat.,* vol. 55, no. 2, pp. 265–274, 2012.

[12]   Sunohara T., Hirata A., Laakso I., Santis V.D. and Onishi T.: "Evaluation of non-uniform field exposures with coupling factors," *Phys. Med. Biol.*, vol. 60, pp. 8129–8140, 2015.

[13]   International Electrotechnical Commission – IEC 61980-1:2020: "Electric vehicle wireless power transfer (WPT) systems – Part 1: General requirements", 2020.

[14]   International Organization for Standardization – ISO 19363:2020: "Electrically propelled road vehicles—magnetic field wireless power transfer—safety and interoperability requirements", 2020.

[15]   Irnich W., Bakker J.M.T. and Bisping H.J.: "Electromagnetic interference in implantable pacemakers," *J. Pac. Clin. Electrophysiol.*, vol. 1, no. 1, pp. 52–61, 1978.

[16] https://www.fda.gov/radiation-emitting-products/electromagnetic-compat-ibility-emc/electromagnetic-compatibility-cellular-phone-interference

[17] https://www.gsma.com/publicpolicy/emf-and-health

[18] Ministry of Internal Affairs and Communication, Japan: "Guidelines on the use of radio communications equipment for implanted medical devices," Aug. 2005.

[19] Ministry of Internal Affairs and Communication, Japan: "Investigation of the effects of the radio emissions from mobile phone handsets and RFID (electronic tag) equipment," Mar. 2007.

[20] Ministry of Internal Affairs and Communications, Japan: "Results of a study on the effects of electromagnetic waves on medical equipment," *MIC Commun. News*, vol. 18, no. 4, Jun. 8, 2007.

[21] Irnich W.: "Interference in pacemakers," *PACE*, vol. 7, no. 6, pp. 1021–1048, 1984.

[22] Irnich W.: "Electronic security systems and active implantable medical devices," *Europace*, vol. 8, no. 5, pp. 377–384, 2006.

[23] Irnich W., Batz L., Muller R. and Tobisch R.: "Electromagnetic interference of pacemakers by mobile phones," *PACE*, vol. 19, no. 10, pp. 1431–1446, 1996.

[24] Barbaro V., Bartolini P., Donato A. and Militello C.: "Electromagnetic interference of analog cellular telephone with pacemakers," *J. Pac. Clin. Electrophysiol.*, vol. 19, no. 10, pp. 1410–1418, 1996.

[25] Irnich W.: "Mobile telephones and pacemakers," *PACE*, vol. 19, no. 10, pp. 1407–1409, 1996.

[26] Irnich W.: "Electronic security systems and active implantable medical devices," *PACE*, vol. 25, no. 8, pp. 1235–1258, 2002.

[27] Toyoshima T., Tsumura M., Nojima T. and Tarusawa Y.: "Electromagnetic interference of implantable cardiac pacemakers by portable telephones," *Jpn. J. Cardiac Pac. Electrophysiol.*, vol. 12, no. 5, pp. 488–497, 1996.

[28] Electromagnetic Compatibility Conference Japan: "Guidelines on the use of radio communication equipment such as cellular telephones – safeguards for electric medical equipment, presented at the EMC Conf. Japan, Electromagnetic Medical Equipment Study Group," 1997.

[29] Hayes D.L., Wang P.J., Reynolds D.W., *et al.*: "Interference with cardiac pacemakers by cellular telephones," *New Engl. J. Med.*, vol. 336, no. 21, pp. 1473–1479, 1997.

[30] Stevenson R.A.: EMI filters for cardiac pacemakers and implantable defibrillators. in Proceedings of the 19th Annual International Conference of the IEEE Engineering in Medicine and Biology Society, October 30–November 2, 1997, Chicago.

[31] Barbaro V., Bartolini P., Calcagnini G., *et al.*: "On the mechanisms of interference between mobile phones and pacemakers: parasitic demodulation of GSM signal by the sensing amplifier," *Phys. Med. Biol.*, vol. 48, no. 11, pp. 1661–1671, 2003.

[32] Tarusawa Y., Ohshita K., Suzuki Y., Nojima T. and Toyoshima T.: "Experimental estimation of EMI from cellular base-station antennas on

implantable cardiac pacemakers," *IEEE trans. Electromagn. Compat.*, vol. 47, no. 4, pp. 938–950, 2005.

[33] Futatsumori S., Hikage T., Nojima T., Koike B., Fujimoto H. and Toyoshima T.: "In vitro experiments to assess electromagnetic fields exposure effects from RFID reader/writer for pacemaker patients," in Proceedings of the Biological Effects of EMFs 4th International Workshop, vol. 1, pp. 494–500, 2006.

[34] Futatsumori S., Hikage T., Nojima T., Koike B., Fujimoto H. and Toyoshima T.: "A novel assessment methodology for the EMI occurrence in implantable medical devices based upon magnetic flux distribution of RFID reader/writers," in Proceedings of the IEEE International Symposium on Electromagnetic Compatibility, pp. 1–6, Jul. 2007.

[35] Futatsumori S., Toyama N., Hikage T., *et al.*: "An experimental validation of a detailed numerical model for predicting implantable medical devices EMI due to low-band RFID reader/writers," in Proceedings of the 2008 Asia-Pacific Microwave Conference, Dec. 2008.

[36] Futatsumori S., Kawamura Y., Hikage T., *et al.*: "In vitro assessment of electromagnetic interference due to low-band RFID reader/writers on active implantable medical devices," *J. Arrhyth.*, vol. 25, no. 3, pp. 142–152, 2009.

[37] Kawamura Y., Futatsumori S., Hikage T., *et al.*: "A novel method of mitigating EMI on implantable medical devices: experimental validation for UHF RFID reader/writers," in Proceedings of the IEEE EMC 2009 Symposium, pp. 197–202, Aug. 2009.

[38] Seidman S., Ruggera P., Brockman R., Lewis B. and Shein M.: "Electromagnetic compatibility of pacemakers and implantable cardiac defibrillators exposed to RFID readers," *Int. J. Radio Freq. Ident. Technol. Appl.*, vol. 1, no. 3, pp. 237–246, 2007.

[39] Seidman S., Brockman R., Lewis B., *et al.*: "In vitro tests reveal sample radiofrequency identification readers inducting clinically significant electromagnetic interference to implantable pacemakers and implantable cardioverter-defibrillators," *Heart Rhyth.*, vol. 7, no. 1, pp. 99–107, 2010.

[40] Gwechenberger M., Rauscha F., Stix G., Schmid G. and Strametz J.S.: "Interference of programmed electromagnetic stimulation with pacemakers and automatic implantable cardioverter defibrillators," *Bioelectromagnetics*, vol. 27, no. 5, pp. 365–377, 2005.

[41] Augello A., Chiara G.D., Primiani V.M. and Moglie F.: "Immunity tests of implantable cardiac pacemaker against CW and pulsed ELF fields: experimental and numerical results," *IEEE Trans. EMC*, vol. 48, no. 3, pp. 502–515, 2006.

[42] Fujimoto H. and Toyoshima T.: "An EMI evaluation mode that can simultaneously simulate the atrium and ventricle," *J. Arrhyth.*, vol. 16, no. 5, pp. 534–540, 2000.

[43] AAMI TIR No.18-1997: "Guidance on electromagnetic compatibility of medical devices for clinical/biomedical engineers—part 1: radiated radio-frequency electromagnetic energy, 1997."

[44] ANSI/AAMI PC69:2007: "Active implantable medical devices—electromagnetic compatibility—EMC test protocols for implantable cardiac pacemakers and implantable cardioverter defibrillators, 2007."

[45] ANSI/AAMI/IEC 60601-1-2:2001: "Medical electrical equipment—part 1–2: general requirements for safety—collateral standard: electromagnetic compatibility—requirements and tests, 2001."

[46] ISO14708-1:2000: "Implants for surgery—active implantable medical devices—part 1: general requirements for safety, marking and for information to be provided by the manufacturer, 2000."

[47] ISO14708-2:2005: "Implants for surgery—active implantable medical devices—part 2: cardiac pacemakers, 2005."

[48] ISO14117:2019: "Active implantable medical devices—electromagnetic compatibility—EMC test protocols for implantable cardiac pacemakers, implantable cardioverter defibrillators and cardiac resynchronization devices, 2019."

[49] EN50061:1988: "Safety of implantable pacemakers, 1988."

[50] Hikage, T., Nojima T. and Fujimoto H.: "Active implantable medical device EMI assessment for wireless power transfer operating in LF and HF bands," *Phys. Med. Biol.*, vol. 61, no. 12, pp. 4522–4536, 2016.

[51] Hikage T., Ito S. and Ohtsuka A.: "Novel interference voltage measurement for beam-type wireless power transfer using an electro-optical converter for EMI assessment of active implantable medical devices," *URSI RADIO SCIENCE LETTERS*, vol. 2, pp.1–3, 2020.

[52] McIntosh R.L., Anderson V. and McKenzie R.J.: "A numerical evaluation of SAR distribution and temperature changes around a metallic plate in the head of a RF exposed worker," *Bioelectromagnetics*, vol. 26, no. 5, pp. 377–388, 2005.

[53] Virtanen H., Keshvari J. and Lappalainen R.: "Interaction of radio frequency electromagnetic fields and passive metallic implants—a brief review," *Bioelectromagnetics*, vol. 27, pp. 431–439, 2006.

[54] Virtanen H., Keshvari J. and Lappalainen R.: "The effect of authentic metallic implants on the SAR distribution of the head exposed to 900, 1800 and 2450 MHz dipole near field," *Phys. Med. Biol.*, vol. 52, pp. 1221–1236, 2007.

[55] Kyriakou A., Chris, A., Neufeld, E. and Kuster N.: "Local tissue temperature increase of a generic implant compared to the basic restrictions defined in safety guidelines," *Bioelectromagnetics*, vol. 33, no. 5, pp. 366–374, 2012.

[56] Corcoles J., Zastrow E. and Kuster N.: "Convex optimization of MRI exposure for mitigation of RF-heating from active medical implants," *Phys. Med. Biol.*, vol. 60, pp. 7293–7308, 2015.

[57] Hikage T., Nagaoka T., Watanabe W., Tanaka N. and Nojima T.: "Experimental estimation of SAR enhancement due to two parallel implanted metal plates using a head phantom and thermograph," in Proceedings of the 5th

EAI International Conference on Wireless Mobile Communication and Healthcare, FA3-3, pp. 1–4, Oct. 2015.

[58]   Zastrow E., Yao A. and Kuster N.: "Practical considerations in experimental evaluations of RF-induced heating of leaded implants," in Proceedings of the 32nd International Union of Radio Science General Assembly and Scientific Symposium (URSI GASS), Montreal, Canada, Aug. 16–26, 2017.

Index

www.ingramcontent com/pod-product-compliance
Lightning Source LLC
Chambersburg PA
CBHW060247230326
41458CB00094B/1478

9 7 8 1 8 3 9 5 3 8 9 2 6